So You Want to Take Physics	Topic	*College Physics* by R. Serway, J. Faughn, 3rd edition	*Physics* by J. Cutnell, K. Johnson, 2nd edition
Section 7.6 pp. 177–183	Relative motion.	3.6 pp. 64–67	3.4 pp 67–71
Section 8.1 pp. 187–192	Circular motion with vectors.	Chapter 7 pp. 170–204, 8.4–6 pp. 219–229, 20.7 pp. 639–641	Chapter 8 pp. 194–208, 9.4 pp. 226–236, 21.3 pp. 601–603
Section 8.2–8.4 pp. 203–210	Practice with polar coordinate systems. These sections complement 8.5	supplemental material	supplemental material
Section 8.5 pp. 201–205	Area and Volume, elements of symmetry. Used in moments of inertia, fluids, heat transfer, waves, electricity and magnetism.	8.4 pp. 219–224, 10.2 pp. 279–283, 12.5–8 pp. 353–361, 15.4–5 pp. 461–466, 16.10 pp. 521–526, 21.1 pp. 660–662	9.4 pp. 226–229, 11.8 pp. 298–308, 13.2–4 pp. 365–376, 16.7 pp. 451–452, 22.3 pp. 640–643
Chapter 9 pp. 209–228	Force, Newton's laws, and free-body diagrams.	Chapter 4 pp. 74–107	Chapter 4 pp. 78–124
Sections 10.1–10.3 pp. 229–236	Vector projections—used throughout the textbooks, but are used heavily in work, torque, and the magnetic force and flux.	5.2 pp. 109–111, 8.1 pp. 212–218, 20.3 p. 631, 21.1 p. 661	6.1 p. 146–148, 9.1 p. 216–223, 21.2 p. 598–601, 22.3 p. 642
Section 10.4 pp. 236–239	Work and the dot product.	5.3 p. 111–113	6.2 p. 148–152
Section 10.5 pp. 239–245	Directed area and flux—used in fluids, heat transfer, and electromagnetism, and waves.	10.2 pp. 279–283, 12.5–8 pp. 353–361, 15.4–5 pp. 461–466, 16.10 pp. 521–526, 21.1 pp. 660–662	11.8 pp. 298–308, 13.2–4 pp. 365–376, 16.7 pp. 451–452, 22.3 pp. 640–643

So You Want to Take Physics

A Preparatory Course with Algebra and Trigonometry

Rodney Cole

University of California, Davis

SAUNDERS COLLEGE PUBLISHING

A Harcourt Brace Jovanovich College Publisher

FORT WORTH PHILADELPHIA SAN DIEGO
NEW YORK ORLANDO AUSTIN SAN ANTONIO
TORONTO MONTREAL LONDON SYDNEY TOKYO

Text Typeface: Times Roman
Compositor: York Production Services
Acquisitions Editor: John Vondeling
Developmental Editor: Jennifer Bortel
Managing Editor: Carol Field
Production Service: York Production Services
Copy Editor: York Production Services
Manager of Art and Design: Carol Bleistine
Text Designer: York Production Services
Cover Designer: Lawrence R. Didona
Text Artwork: York Production Services
Director of EDP: Tim Frelick
Production Manager: Joanne M. Cassetti
Marketing Manager: Marjorie Waldron

Cover Credit: David Rogers

Library of Congress Cataloging-in-Publication Data

Cole, Rodney.
So you want to take physics : a preparatory course with algebra and trigonometry / Rodney Cole.
p. cm.
Includes index.
ISBN 0-03-097215-9
1. Mathematical physics. 2. Algebra. 3. Trigonometry.
I. Title.
QC20.C615 1993
512′.13′02453—dc20 92-41426
CIP

234 082 987654321

Contents

Preface

So You Want To Take Physics: A Preparatory Course with Algebra and Trigonometry is designed for a preparatory course for science students who will subsequently take algebra and trigonometry-based college physics. There is also a version designed for a calculus-based course: *So You Want To Take Physics: A Preparatory Course with Calculus.* Both provide help for students whose mathematical skills need improvement in specific areas, such as logarithms, vectors, and polar coordinates, where the mathematics often become a stumbling block that prevents students from understanding the physics. Because the discussions in the book focus on why a given mathematical technique is important and on how the mathematics relates to physical examples, they can also benefit the student who is currently taking physics and who needs extra help with the mathematics. The main goal of the book is to improve the mathematical skills and problem-solving capabilities of the reader. It is not designed as a short course in physics nor as a comprehensive study guide for the standard physics texts, such as Serway and Faughn's *College Physics* and Cutnell and Johnson's *Physics.* However, the book can be used as a self-help workbook for students taking physics, so a map showing where the topics treated in this book are used in these standard physics texts has been provided on the inside of the front cover.

FEATURES

In writing the book, the guiding principles have been to help the reader build skills and to overcome the anxiety that often accompanies studying physics by using carefully crafted examples and problems and by linking the illustrations and mathematics with clear discussions. The book's direct conversational style makes both the concepts and theory accessible to the reader, and its many illustrations link the mathematics to physical processes, which appeals to a visually oriented reader. The important features include:

Study Skills

Often after a midterm, students who have performed poorly ask how they can improve their grade next time. I usually start by asking how they prepared for the test. Invariably, I hear they studied at length, reading the book several times over and perhaps reviewing the assigned homework problems. Of course, many students do not understand that reading the book alone is not an effective way to study physics. Physics requires practice. Therefore, the first chapter of this book starts by discussing an effective study regimen and great care is taken throughout the book to inform the reader as to what he or she should be doing to be successful. An appendix discusses how to prepare both homework and exam solutions.

Problem-Solving Skills

There is little doubt that over the years the skills of the students have changed dramatically. In a recent article, one Worchester Polytechnic Institute professor commented that, "The United States is no longer a nation of tinkerers." Students don't come to class knowing what a transistor or an electromagnetic relay looks like. They don't take things apart and put them back together anymore." This lack of basic skills is nowhere more apparent than in the arena of problem solving. Increasing the problem-solving skills of the reader is a major goal of the book and is implemented through a variety of mechanisms. The first chapter introduces techniques for estimating the answer, and examples are used throughout the text. The first chapter also establishes a problem-solving paradigm that is stressed in the second chapter and reinforced in subsequent chapters.

The Laboratory

Although the lecture portion of a physics course receives most of the attention, physics courses have an equally important laboratory component that requires data analysis. Chapter 5 discusses probability and error analysis and covers the standard statistical techniques with which the student is often expected to be conversant, but which are seldom covered comprehensively. The laboratory part of a course can pose a major hurdle to many students, and this chapter is important to surmounting it.

Learning Style

In writing the book, I have attempted to be conscious of the different learning styles that the readers may have. Current research indicates that some people are analytic learners, and the analytic treatment of equations will appeal to more to these readers. Others are visually oriented, and the way the illustrations are used to help the reader understand the equations will appeal more to them. Still others may be oriented toward the descriptive passages that link the drawings to the equations. Extensive discussions detail what the equations mean and show how they describe the physics.

Mathematical Level

The book starts gently, reviewing algebra and problem-solving techniques in Chapter 1 and waiting until Chapter 3 to introduce logarithms and trigonometry. The trigonometry is then used and expanded throughout the rest of the book, so the student has ample opportunity to practice. Chapter 6 provides extensive work with graphs of functions illustrating the geometrical connections between the slope, and the area under a curve. These visualization exercises help readers gain an understanding of the concepts of kinematics and develop skills for interpreting graphs. The text builds on these basic ideas, taking the reader to successively more difficult material through Chapter 10 which presents the projections of vectors. Although vector products are not explicitly introduced, Chapter 10 contains the basic techniques which are important to so many aspects of physics. These topics often cause students difficulty, but might be considered too advanced for some preparatory courses, and therefore, are placed at the end of the book so that they may be easily omitted. Other advanced topics that might be regarded as optional are signaled by a bullet (•) before the section number, for example (• **6.4 Density**). The mathematical level of the book is meant to challenge the reader, but at the same time provide the insight to prevent it from becoming frustrating.

Problems

Many of the problems in the book are the result of years of research with and questions raised by students. These problems are new and will not be found in other physics texts. In addition, there are a few of the old favorites, some of which have been given new twists. I have attempted to write problems that use mathematics as it applies to physics. For example, in vector addition and subtraction, students have difficulty seeing how the vectors fit together geometrically. The problems in this chapter are weighted toward relative velocity problems and visualization problems on writing an equation given a drawing of three vectors. This focus provides the reader with essential practice in relating the geometry to the mathematical equations, which is a skill the reader will use throughout all of physics. Readers will find most of the problems challenging. There are a small number of especially difficult problems, which are preceded by a bullet. These problems are included because they illustrate important points, and they also provide practice for readers who find their skills developing and want practice working more difficult problems.

One criticism students always levy at a physics textbook is that there are never enough worked examples. Because this book functions as a self-help workbook in addition to a course textbook, it is critical that there be an abundance of examples. Therefore, solutions to a majority of the odd-numbered problems are included in an appendix. These solved problems serve as further examples to help students gain problem-solving skills without hindering the readability of the text. Students can only gain true problem-solving skills, however, by attempting problems that are unknown to them. The even-numbered problems serve this purpose, and they do not have solutions or answers in the text.

Supplemental Materials

An *Instructor's Manual* prepared by the author provides solutions to the unsolved problems and discusses a course design that has been successful. It contains a chapter-by-chapter discussion of interesting class projects and the major stumbling blocks encountered by students.

Acknowledgments

I am indebted to the people who have provided many wonderful suggestions for improving the text and I send them my heartfelt thanks. To my students at the University of California, Davis, who suffered the early drafts, and without whose ideas and critical responses this text could not have been written, I thank you. I wish to thank the reviewers who have taken the time to read the manuscript carefully at various stages of its production and have offered valuable suggestions, criticisms, and encouragement:

Helena Dedic
Vanier College

Roger Freedman
University of California
at Santa Barbara

Joseph M. Hoffman
Frostburg State University

Brian Holton
Rutgers University

John L. Hubisz
College of the Mainland

Eugene V. Ivash
University of Texas at
Austin

Robert Kadesch
University of Utah

Fritz Kissner
State University of New York,
College at Plattsburgh

Donna MacDuff
British Columbia Institute of
Technology

Raymond Serway
James Madison University

Harold Skelton
West Chester University

In particular, I wish to thank Steve Van Wyk for his detailed reading of the final manuscript, for providing additional problems, and for performing the thankless task of corroborating the solutions to the problems. I also wish to thank Neal Peek for his improvements on the problem-solving aspects of the text and his wonderful sense of where the manuscript was needlessly difficult. Throughout the production of the text, my publisher, John Vondeling, provided inspirational support and encouragement, and Developmental Editor Jennifer Bortel kept all the innumerable tasks rolling smoothly toward a conclusion. I wish to thank them and the rest of the professional staff at Saunders College Publishing for their outstanding work, including Production Manager Joanne Cassetti and Manager of Art and Design Carol Bleistine.

I wish to thank the Learning Skills Center, in particular Virginia Martucci and Ward Stewart, and the Physics Department, in particular Doug McColm, at the University of California, Davis, for providing the supportive and nourishing environment that allowed this project to develop. Finally, I am indebted to my wife, Norma, for her long nights of help with the editing of the manuscript, and to my family, to whom this book is dedicated, without whose love and understanding this project would not have been completed.

Rod Cole
University of California
Davis, California
October 1992

CHAPTER 1
Keeping It in the Ballpark

1.1 STUDYING PHYSICS

How do you study for physics? Do you read your physics book the same way you read a book for literature class? Although physics and English literature are both intellectual disciplines, the way you study for them is not the same. You can usually read a novel for English class in one or two nights. You can scribble any notes in the margins, provided the book is yours and not the library's. In contrast, you cannot sit down and read a physics book in one or two evenings and expect to get anything out of it. Think of a chapter in a physics book as a meal that will take many hours to consume. You will enjoy it far more if you take small bites and savor each one, allowing the full flavor to develop rather than quickly devouring each course. The most important thing you can learn from this book is to study physics attentively, and remember that learning is more thorough over the course of many small sessions than over a few long nights of cramming.

One reason that physics must be taken in small bites is that the theories are very intricate, and the number of different problems to which any given theory can be applied is often astronomical. This means that at most we can pick a set of important examples illuminating critical points that otherwise might be missed. The only way in which you can see these points and learn these techniques is by doing problems yourself, one at a time. You can liken it to the difference between looking at a map of a forest and actually going to visit a forest. A map points out the major landmarks and gives you a feel for the geography, but it takes an actual hike to see the beauty. For example, nowhere in the map does it show you the exquisite detail of the veins and textures that makes up a leaf or the serenity of a gurgling brook sheltered deep within the forest. These are discoveries that can only be made by exploring for yourself, and on your journey through physics, you will acquire many useful skills that will serve you the rest of your career.

A physics textbook is much like a map: It derives the theory and attempts to give you a sense of geography. However, to learn the beauty and secret places of physics you must explore for yourself; anything less would be an injustice to the theory and to yourself. To explore the theory, you must apply it to specific problems. In the course of working those problems, you will learn many professional tricks and find many beautiful ideas not present in any textbooks. This process will greatly enhance your understanding of physics if your studying is more like a hike than like a passive reading session. It should be done with a pad of paper and pencil so you can work through the various examples in the text as you read. In fact, when you come to an example, you should try to solve it yourself before looking at the solution. On the other hand, don't spend too much time trying to solve the example. Work until you get stuck, and then consult the solution. Always be alert for interesting ideas and useful techniques in the examples and problems. To understand physics requires active participation on your part and not passive reading in the easy chair. Physics is not a story but a very powerful tool that will give you a deeper understanding of the universe than you now possess.

Because physics is a tool, it takes practice to learn to use it. Like mastering the violin, mastery of physics doesn't come by reading an instruction manual, but takes hours and hours of practice. To do well in physics you must spend those hours practicing. The problems are designed to give you practice in physics. It is extremely important that you work not only the assigned problems, but others that have not been assigned as well. It is your journey. Your prowess in physics will be directly related to your time spent practicing. You would not expect someone to play Mozart without practicing for hours and hours each day, and so it is with a scientist. Scientists must practice their trade, and physics is one of the key elements in any scientist's toolkit.

The frequency of the practice is also important because to absorb the physics and become competent, you need constant exposure. You should practice physics at least one hour every day, and not in large stretches just before midterms. Neither should you spend five hours studying physics Thursday because you skipped Sunday through Wednesday, nor can you get by reading one hour a day. You must read a section and work the problems on the material in that section, repeating the process every day. It does no good to just sit and read; you must get out the pencil and work the problems. As a physicist once said, "Learning starts at the pencil and works its way up my arm." Let's say you have a problem set due each Friday. After lecture on Friday afternoon, spend half an hour or so reading the next week's assignment and attempting one of the easy problems. If you spend just a little bit of time each day reading and doing the problems, then by Sunday night you'll probably have four or five problems done for your next week's assignment. If you get stuck, you still have the whole week to get help from a teaching assistant or your instructor. On the other hand, if you wait until Thursday night, you'll only have yourself to consult—and to blame for the ensuing disaster.

You will find that your learning will be more thorough and easily retained if you work a little bit each day. Set a goal to work two problems a day, and take them one at a time. It is not hard to motivate yourself to work a problem or two, but it is surely hard to motivate yourself to work the whole set. Trying to bite off more than you can chew creates problems. A half an hour here and half an hour there, and you'll be much happier than facing that mountain of a problem set Thursday night—alone.

Study Technique

1. *Start early.*
   After class on the day your problem set is due, start on the next set.
2. *Tackle the reading and the problems together.*
   Work with pencil and paper. Work all the examples.
3. *Work with small chunks at a time.* You should work one or two problems a day. This gives you
   - time to get help if you are stuck
   - more time for your subconscious to mull over the problem
   - a more thorough learning experience due to the regular exposure to the material

1.2 SCIENTIFIC NOTATION AND SIGNIFICANT FIGURES

Although science is considered to be very precise, our knowledge of the universe is often incomplete, based on finite measurements—i.e., limited to some extent, in a real world. This has been parodied by the character Mr. Spock in the *Star Trek* episodes. When queried, he knows any piece of data to many decimal places. Being precise, though, is not the same as giving your results to as many decimal places as your calculator can produce. Precision is not in the number of decimal places. It is in the careful reporting of your results in a manner that does not lead others astray.

For instance, if I divide 2.54 by 1.3, my calculator produces 1.9538462. However, it would be very imprecise to report 1.9538462 as my answer because I would be leading others astray. I only know the 1.3 (the least accurate number) to two significant figures; therefore, I should report my answer to two significant figures: 2.0 where I have rounded the 1.95 up to 2.0. The number of decimal places that are important is the **number of significant figures.** 2.54 has three significant figures, and 1.3 has two. A general rule is that your answer from the product or division of two numbers should never have more significant figures than the number possessing the least number of significant figures.

EXAMPLE 1-1

Find the product of 2.54 and 1.3.

Solution:

$$(2.54)(1.3) = 3.3$$

Note that even though the answer produced on a calculator is 3.302, the least accurate number is 1.3, which has two significant figures, so the result should be rounded to two significant figures.

Some points require special notice. For instance, 0.0000356 has only three significant figures. The four zeroes between the decimal point and the 3 show only the position of the decimal point and are not considered significant. Addition and subtraction demand special care when involving decimals because both are sensitive to the placement of the decimal point. We can see this by considering two examples.

EXAMPLE 1-2

Find the sum of 0.0235 and 0.0021.

Solution:

$$\begin{array}{r} 0.0235 \\ +\ 0.0021 \\ \hline 0.0256 \end{array}$$

Note that even though 0.0021 has only two significant figures, the answer has three. In this case, you do not round off the answer to two because the two numbers are known to the same number of decimal places.

EXAMPLE 1-3

Find the sum of 0.0236 and 0.321.

Solution:

$$\begin{array}{r} 0.0236 \\ +\ 0.321 \\ \hline 0.3446 \end{array} = 0.345$$

Here, the last 4 is rounded up to a 5 because of the 6. Even though the two numbers have the same number of significant figures, the sum must be truncated because one is not known to the same number of decimal places.

Usually, the formulas you will need to evaluate will not be as simple as the preceding examples, and they will contain many numbers, with few—if any—having the same number of significant figures. You could round off at each step, which would involve quite a bit more time, or you could wait until the end of the calculation and then round off the answer to the required significant figures. The latter approach will save significant amounts of time (no pun intended), and it will lead to a more precise answer. The reason can be seen by looking at the way modern computers add and multiply. A computer has a finite word size that represents the most digits, actually the largest integer, it can hold in a memory unit. If the result of floating-point calculation exceeds this value, it will truncate the result. Normally, the number of digits is so large that this limit imposes no problems. If, however, a calculation involves a large number of steps, and the numbers are truncated at each step, then over the length of the calculation, the "round-off" error, as it is called, can build up to give a result with no significant figures! This means that the number produced by the computer has nothing to do with reality. In the words of a renowned physicist, Hans Bethe, "The trouble with computers is you can believe them or not." They always give you an answer, but it may have little to do with what you want. To get around the round-off problem, most computers have an extended precision mode that allows you to do the calculation using numbers with more digits, and at the end of the calculation, you can round the answer off to the correct number of significant figures.

The point of this is that to avoid introducing a compounding round-off error into your calculations, you should wait until the end of your calculation to round the result to the correct number of significant figures.

Another problem with significant figures arises when the decimal point is left off. How many significant figures does 7200 have? There is no way to tell whether the zeroes are significant. A way to avoid this is to use scientific notation. A number is written as the first digit, followed by the decimal point, with all remaining digits multiplied by a power of ten. For example, 7200 could be written 7.2×10^3. If some of the zeroes are significant, they can be included explicitly. 7.20×10^3 has three significant figures, 7.2000×10^3 has five.

Scientific notation is also handy for representing either very large or very small numbers. For example, 2,300,000,000,000,000,000 can be written 2.3×10^{18}, and 0.0000000000000000023 as 2.3×10^{-16}. However, the real power of scientific notation lies in multiplication and division, for which such notation is a matter of adding or subtracting, respectively, the exponents.

EXAMPLE 1-4

Find **a)** $(2.0 \times 10^6)(3.0 \times 10^3)$, and **b)** $(3.0 \times 10^3)/(2.0 \times 10^6)$.

Solution:

a) $(2.0 \times 10^6)(3.0 \times 10^3) = 6.0 \times 10^{6+3} = 6.0 \times 10^9$

b) $(3.0 \times 10^3)/(2.0 \times 10^6) = 1.5 \times 10^{3-6} = 1.5 \times 10^3$

When adding or subtracting numbers in scientific notation, care must be taken to align the decimal places. This does not mean that you have to write out the numbers in decimal form, but you must write the numbers to the same power.

EXAMPLE 1-5

Find the sum of 7.34×10^9 and 2.4×10^7.

Solution:

First write them to the same power of ten, say 10^9. $2.4 \times 10^7 = 0.024 \times 10^9$.

$$\begin{array}{r} 7.34 \times 10^9 \\ +\ 0.024 \times 10^9 \\ \hline 7.36 \times 10^9 \end{array}$$

Note how the larger exponent (10^9) was used, so that the final answer was automatically in the correct form.

Now let's apply the rules to a real example. The units used are kilograms (kg), meters (m), seconds (s), and degrees Kelvin (K) and are explained in the next section.

EXAMPLE 1-6

Given the formula

$$v_{rms} = \sqrt{\frac{3kT}{m}}$$

and the data $k = 1.3807 \times 10^{-23}$ kg•m2/(s^2•K), room temperature T ≐ 293 K, and m = 4.7×10^{-26} kg for diatomic nitrogen molecules, find the root mean square speed of nitrogen molecules at room temperature.

Solution:

The three is an integer, so it is exact. We won't bother rounding until the end, when we will keep two significant digits because the mass m = 4.7×10^{-26}kg is only known to two significant figures. Taking the calculation step by step and showing the intermediate numerical results obtained on the calculator,

$$3(1.3807 \times 10^{-23}) = 4.1421 \times 10^{-23}.$$

$$4.1421 \times 10^{-23} \times 293 = 1.2136353 \times 10^{-20}$$

$$\frac{1.2136353\times10^{-20}}{4.7\times10^{-26}} = 258220.28$$

Now take the square root and round

$$v_{rms} = 508.15379 = 510 \text{ m/s}$$

The answer must be rounded to two significant figures because the mass is only known to two. Note that the units also follow the rules of algebra.

$$\sqrt{\frac{\frac{\text{kg}\cdot\text{m}^2}{\text{s}^2\cdot\text{K}}\times\text{K}}{\text{kg}}} = \sqrt{\frac{\text{m}^2}{\text{s}^2}} = \frac{\text{m}}{\text{s}}.$$

Following your units through the entire calculation is a very good habit to get into. It will save you from making many errors. For instance, in this example, if we had not obtained m/s for the speed, we would know we had made a mistake somewhere.

1.3 UNITS

The international system of units used in physics is the SI system *(Système Internationale d'Unités)*, which is also called the metric system. Within this system, there are fundamental quantities, such as length, mass, charge, and time, that cannot be written as combinations of other units. The SI units for these fundamental quantities are the meter (m) for length, the kilogram (kg) for mass, the coulomb (C) for charge, and the second (s) for time. Other units can be constructed from these fundamental units. For instance, the unit of speed is meter/second (m/s). The unit of force, the newton, is kilogram • meter/second squared (kg • m/s^2). Thus, the units of speed and force, being constructed from other quantities, are not fundamental.

The main advantage to the SI system is that it is a base-ten system of units, so that to convert units **within the system** requires only a movement of the decimal point. For example, a centimeter is 1/100 of a meter, so to convert 51.3 cm to meters requires movement of the decimal two places to the right, i.e., 51.3 cm = 0.513 m. This is much easier and more sensible than the English system, where you go from inches to feet to yards to miles, using conversions that bear not the slightest resemblance to one another. Unfortunately, the metric system is not the only system in the world. Many machinists must work on machines that use the English system of inches and feet. Thus, in order to get a device built, it is often necessary to use one system to do the calculations and to convert the results to another system. Also, you are probably more familiar with the English system than with the SI system, so it is a useful exercise to convert the metric units into English units to compare them.

The SI system is particularly well suited for scientific notation, and to exploit this property there are certain corresponding prefixes, such as *centi*, which means 1/100 or 10^{-2}) of whatever it precedes. The most commonly used prefixes are listed in Table 1.1, and you should memorize them because you will use them all the time.

Table 1.1 Metric Prefixes

Power	Prefix	Symbol
10^{-15}	femto	f
10^{-12}	pico	p
10^{-9}	nano	n
10^{-6}	micro	μ
10^{-3}	milli	m
10^{-2}	centi	c
10^{-1}	deci	d
10^{3}	kilo	k
10^{6}	mega	M
10^{9}	giga	G
10^{12}	tera	T

The easiest way to convert a unit from one system to another is to multiply and divide by 1 because this will not change the result and will only convert it to new units.

EXAMPLE 1-7

a) Convert 36.0 in. into centimeters. **b)** Convert 36.0 cm into in.

Solution:

a) In Table 1.2 we find 1 in. = 2.54 cm. We can convert this to 1 by trivially dividing through by 1 inch, so that

$$1 = \frac{2.54\text{cm}}{\text{in.}}$$

then multiply the 36 in. by 1:

$$(36.0 \text{ in.})(2.54 \text{ cm/in.}) = 91.4 \text{ cm}$$

b) This time, we want the centimeters to cancel, so we write 1 as:

$$1 = 1/(2.54 \text{ cm/in.}) = \frac{1\text{in.}}{2.54\text{cm}}$$

and then multiply by the 36 cm:

$$(36.0 \text{ cm})\frac{1\text{in.}}{2.54\text{cm}} = 14.2 \text{ in.}$$

The best way to become familiar with the metric system is to convert various units into something familiar. Following are some approximate sizes. You should extend the table using examples that are familiar and accessible to you.

Table 1.2 Conversions

Length	1 m = 3.23 ft 1 km = 0.62 mi 1 in. = 2.54 cm (exact) 1 mi = 5280 ft
Velocity	1 m/s = 2.24 mi/hr 30 mi/hr = 44 ft/s
Mass & Weight	1 kg = 0.0685 slugs 1 kg weighs 2.2 lb at sea level 1 N = 0.22 lb

Approximate SI Sizes

1 cm ≃ width of the nail on my small finger.
1 meter ≃ very large step.
Vertically, 1 meter comes up to my hip pocket.
1 pound ≃ half a kilogram.
1 gram ≃ the mass of two standard paper clips.
1 newton ≃ one quarter pound (the weight of 1 stick of butter).

EXAMPLE 1-8

Estimate the conversion factor between miles and kilometers.

Solution:

A meter is a bit larger than a yard, and there are approximately 1,700 yards in a mile. So a good guess would be 1600 meters to a mile. Thus,

$$1 \text{ mile} \simeq 1.6 \text{ kilometers}$$
or
$$3 \text{ miles} \simeq 5 \text{ kilometers}$$
or
$$1 \text{ kilometer} \simeq 5/8 \text{ mile}$$

The preceding example might appear to be very imprecise, and of course it is, but that does not mean that it is not science. In order to succeed, it is not necessary to know, say, π to 22 places, but it is important to know $\pi \simeq 3.1$ or 22/7. That is, it is necessary to know the "ballpark" figure for π. Likewise, it is good to know the ballpark conversions between the English and the metric systems. You must convert these new measures into something that has meaning to you, which brings up the next point.

1.4 A BIT OF ALGEBRA

In this section we will briefly review some of the troublesome aspects of algebra. This review is not a comprehensive treatment, but serves only to refresh your memory. The basic tenet of algebra is that you may do anything to an equation (except divide by zero) so long as you do the same thing to both sides of the equation. For instance, to solve the equation

$$x + 3 = 8 \tag{1-1}$$

for the variable x, we can subtract 3 from both sides of the equation, and will thus leave the x alone on the left-hand side giving

$$x + 3 - 3 = 8 - 3 \tag{1-2}$$

$$x = 5 \tag{1-3}$$

Note that to isolate x on the left-hand side of the equation involves using the inverse process, i.e., in equation (1-1) 3 is added to x, so in our solution, we must subtract the 3 in order to cancel it out of the left-hand side. Similarly, if a number multiplies the x, then we must use the inverse of multiplication which is division. In the equation

$$10x + 3 = 8 \tag{1-4}$$

first clear the 3

$$10x = 8 - 3 \tag{1-5}$$

and next divide both sides by 10

$$\frac{10x}{10} = \frac{5}{10} \tag{1-6}$$

$$x = \frac{1}{2} \tag{1-7}$$

The previous two examples represent very simple examples of basic algebra. Your skills should be sufficiently polished that you can solve equations like these without thinking. If you cannot, then you will need to practice. It is very important that you have a good grasp of the techniques algebra because we will use it all the time.

Fractions often pose a formidable hurdle. The top number in a fraction is the numerator, and the bottom number is the denominator. Your must remember that the denominators cannot be simply added, i.e.,

$$\frac{1}{3} + \frac{1}{5} \neq \frac{1}{8}$$

Don't make this mistake!

but must be converted to the lowest common denominator (LCD). 3 and 5 are both integer multiples of 15 which is the LCD in this case. Had the denominators been 3 and 9, then the LCD would be 9 because 9 is the lowest integer multiple of both 9 and 3.

Another major technique in algebra is that you can always multiply any quantity by a number divided by itself because it is 1. If we multiply both the numerator and denominator of the ⅓ by 5, we will not have changed the fraction, but we will have made the denominator 15, i.e.,

$$\frac{1}{3}+\frac{1}{5}=\frac{5}{5}\bullet\frac{1}{3}+\frac{3}{3}\bullet\frac{1}{5}$$

$$\frac{1}{3}+\frac{1}{5}=\frac{5}{15}+\frac{3}{15},$$

and similarly for the $1/5$, we have multiplied and divided by 3. Because both denominators are now the same, namely 15, the numerators can be added.

$$\frac{1}{3}+\frac{1}{5}=\frac{5+3}{15}=\frac{8}{15}.$$

EXAMPLE 1-9

Evaluate $\frac{a}{x^2y}+\frac{b}{y}$ by using the lowest common denominator.

Solution:

The LCD is x^2y, so we must convert the b/y to the proper form by multiplying the numerator and denominator by x^2.

$$\frac{a}{x^2y}+\frac{b}{y}=\frac{a}{x^2y}+\frac{b}{y}\bullet\frac{x^2}{x^2}$$

$$\frac{a}{x^2y}+\frac{b}{y}=\frac{a}{x^2y}+\frac{bx^2}{x^2y}$$

Next add the numerators

$$\frac{a}{x^2y}+\frac{b}{y}=\frac{a+bx^2}{x^2y}$$

This result is as far as the expression can be evaluated because we don't know the values of the variables. Note how complete equations were written at each step in this example. Always write complete equations. It makes your solution easier to follow, especially tomorrow, when you will have forgotten what you were doing today.

Another troublesome topic is square roots which are frequently used in geometry, vectors, and statistics, so we will review these next. Taking the square root is the inverse of squaring, that is, multiplying a quantity by itself.

$$x = \sqrt{x^2}$$

Problems arise when trying to simplify square roots. Sums cannot be pulled out of square roots, i.e.,

$$\sqrt{x + y} \neq \sqrt{x} + \sqrt{y}$$ **Don't make this mistake!**

but products can be simplified.

$$\sqrt{x \cdot y} = \sqrt{x} \cdot \sqrt{y} \tag{1-8}$$

When pulling variables out from under the square root, don't forget to take the square root of the variable being removed.

$$\sqrt{(x^3 + x^2)} = x\sqrt{x + 1} \tag{1-9}$$

because

$$\sqrt{(x^3 + x^2)} = \sqrt{x^2} \cdot \sqrt{x + 1} \tag{1-10}$$

When square roots appear in the denominator they should be "rationalized" so that the square root only appears in the numerator. For simple expressions this may be accomplished by multiplying and dividing by the square root of the expression. For instance,

$$\frac{1}{\sqrt{3}} = \frac{1}{\sqrt{3}} \cdot \frac{\sqrt{3}}{\sqrt{3}}$$

$$\frac{1}{\sqrt{3}} = \frac{\sqrt{3}}{3}$$

For more complex expressions of the form

$$\frac{1}{a + \sqrt{x + a}}$$

the expression in the denominator and numerator must be multiplied by the conjugate in which the sign of the square root has been changed.

$$\frac{1}{a + \sqrt{x + a}} = \frac{1}{a + \sqrt{x + a}} \cdot \frac{a - \sqrt{x + a}}{a - \sqrt{x + a}} \tag{1-11}$$

Multiplying out the denominator leaves

$$\frac{1}{a + \sqrt{x + a}} = \frac{a - \sqrt{x + a}}{a^2 - (x + a)} \tag{1-12}$$

Note that there is no longer any root in the denominator so the expression has been rationalized.

Heretofore, we have only solved linear equations, i.e., equations involving only the first power of x. We have also learned how to rationalize nonlinear expressions involving square roots; however, you will often run into quadratic equations, particularly in mechanics where the path followed by an object is often a parabola. As you will see in the next chapter, the position as a function of time might be given by an expression like

$$\boldsymbol{x} = \boldsymbol{x}_o + \boldsymbol{v}_o t + \frac{1}{2}\boldsymbol{a}t^2. \tag{1-13}$$

At this point if you have not had any physics, it is not important to know what the various letters represent other than to note that it is the equation of a parabola. For such an equation, you might know a particular value for $\boldsymbol{x}$ and want to find the t for that value of $\boldsymbol{x}$, which involves solving the quadratic equation. The simplest solution to a quadratic occurs when the quadratic equation can be factored. For instance, in the equation

$$x^2 + 5x + 6 = 0 \tag{1-14}$$

the 6 can be factored into $6 = 2 \cdot 3$, and the sum $2 + 3 = 5$ which is the coefficient middle term. So you would try

$$(x + 2)(x + 3) = 0 \tag{1-15}$$

Checking:
$$x^2 + 2x + 3x + 2 \cdot 3 = x^2 + 5x + 6. \tag{1-16}$$

We can make the left-hand side of equation (1-15) zero by letting x be

$$x = -2 \text{ or } -3$$

These values are termed the roots of the quadratic equation. If you cannot quickly factor the quadratic equation, it is not generally profitable to pursue that route further; rather, it is generally better to use the quadratic formula.

For the quadratic equation of the form

$$ax^2 + bx + c = 0 \tag{1-17}$$

the roots are:

$$x = \frac{-b \pm \sqrt{b^2 - 4ac}}{2a} \tag{1-18}$$

This formula works for finding the roots of any quadratic equation, even one that can be factored as illustrated in the next example.

EXAMPLE 1-10

Use the quadratic equation to find the roots of $x^2 + 5x + 6 = 0$.

Solution:

Start by comparing the equation to $ax^2 + bx + c = 0$. Then $a = 1$, $b = 5$, and $c = 6$. Substituting into the quadratic formula,

$$x = \frac{-5 \pm \sqrt{25 - 4 \cdot 1 \cdot 6}}{2 \cdot 1} = \frac{-5 \pm 1}{2}$$

Using the plus signed solution gives

$$x = -2$$

and the negative signed solution gives

$$x = -3$$

Solving quadratic equations is not always straight forward. Sometimes not all of the solutions given by the quadratic equation actually solve the original problem, and this stems from the $\pm\sqrt{}$ operation. Note that the equation $x^2 = 25$ has two roots, $x = +5$ and -5. Both may be admissible solutions, or only one may be admissible as shown by the next example.

EXAMPLE 1-11

Find the roots of the equation $x + \sqrt{x + 2} = 0$.

Solution:

At first this equation may not look like a quadratic equation at all, but the solution to this equation does involve a quadratic because we must clear the square root in order to solve the equation. There are a couple of ways to do this. You can multiply both sides of the equation by the conjugate, or you can isolate the square root on one side of the equation and square both sides. Following the second approach:

$$x = -\sqrt{x + 2}$$

Squaring both sides,

$$x^2 = x + 2$$

where you should note that we have lost the minus sign in the squaring of the right-hand side. Writing the equation in quadratic form,

$$x^2 - x - 2 = 0$$

This can be factored yielding

$$(x - 2)(x + 1) = 0$$

and producing the two roots

$$x = 2 \text{ and } -1$$

However, note that one of the roots does not solve our original equation, $x + \sqrt{x + 2} = 0$. Only $x = -1$ solves this equation. The other root solves $x - \sqrt{x + 2} = 0$, and was introduced by squaring the original equation.

When solving quadratic equations always check that both solutions are consistent with the original conditions of the problem.

1.5 THE BALLPARK AND THE BACK-OF-THE-ENVELOPE

Invariably, science progresses because someone gets an idea. Ideas often coalesce in a part of the brain not noted for its analytic abilities. When first created, ideas are not in a detailed form that you know will work; rather, you may have a strong "feeling" it will work. Your next step is to decide whether to spend the time pursuing the idea. You will find that time is your most precious quantity and you will not want to waste any of it. Therefore, this is a very important question, one that pervades all of science, particularly physics, where it is easy to be lured down the wrong path. The way to proceed is to do a "back-of-the-envelope-calculation," i.e., a calculation short enough to fit on the back of an envelope, one dealing with orders of magnitude, using only the major contributing factors, and ignoring small details. It is an educated guess, nothing more. The result is not precise, but if it is in the ballpark, there is a possibility the idea might work and should be pursued further with a detailed calculation.

Of course, in order for a back-of-the-envelope calculation to succeed, you must know what the ballpark is. At worst, the ballpark is within an order of magnitude of the result that a detailed calculation would produce, but this is not necessarily a hard-and-fast rule. *Order of magnitude* means within a factor of ten. Sometimes you will be able to define the ballpark more strictly. If we wanted to be within a factor of ten of 4,000, it would mean between a factor of ten greater and a factor of ten less, so the ballpark would be between 400 and 40,000. The limits are meant to be taken loosely. A value of 40,001 would still be considered in the ballpark, but a figure of 75,000 would not.

Having a good idea what the ballpark is can save you from making embarrassing mistakes, as shown in the next example.

EXAMPLE 1-12

On an exam, you are given the following problem:

A deuteron and a neutron collide head on. If the magnitude of the change in velocity of the deuteron is 5.67×10^3 m/s and that of the neutron is 1.13×10^4 m/s, then find the mass of the deuteron. The mass of the neutron is 1.68×10^{-27} kg.

Solution:

The actual solution involves conservation of momentum, but from the context of the problem it should be obvious that the mass of the deuteron must be in the same ballpark as that of the neutron. Because the change in velocity for the deuteron is roughly half that of the neutron, the deuteron must be about twice as massive. The important point is that you are looking for a mass about 10^{-27} kg. You would not write 5.64 kg for your answer because it is ridiculously outside the ballpark! Yet students often do just that; they put down a ridiculously wrong answer, and this is the result of not taking the time to estimate the ballpark for the answer.

Always estimate the ballpark!

Estimation is a fine art, and to be good at it requires long hours of practice and years of experience. So it is very important that you start practicing. In science, Enrico Fermi is considered to have been the king of estimation. One apocryphal story occurred during the Manhattan Project to develop the atomic bomb during World War II. One of the most difficult aspects of testing the first bomb was to measure the amount of energy produced so that the theories used to create the bomb could be verified, a difficult problem because any detector close to the bomb would be immediately destroyed by the terrific energy released. Two methods of detection were developed: fast detectors that could relay data before they were destroyed, and remote sensing techniques. These experimental aspects delayed the actual test and added considerably to the cost of the project, but they were very necessary. In the moments before the test, Enrico Fermi ripped a piece of paper into small bits, and fixed a meter stick to the bunker itself. Just before the shock from the blast hit the bunker, he released the paper bits, which fluttered downward in the still air. When the shock hit, he measured the relative displacement between the bunker and the fluttering paper bits. After a brief calculation, he had estimated the energy released to within 10% of the value obtained after months of analyzing the experimental data.

Whether the preceding story is true or just part of the legend is immaterial. Enrico Fermi certainly was good at estimating almost anything. People used to challenge him at cocktail parties. On one such occasion someone asked him to estimate the number of piano tuners in New York City. He produced a result remarkably close to the actual value. Consider in the next example how such an estimate might be made.

EXAMPLE 1-13

Estimate the number of piano tuners listed in the San Francisco city phone book.

Solution:

The number of piano tuners will, of course, be limited by the amount of work available, i.e., by the number of pianos that need tuning, so let us start by first estimating how many pianos need tuning in a year. There are approximately 2 million people in San Francisco. Individuals don't tend to own pianos; rather, families own pianos. (We can ignore the few single music majors). There might be five individuals per family, which leaves 4×10^5 families. Out of these, perhaps only one in ten owns a piano. We are now down to 4×10^4 pianos, but not all of the pianos will be tuned during the same interval of time. Some will be tuned every few months; some will go years without tuning. On the average, we can assume that a piano will be tuned once a year.

Next, estimate how many pianos a tuner can tune in a year. Assuming that a piano tuner can tune 4 per day, because there are approximately 200 working days per year, a piano tuner can tune 800 pianos a year. This means there is work for about $(4 \times 10^4$ pianos/year$) \div (8 \times 10^2$ pianos/tuner • year$) \simeq 50$ tuners. While this figure is not meant to be exact, we are probably within a factor of 10 (there are certainly more than 5 and fewer than 500).

Note the logic in this example. The number of piano tuners will be limited by the amount of work available, so we first need to find out how many pianos need tuning in a year. We then divide this figure by the number of pianos a tuner can tune in a year. This leaves the number of tuners needed to service the pianos in San Francisco. The actual count was 35 piano tuners. Actually, our estimate was probably closer than it appears because individual piano tuners are no longer listed. Businesses that do piano tuning are listed, and many of the larger businesses will probably have more than one piano tuner. As you can see, estimation involves experience, and that only comes from practice—lots of it! Don't be discouraged. Keep practicing. You will get better.

Another extremely useful tactic is to keep track of the units; they can be a great help in keeping the steps straight. Consider the following scientific examples.

EXAMPLE 1-14

Estimate the number of electrons in your body. *(Come on! How long have you had this body? And you don't know the number of electrons in it?)* Remember to try this example before looking at the solution.

Solution:

Avogadro's number gives $N_A \simeq 6 \times 10^{23}$ molecules/mole. This number is the place to start to estimate the number of atomic particles in any macroscopic chunk (*macroscopic* means stuff $\simeq$ our size, as contrasted with *microscopic*—much smaller than we are). By looking at the units, if I could figure the number of moles in my body, then I could get the number of molecules by multiplying N_A by the number of moles. From there, it is an easy step to get the number of electrons. We are mostly water, so let's find the mass of 1 mole of water. The chemical formula for water is H_2O. The atomic weight of oxygen is 16, so a mole of oxygen has a mass of 16 grams. The atomic mass of hydrogen is 1.0, so a mole of H_2O must have a mass of 18 $g = 18 \times 10^{-3}$ kg. If I have a mass of 70 kg, then I have

$$N = \frac{70000\text{g}}{18g/\text{mole}} \simeq 4 \times 10^3 \text{ mole}$$

in my body (note how the *g* units cancel). Next, get the number of electrons per mole. Oxygen has 8 electrons (8 protons and 8 neutrons in the nucleus give it an atomic mass of 16), so a water molecule must have 10 electrons. Thus,

$$(10 \text{ electrons/molecule})(6 \times 10^{23} \text{ molecules/mole}) = 6 \times 10^{24} \text{ electrons/mole.}$$

Now multiply by N to get the number of electrons:

$$N_e = (4 \times 10^3 \text{ moles})(6 \times 10^{24} \text{ electrons/mole}) \simeq 2 \times 10^{28} \text{ electrons}$$

Physical insight often comes from asking a very simple question. Back-of-the-envelope calculations, although relatively simple, often depend on your past experience. The greater the range of your past experiences, the easier back-of-the-envelope calculations become. For instance, you have lived all your life on this planet. What is its mass? Would you guess a hundred thousand kilograms, a million, a billion, 10^{20}, 10^{100}? Just how massive is the earth?

EXAMPLE 1-15

Estimate the mass of the earth.

Solution:

Where you start depends on your background knowledge. If you know some geography, you might start by estimating the size of the earth and guessing at its density. The first measurement of the earth's radius was done about the third century B.C. and is discussed in chapter 8. If you don't have any idea how big the earth is, it becomes exceedingly difficult to estimate the mass. Let's assume we know the earth's circumference. The earth is about 25,000 miles in circumference, and since $(2\pi r)$ = 25,000 mi, this gives a radius of about

$$r = \frac{C}{2\pi} = \frac{25000 \text{ mi}}{2\pi}$$

We are approximating, so let's make the arithmetic easy.

$$r \sim \frac{24000 \text{ mi}}{6} = 4{,}000 \text{ mi}$$

Or, in meters

$$r = 4{,}000 \text{ mi} \frac{5 \text{ km}}{3 \text{ mi}} = 6{,}000 \text{ km} = 6 \times 10^6 \text{ m}$$

Mass is density × volume:

$$M = \rho\, V$$

The volume of the earth is

$$\frac{4}{3}\pi\, r^3 = \frac{4}{3} 3\,(6 \times 10^6 \text{ m})^3$$

$$V \sim 4\,(200 \times 10^{18} \text{ m}^3) = 8 \times 10^{20} \text{ m}^3$$

What should we use for the density of the earth? Water has a density of 1000 kg/m^3. Iron has a density of about seven times that of water. Because the core of the earth is mostly iron, we would assume that the actual density of the earth is closer to iron than to water. We could try something like 6,000 kg/m^3:

$$M \simeq (6{,}000 \text{ kg/m}^3)(8 \times 10^{20} \text{ m}^3) = 5.4 \times 10^{24} \text{kg}$$

The actual mass of the earth is 5.98×10^{24} kg. Not bad!

On the other hand, had we known some elementary physics, we might have taken a different tack. You will learn later that the gravitational force of an object of mass M on another object of mass m is given by Newton's law of gravitation:

$$F = \frac{GM\,m}{r^2}$$

where G is a constant, 6.67×10^{-11} N•m^2/kg^2. Let M be the mass of the earth and m be the mass of an object attracted to the earth. This force will cause an acceleration of the mass m of 9.8 m/s^2. Newton's second law states that the net force will be equal to the mass times the acceleration:

$$\frac{GMm}{r^2} = m(9.8 \text{ m/s}^2)$$

Note that the mass of the object drops out, giving us

$$M = (9.8 \text{ m/s}^2)\frac{r^2}{G}$$

$$M \simeq (10 \text{ m/s}^2)\frac{(6 \times 10^6 \text{ m})^2}{(6 \times 10^{-11} \text{ N} \cdot \text{m}^2/\text{kg}^2)} = 6 \times 10^{24} \text{kg}$$

We've seen two different methods applied to the same problem. Which you use depends on your past experiences. Usually there are several methods that will yield a good estimate. Keep an open mind and try to expand your experiences as you progress through your physics class.

Note that behind a successful estimate is quite a bit of supplemental information. Being a skillful estimator relies on having a wealth of background information at your fingertips. At this point, you might not have the necessary background information to have worked the preceding two examples requiring detailed knowledge of atomic structure and of the gravitation force, but as you acquire scientific skills and practice the art of estimating, you will improve dramatically. You should always be trying to remember orders of magnitude for the major processes we encounter. For instance, to estimate whether an object will float, we compare the density of the object to that of the fluid. We need to have an idea of the density of various objects and fluids. The density of water is 1000 kg/m^3. It is easiest to remember other densities relative to water (specific gravity). Steel has a specific gravity of 7, meaning that steel is 7 times denser than water. Air has a specific gravity of 0.001. Ice has a specific gravity of 0.9, and wood generally lies in the range of 0.5 for balsa to 0.9 for walnut. However, ironwood is actually denser than water and therefore sinks.

1.6 WORD PROBLEMS!

You may have also noticed that estimation uses mathematics in a different manner than you have been taught. Mathematicians like to teach mathematics in an axiomatic fashion. Each new piece follows logically from the last, with the entire base of logic built upon a few solid axioms. Physics uses mathematics more as what the late Richard Feynman, a famous physicist who served on the Challenger investigation commission, called "Babylonian mathematics," where pieces of information are all networked together. The individual pieces of information follow from the fundamental laws, but if you have forgotten one piece or it has not yet been discovered, there are several alternative paths to get to that piece. For example, in going from point 1 to point 2 in Figure 1.1(b), your thought process might begin at 1, and then you may follow a very different path using the relations among pieces of information. You do not have to reach it by building up from the axioms. Perhaps you can get to that piece from the side or by working down from above.

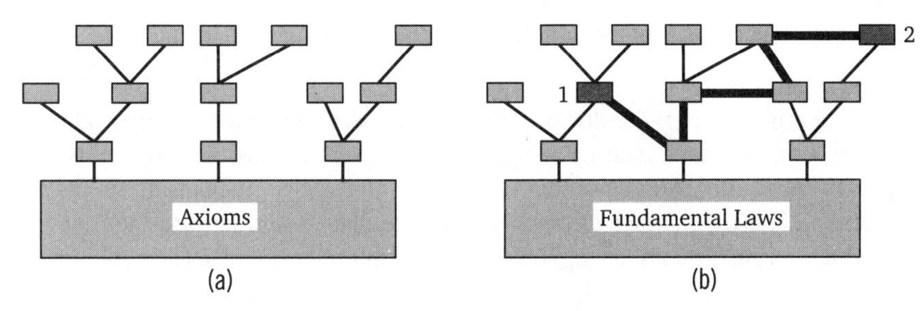

Figure 1.1
(a) Axiomatic mathematics and (b) Babylonian mathematics

Babylonian mathematics demands that you see connections among all sorts of different pieces of information you may possess. Very often, the problems you will encounter in physics are those that require a "Babylonian" approach. This approach makes physics problems different from those you have traditionally worked in mathematics. About the closest a traditional algebra class comes to doing Babylonian math is when doing word problems.

Word problems require a firm knowledge of the rules of algebra and a great deal of common sense. *Common sense* is a way of saying that it takes experience to be adept at working word problems. It takes lots of practice to sharpen those skills. There are some basic rules to solving word problems or physics problems in general. Very often, a student will start by searching for a formula. Don't make this mistake! Constructing or finding an appropriate formula is one of the last steps in problem solving. First, you must understand what is happening in the problem. This may involve figuring out in which direction an object is moving, or even what objects are moving. Are the objects being pulled or pushed? If you first visualize what is happening in the problem, then the mathematics usually follows quite naturally. If, however, you grab a formula and try to beat it into the problem, you will produce nothing worthwhile.

Following are some rules for problem solving: 1) In physics, always try to start with a general principle, such as Newton's second law or conservation of energy. You will learn about these principles as you continue on in physics. 2) Try to draw a picture to organize your thoughts and the given information. Write any given information on your picture. This helps to organize the given information into a readily identifiable state. 3) Try to identify all the major events in the drawing. Seeing where the major events occur in the problem can help you to break the problem into smaller parts in a natural manner. 4) Write down what you know about the problem, identifying unknowns. Unknowns are variables, usually able to assume a range of values, so use letters typically used for variables, such as x, y, or z. 5) Write an equation relating the variables. Use the general principle you identified in step 1. Use common sense. Sometimes this will require all of your mathematical skills. 6) Keep track of your unknowns and equations. You need one equation for each unknown, or you will wind up like a dog chasing its tail, wasting a lot of time.

Problem solving requires practice, so let's carefully work through some word problems and apply the technique we have just covered.

EXAMPLE 1-16

Some horses and chickens are in a barn. The total number of heads and wings equals the number of feet. However, there are 12 fewer wings than feet. How many horses and chickens are in the barn?

Solution:

Here, there is not really a general principle to write down, nor is it necessary to make a drawing other than to identify that a horse has four legs, one head, and no wings, but a chicken has two feet, two wings, and one head. What are the variables? We don't know how many horses or how many chickens are in the barn, so let's call the number of horses x and the number of chickens y. Note that we did not use the number of heads or wings or feet as variables because we are trying to solve for the number of horses and chickens.

The first information we are given is that the total number of heads and wings equal the number of feet. This we can organize into an equation:

$$\text{\# of heads} + \text{\# of wings} = \text{\# of feet}$$

$$(x + y) + 2y = 4x + 2y$$

Simplifying

$$y = 3x$$

This one equation has two unknowns, so we must obtain another equation to solve the problem. Is there any more information given? Yes, there are 12 fewer wings than feet.

$$\text{\# of wings} = \text{\# of feet} - 12$$

$$2y = (4x + 2y) - 12$$

Solving for x,

$$4x = 12$$

$$x = 3$$

So there are 3 horses. Now we can substitute this into the first result to find the number of chickens:

$$y = 3(3) = 9$$

EXAMPLE 1-17

A tank of water is to be emptied by siphoning using 3 hoses. When hose A is used alone, it takes 8 minutes to empty the tank; however, if all three hoses are used at once, it takes 2 minutes. Hose C has half the capacity of hose B. How long will it take to drain the tank with hose B alone?

Solution:

How fast the tank is emptied involves rates. The rate that water leaves the tank when three hoses are siphoning will be the sum of the individual rates. Although you cannot be expected to know it yet, the general principle involved is conservation of mass—which is a way of saying that mass cannot be created or destroyed, so the mass that disappears from the tank must be the sum of the mass that passes through the three hoses.

For a rate, we might try considering something like volume/time. We don't know the volume of the tank, so we can represent it with a variable, V. Hose A has a rate of $r_A = V/8$. We want to find the time it will take for hose B to empty the tank, so that time is represented by the symbol t. The rate B empties the tank is $r_B = V/t$. C has half the capacity of B, so its rate is $r_C = V/2t$. The net rate is then

$$r_{net} = \frac{V}{2 \text{ min}}$$

From the general principle

$$r_{net} = r_A + r_B + r_C.$$

$$\frac{V}{2} = \frac{V}{8} + \frac{V}{t} + \frac{V}{2t}$$

Note that the volume of the tank divides out of the problem, which is a lucky break for us because we were not given the volume. This simplification frequently occurs in physics problems. Very often, if you are not given a quantity you need, just use a variable to label it and continue on with the problem. It may very well cancel itself out. Now we are left with algebra. First clear the denominator by multiplying through by the lowest common denominator, $8t$, and then solve for t.

$$4t = t + 8 + 4$$

$$t = 4 \text{ min}$$

There is more to physics than just the mathematics. This means that often you will be asked to draw conclusions about the underlying physical principles, and these go beyond the mathematical formulas involved.

EXAMPLE 1-18

A hiker starting at point P wants to get to a forest cabin, which is 2 miles from the road at point Q. A straight road, 3 miles long, connects P and Q, as shown in Figure 1.2. The hiker can walk at 8 mi/hr on the road and 3 mi/hr through the forest. Suppose that the hiker walks a distance x down the road before cutting through the woods. **a)** Find an expression for the time it takes to make the trip in terms of x. **b)** Find the time it takes if the hiker hikes only through the forest. **c)** Find the time it takes if the hiker hikes to Q and then to the cabin. **d)** Why does the shortest distance not take the least amount of time?

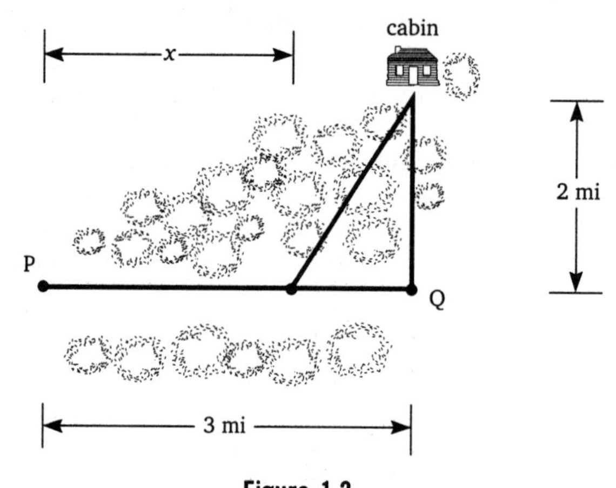

Figure 1.2

Solution:

The drawing should include the major events. In this problem, that includes where the hiker leaves the road. This spot is variable, so it can be placed anywhere on the diagram. You want a general solution, so don't pick the easiest place to solve the problem. Have the hiker leave the road at a place that is hard for you to calculate. In Figure 1.2 that is some place between P and Q, but not midway or at the end points. Don't give the picture "false" symmetry.

Because of the drawing, you can see that the problem breaks into two parts, so the time the hiker walks can be broken into the time hiking along the road and the time hiking through the forest. Realizing that the time must be broken into two parts is a key part of the solution for this problem.

$$t = t_r + t_f$$

a) We are given the distances, so we can use distance = rate × time. The hiker walks x along the road

$$t_r = \frac{x}{8}$$

and the distance hiked through the forest is the hypotenuse on the triangle in Figure 1.2, so we can use the Pythagorean theorem. One leg is 2 mi, and the other we don't know, but we can write it in terms of x as $3 - x$. The hypotenuse is then

$$r = \sqrt{(3 - x)^2 + 2^2}$$

$$t_f = \frac{\sqrt{(3 - x)^2 + 2^2}}{3}$$

The total time is

$$t = \frac{x}{8} + \frac{\sqrt{(3 - x)^2 + 2^2}}{3}$$

where x is in miles and t is in hours.

b) If the hiker hikes only through the forest, $x = 0$ in the formula we derived in step a.

$$t = \frac{\sqrt{9 + 4}}{3}\,\text{hr} = 1.2\,\text{hr}$$

c) If the hiker walks to point Q, then $x = 3$, and

$$t = \frac{3}{8} + \frac{2}{3} = 1.0\,\text{hr}$$

d) Because the hiker can walk more quickly along the road than through the forest, the hiker can get to the cabin in a shorter time by staying on the road until nearing the cabin.

Because of Fermat's principle, this same effect is responsible for refraction of light at a boundary in optics and leads to Snell's law. Fermat's principle states that a ray of light will follow the path connecting two points that takes the least amount of time. The path that takes the least time is not the shortest distance; rather, it is the path where the light spends extra time in the material in which it can travel fastest.

To review the rules for problem solving:

Problem–Solving Technique

1. Try to identify a general principle that applies to the problem.

2. Draw a picture and mark down on the picture the given information.

3. Your picture should include the major events in the problem.

4. Write down what you know about the problem, identifing relevant variables and unknowns. Use an appropriate symbol, such as an x or a y. For velocity, for instance, you might use a v.

5. Try to relate the given information and the variables in equations. Use the general principle for this step.

6. Remember, you need one equation for each unknown! Otherwise you can waste a lot of time in a loop, much like a dog chasing its own tail.

The rest of this book builds on the ideas in this chapter. You will be asked to see connections among ideas that you may not have realized are related, to use mathematics in a new and different way. Try your luck on the following problems, and remember that there will often be several different ways to reach any result. You will find Tables 1.1 and 1.2 useful at various times in the rest of the book.

Chapter 1 Problems

[1.1] Solve the following equations for x. All may be converted to linear equations.

a) $8x - 3 = 0$ **b)** $2x + 4 = 3x - 1$ **c)** $7x = \frac{7}{3} + \frac{14}{7}$ **d)** $\frac{1}{x} + 3 = 5$

e) $\frac{2x}{3} - \frac{3}{4} = \frac{x}{4}$ **f)** $\frac{1}{x} + \frac{1}{3x} = \frac{1}{9}$ **g)** $\frac{3}{\sqrt{x}} = 2$ **h)** $\sqrt{\frac{x+1}{3}} = 1$

[1.2] Solve the following equations for x.

a) $x^2 + 8x + 7 = 0$ **b)** $3x^2 - 5x = 2$ **c)** $x^2 - 7x + 3 = 0$ **d)** $\frac{3}{x} = \frac{x}{3}$

e) $\frac{1}{x} + x = 2$ **f)** $\sqrt{x+2} + 2 = x$ **g)** $\sqrt{x^2 - 1} + \sqrt{x^2 - 1} = x$ **h)** $\frac{x^2 + 2x + 1}{x^2 + x} = 3$

[1.3] Solve the following equations for x in terms of a, b and c.

a) $\frac{1}{x} + \frac{1}{3x} = \frac{1}{9a}$ **b)** $\frac{a}{x} - b = c$ **c)** $\frac{a(1 + \sqrt{x})}{x - 1} = b$ **d)** $\frac{\frac{1}{x} + \frac{1}{a}}{x + a} = 1$

[1.4] Estimate the amount of residential garbage produced in the United States each week.

[1.5] What is the total mass (in kilograms) of all the bicycles in your town?

[1.6] a) How many ping–pong balls will fit into your dresser drawer? Assume that each ping-pong ball is a 1 in. diameter sphere. The volume of a sphere is $(4/3)\pi r^3 \approx 4r^3$; however, we cannot pack the ping-pong balls into the cubic meter without leaving air spaces. If we include the air spaces, a ball + air space would form a cube approximately 1 in. on a side. So assume cubic ping-pong balls, 1 in. on each side. **b)** Using the same type of analysis as in step **a**, how many protons would it take to fill a hydrogen atom with protons? A proton has an effective radius of 10^{-15} m, and a hydrogen atom has an average radius of 10^{-10} m. What does your result tell you about atoms?

[1.7] How many hairs are on **a)** your arm, **b)** your head?

[1.8] How long in meters is a 70-mm feature-length film? Make a guess as to the rate at which the still picture frames are flashed on the screen and the duration of a feature film (time).

[1.9] What is your density in kilograms/cubic meter (kg/m^3)? If you know your weight, you can find your mass in kilograms. Next, find your volume. What size of cylinder could you pack your body into? The volume of a cylinder is $\pi r^2 h$. Is your body the same radius all over? How might you estimate the average radius of your body? A different path you might consider is that, because you float in water about like an ice cube does, your density should be about the same as that of ice.

• **[1.10]** What is the density of the ocean liner Queen Mary II? (See the hint at the end of Problem 1.8.)

[1.11] How many grains are in a handful of sand?

[1.12] What is the volume of the earth's atmosphere in m^3?

• **[1.13]** How many molecules of air are in a standard classroom?

[1.14] What volume of air in m^3 do you exhale from your lungs in a day?

[1.15] **a)** When sitting in a relaxed state, how many molecules of air do you breathe in 1 minute? **b)** After running a 50—meter wind sprint, how many molecules do you breathe in a minute?

[1.16] Each winter, an egg-drop contest is held at the physics building, in which eggs are dropped from the top of the five-story building. The object is to use materials to package the egg so that it will not break when it hits the ground. Estimate how fast in m/s an egg will be going just before it hits the ground when dropped from the top of such a building. For how long will the egg fall?

[1.17] How many standard paper clips are in 1 pound of clips?

• **[1.18]** Estimate the distance to the moon in the following manner. When the moon is well up in the sky, hold a ruler at arm's length in front of you and measure the diameter of the moon. Referring to Figure 1.3, use the similar triangles formed by your eye and the ruler and your eye and the moon's diameter to estimate the distance to the moon. Use the information in Example 1-15. *Hint: To complete the solution you will need to estimate the diameter of the moon. Is there any way to loosely estimate this from the radius of the earth?*

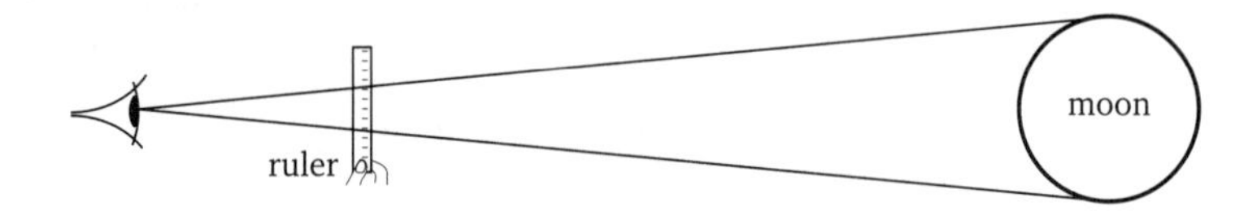

Figure 1.3
Problem 1.17. Estimate the distance to the moon.

• **[1.19]** Estimate the average density of a water molecule in kg/m^3. How does it compare to the density of water? From your estimate of the average density of the water molecule, estimate the mean diameter of the water molecule.

[1.20] Write an equation for the volume of a rectangular solid formed by cutting squares from the corners of a 10 in. × 20 in. rectangular piece of metal and folding up the sides.

[1.21] What two numbers whose sum is 8 have the maximum product?

• **[1.22]** A rectangular building is to be constructed on a right triangular lot, where the legs of the triangle are 50 ft and 100 ft long. Assume that one corner of the building is to be at the vertex of the right angle. **a)** Find an expression in one variable for the area of the largest rectangle that will fit on the lot. **b)** Graph the area function vs the variable locating the maximum area and the value of the variable where the maximum area occurs.

[1.23] Water freezes at 0° C or 32° F, and it boils at 100° C or 212° F. **a)** Derive a function to convert degrees Fahrenheit to degrees Celsius. Check it by converting 32° F and 212° F to Celsius. **b)** Use the formula to convert normal body temperature 98.6° F to Celsius.

CHAPTER 2
A Visit to Lineland

(What this place needs is a little dimension. Get the point?)

Submitted for your approval: one reader, a fun-loving math addict to be transported from an easy chair on Main Street to a one-dimensional land where all creatures are points. Here, you will investigate the motion of point objects along a line. You will meet the kinematic quantities of position, velocity, and acceleration. You will explore their geometrical basis and the relationships among these quantities. It is very important that you learn to interpret the motion through space and time in terms of graphs. You will plot quite a few graphs, but you must concentrate on interpreting them as well—looking at a graph and visualizing the physical path traveled by an object.

2.1 POSITION AND VELOCITY

A fundamental part of the description of any system is to describe what part of it is where, and when. A simple example is to drive between two cities in your car. If it takes you 2 hours to drive from one city to another, and you are driving with an average speed of 60 mph, how far did you travel between the two cities? From a dimensional analysis, we can see that multiplying the time by the average speed yields a distance of 120 miles.

$$x = 60\,\frac{\text{mi}}{\text{hr}} \bullet (2\text{ hr}) = 120\text{ mi}$$

where the hour in the numerator cancels the hour in the denominator, leaving dimensions of miles. We cannot, however, always rely on dimensional analysis for problems like this one. We need a theory that will predict the position at each moment in time. What quantities will be important? Obviously, the speed will play some sort of role, so we start by defining the average speed to be

$$\overline{v} = \frac{\text{total distance}}{\text{total time}} \qquad \text{average speed} \qquad (2\text{-}1)$$

The speed is always an unsigned number. Such a quantity is a *scalar*. The total distance and total time mean just that. If you drove at 60 mi/hr for 1 hour then rested for 1/2 hour, then drove at 60 for 1/2 hour, your average speed would not be 60 mi/hr; it would be

$$\bar{v} = \frac{60\text{mi} + 30\text{mi}}{2\text{hr}} = 45 \text{ mi/hr}$$

You must include the half-hour rest in the total time.

We can relate average speed to the position by introducing the concept of *displacement.* If $\mathbf{x}$ represents the position of an object at any instant, then the net distance traveled is the final position x minus the initial position $\mathbf{x}_0$, which we call the displacement $\Delta\mathbf{x}$:

$$\Delta\mathbf{x} = \mathbf{x} - \mathbf{x}_0. \qquad \text{displacement} \qquad (2\text{-}2)$$

Note that we have defined the final position as a variable x because we will find different final positions at different times. The initial position $\mathbf{x}_0$, once given, does not change. It is a constant.

Unlike the speed, the displacement has a direction associated with it. If $\mathbf{x}_0$ is greater than the final position $\mathbf{x}$, then the displacement is in the negative $\mathbf{x}$ direction, opposite to a positive displacement. Thus, the displacement has a length, termed its *magnitude*, and a *direction* (+ = forward, – = backward). Any quantity having a magnitude and a direction is a *vector.* We can represent a vector in 1 dimension using boldface type for the vector; for example x would represent a position vector. The position is a vector because we can be at a positive position or a negative position.

> A Scalar is any quantity that can represented by a number. A scalar has a size but not a direction.
>
> A Vector is any quantity that has both size and direction. The size is the *magnitude,* and is the scaler part. In one dimension, the direction is represented by the sign—positive or negative.

From the displacement we define a velocity, and to distinguish it from s*peed,* which is a scalar, we use bold type. The bar denotes that it is an average.

$$\bar{v} = \frac{\Delta\mathbf{x}}{\Delta t} \qquad \text{velocity} \qquad (2\text{-}3)$$

$\bar{v}$ will also have a direction, so we will have to give it a new name, *velocity.* The *speed* is just the *magnitude* of the *velocity.* It is important to note that there is a direction (+, –) even for one-dimensional motion. *Any* one-dimensional positive-vector quantity (displacement, velocity, acceleration) points in the direction of increasing $\mathbf{x}$. Position is a vector. In Figure 2.1 the gray ball has a position of –2, but because it is moving 3 m/s toward increasing $\mathbf{x}$, both the displacement and the velocity must be positive. Do the displacement and velocity always have the same direction? (After working it out, see the footnote *.) We treat vectors in greater detail in subsequent chapters.

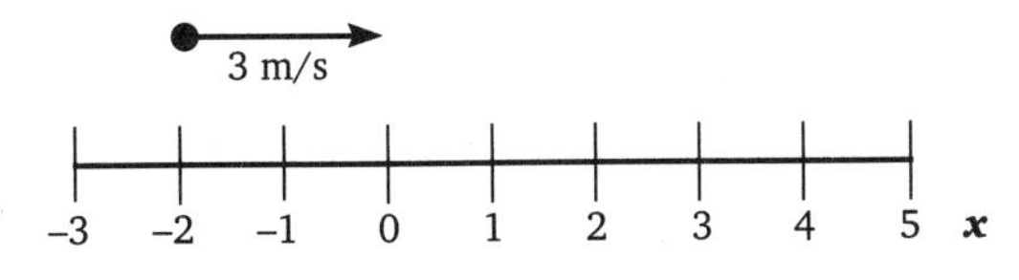

Figure 2.1
A ball with a negative position and a positive velocity

What we are really after is our position as a function of time. We can solve the velocity formula for the final position by using a very simple technique that will turn out to be very powerful when we use calculus. You have learned very early that any result will be unchanged if you multiply and divide by the same number. Hence, let's multiply and divide the displacement by the time, such that,

$$\Delta \mathbf{x} = \frac{\Delta \mathbf{x}}{t}\, t$$

where $\Delta x/t$ is the average velocity so that

$$\mathbf{x} - \mathbf{x}_0 = \mathbf{v}t$$

$$\mathbf{x} = \mathbf{x}_0 + \mathbf{v}t \qquad \text{position as a function of time} \qquad (2\text{-}4)$$

where $\mathbf{x}_0$ is the position at $t = 0$. We see from our simple trick $\Delta \mathbf{x} = (\Delta \mathbf{x}/t)\, t$ that to find the displacement $\Delta \mathbf{x}$ we need to know the velocity $\mathbf{v} = \Delta \mathbf{x}/\Delta t$. Perhaps it is not such a simple trick after all, but, is something quite powerful.

2.2 VELOCITY AS THE SLOPE OF THE POSITION-TIME GRAPH

Geometrically, Equation (2-4) is easy to interpret. It is nothing more than the equation for a straight line with intercept $\mathbf{x}_0$ and slope $\mathbf{v}$. Note that if the velocity is not constant the graph of x vs t will not be a straight line.

The velocity $\mathbf{v} = \Delta \mathbf{x}/\Delta t$ has the geometrical interpretation of being the slope of the position–versus–time graph (rise over run). An interesting consequence of our treatment of direction is that slope is a vector, and therefore it can be positive or negative. Constant speed means that during any time interval of duration Δt, the object will travel the same distance interval $\Delta \mathbf{x}$ so that $\dfrac{\Delta \mathbf{x}}{\Delta t}$ is constant. It is important to keep in mind that the graph in Figure 2.2 is a graph of position versus time and is not a trajectory; i.e, Figure 2.2 does not show a path of sequential positions in a two-dimensional position space.

* The direction of your displacement is the direction in which you are moving, which is also the direction of the velocity. Velocity and displacement must always have the same direction.

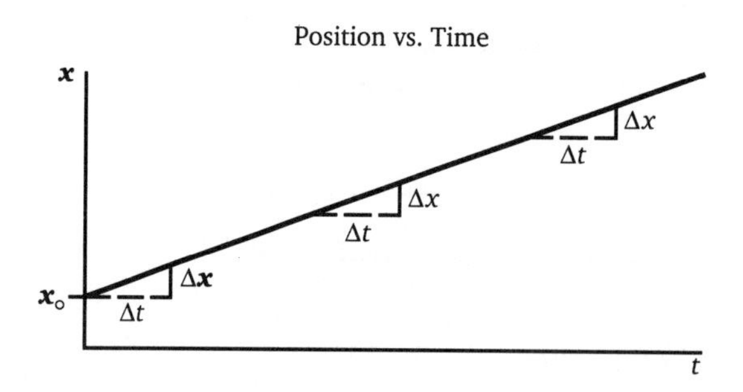

Figure 2.2
For a constant velocity, equal distances are covered in equal time intervals.

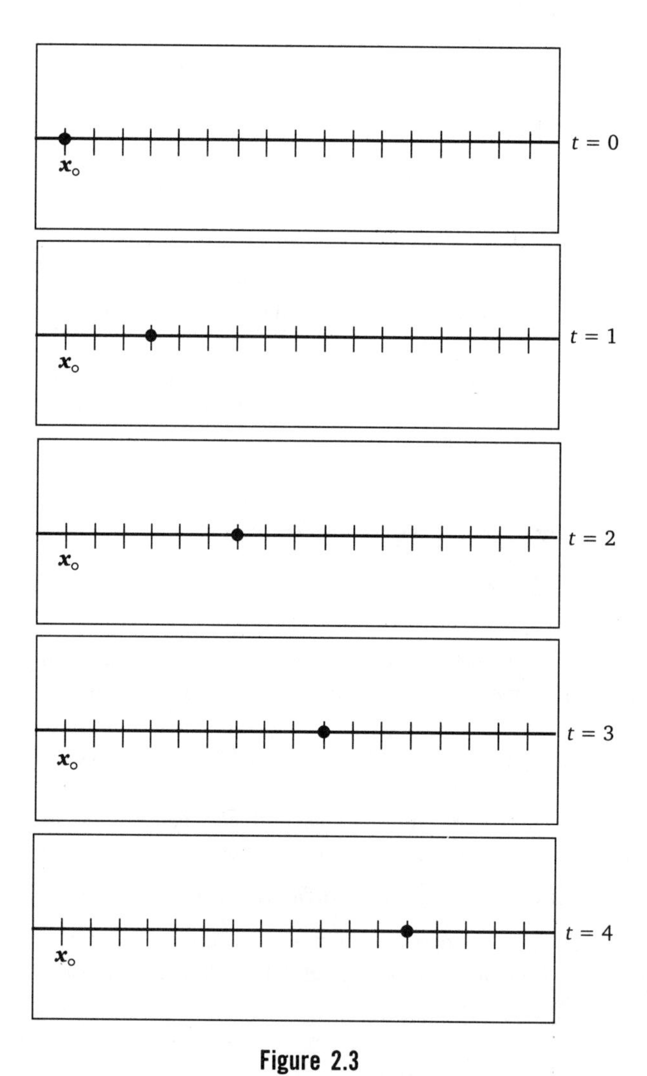

Figure 2.3
Successive snapshots of the displacement of a bead moving with constant speed.

Consider a bead moving with constant velocity along a straight wire. Constant velocity means that it must make equal displacements in equal time intervals, as shown in Figure 2.3. Each successive frame is a view of the bead at a later moment in time, and the intervals between frames are equal. In contrast, the graph in Figure 2.2 is more abstract; it represents the motion of the center point of the bead in time. In this chapter, when we talk of a *bead*, we really mean a one-dimensional pointlike particle that moves along a line. It is easier to visualize, however, if we refer to a bead moving along a wire. It takes some practice, but you must learn how to use a graph of position versus time to visualize the actual motion in space. We work on this visualization extensively in this chapter.

2.3 ACCELERATION

Now suppose that $\boldsymbol{v}$ changes in time. We need something to define the rate at which $\boldsymbol{v}$ changes. This is the *acceleration* $\boldsymbol{a}$:

$$\boldsymbol{a} = \frac{\Delta \boldsymbol{v}}{\Delta t} \qquad \text{acceleration} \qquad (2\text{-}5)$$

Acceleration is also a vector quantity. You are probably accustomed to using "accelerate" strictly to mean "speed up," and "decelerate" to mean "slow down." When acceleration is a vector, however, it can mean to speed up or to slow down, depending on its direction. To see what I mean, consider mathematically how the acceleration affects the velocity.

Suppose that an object is subjected to a constant acceleration and we wish to find the velocity. Again, we use our simple algebra trick of multiplying and dividing the velocity by the time:

$$\boldsymbol{v} - \boldsymbol{v}_0 = \frac{\Delta \boldsymbol{v}}{t}\, t$$

Obviously, to find $\boldsymbol{v}$ we need the acceleration $\boldsymbol{a} = \frac{\Delta \boldsymbol{v}}{t}$ which implies

$$\boldsymbol{v} - \boldsymbol{v}_0 = \boldsymbol{a}t \qquad (2\text{-}6)$$

where the naught on $\boldsymbol{v}_0$ means the velocity at time $t = 0$, and $\boldsymbol{v}$ is a variable velocity. Solving for $\boldsymbol{v}$:

$$\boldsymbol{v} = \boldsymbol{v}_0 + \boldsymbol{a}t \qquad \text{instantaneous velocity} \qquad (2\text{-}7)$$

From Equation (2-7) we see that if the acceleration is positive and the initial velocity $\boldsymbol{v}_0$ is positive, then the velocity $\boldsymbol{v}$ will increase in time, i.e., speed up. On the other hand, if $\boldsymbol{v}_0$ is positive and $\boldsymbol{a}$ is negative, $\boldsymbol{v}$ will decrease. The speed will also decrease if $\boldsymbol{v}_0$ is negative and $\boldsymbol{a}$ is positive, as in Figure 2.4. So what is important is not whether $\boldsymbol{a}$ is positive or negative, but rather, its direction relative to $\boldsymbol{v}$. If the acceleration is in the same direction as the velocity, then the object will speed up; conversely, if the acceleration is in a direction opposite to the velocity, the object will slow down. This can be confusing. Before going on, get up out of your chair and demonstrate your understanding to your roommate by walking across the room with a negative initial velocity and a positive acceleration.

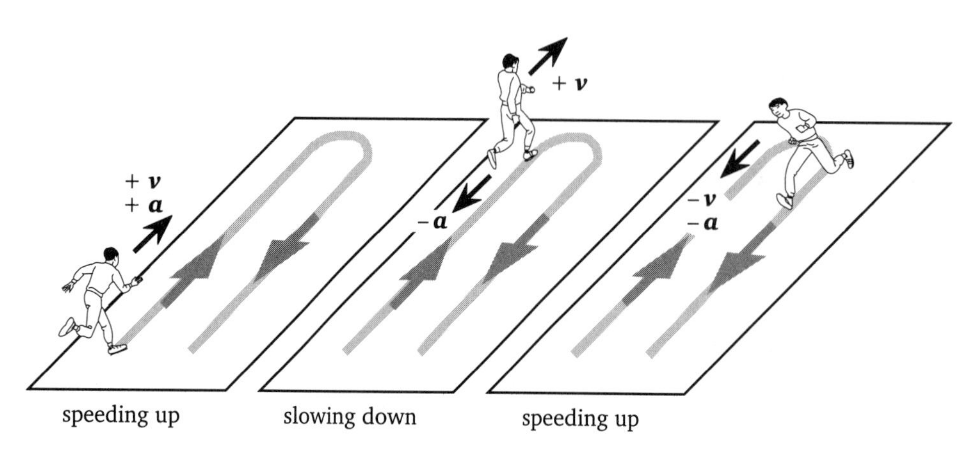

Figure 2.4
When the velocity and acceleration point in the same direction, the person speeds up. When the velocity and acceleration point in opposite directions, the person slows down. Speeding up and slowing down an object are both cases of accelerating the object.

The velocity-versus-time graph for Equation (2-7) looks qualitatively like the graph for position versus time in Figure 2.2. As an exercise, take a few moments to sketch the velocity-versus-time graph. Be sure to label the $\boldsymbol{v}$ intercept. What is the slope of the line?

At this point you might be confused, because we now have two equations for the velocity. We have the definition of the velocity—Equation (2-3), $\overline{v} = \Delta x/\Delta t$, and Equation (2-7). Which do we use? In Equation (2-7), the velocity continuously changes as time evolves, so it must represent the velocity at some particular instant in time. The velocity given in Equation (2-7) is thus termed the *instantaneous velocity*. Equation (2-3) is the *average velocity* and is the total displacement divided by the total time interval. Which equation you use depends on the information you have. If you have only the total displacement and the total time, then you can calculate only the average velocity. On the other hand, if you know the acceleration, the initial velocity, and how long the body accelerated, then you can calculate $v = v_0 + at$. For cases when the velocity is constant, the average velocity and the instantaneous velocity are the same because the acceleration is zero.

The difference between the instantaneous velocity and the average velocity is illustrated in the following example.

EXAMPLE 2-1

From the graph of position vs time in Figure 2.5, calculate the average speed between points A and D, and find the instantaneous speed and the acceleration at points A, B, C, and D.

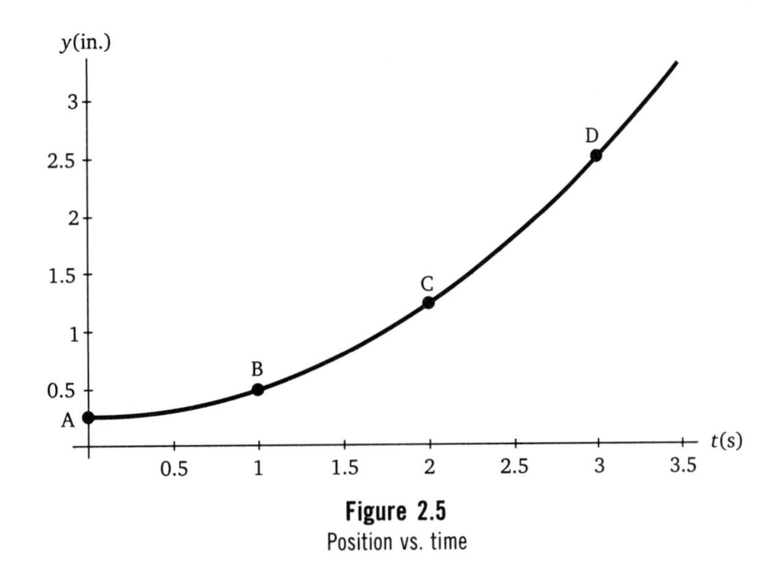

Figure 2.5
Position vs. time

Solution:

The average velocity is the total displacement divided by the total time required to travel that distance. The average velocity for traveling the distance from A to D is

$$\overline{v} = (2.5 - 0.25 \text{ in.})/3 \text{ s} = (9/4) \text{ in.}/3\text{s} = 0.75 \text{ in./s}$$

Now, to find the instantaneous velocity, we need to measure the derivative—i.e., the slope of y vs t. This can best be accomplished by drawing a line tangent to the curve at the point where you want to find the slope and to measure the slope of that tangent line. In Figure 2.6, tangent lines are drawn at the points, A and C and the slope is computed.

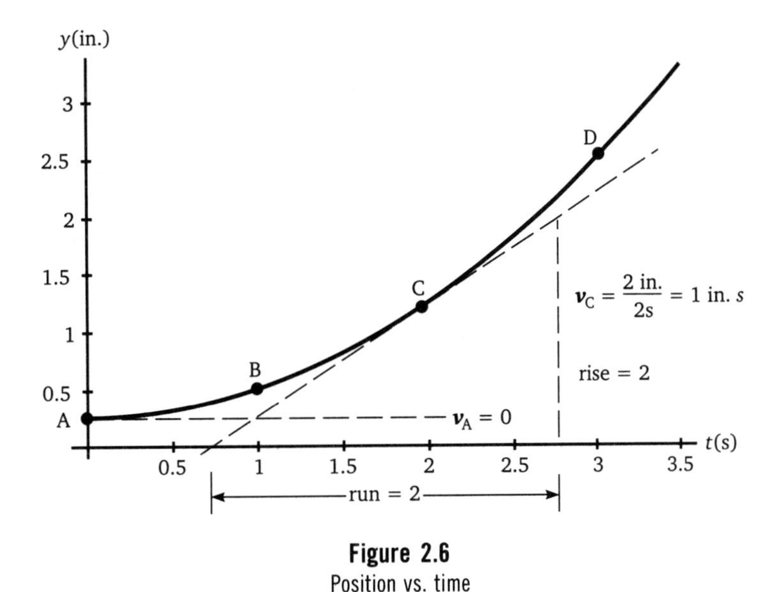

Figure 2.6
Position vs. time

From the slopes: $v_A = 0$, $v_B = ½$ in./s, $v_C = 1$ in./s, $v_D = 3$ in./2 s. The slope of the graph at A is zero. The process of obtaining the slope at C is illustrated in Figure 2.6. To get the slope, you must determine the rise/run of the tangent line. As you can see, reading slopes from a graph is a very inexact science at best. We can get the acceleration next by plotting the instantaneous velocity vs time and computing the slope of that curve.

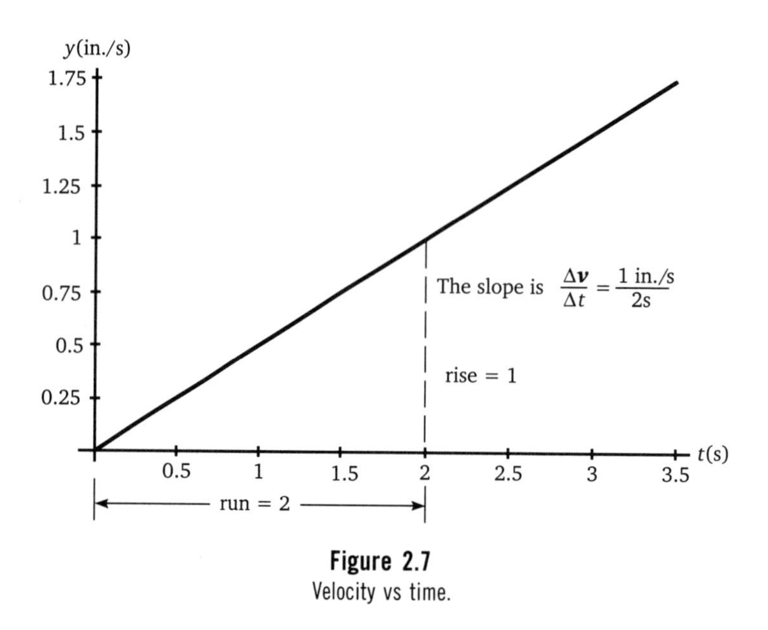

Figure 2.7
Velocity vs time.

From the graph, it can be easily seen that the slope is ½ and constant, so the acceleration is $\boldsymbol{a}$ = ½ in/s^2. Note that the position-vs-time plot (Figure 2.5) curves upward, implying a positive curvature that agrees with our positive acceleration.

It is important to understand the difference between the average velocity and the instantaneous velocity. For instance, suppose we wish to find the position as a function of time. The velocity changes continuously. Which velocity do we use in our formula

$$\boldsymbol{x} = \boldsymbol{x}_o + \boldsymbol{v}t \text{ ?} \tag{2-8}$$

If we use the instantaneous $\boldsymbol{v}$ at time t, we will grossly overestimate the distance traveled because $\boldsymbol{v}$ is the velocity at the end of the interval and the bead did not have that velocity for the whole interval. It started at $\boldsymbol{v}_o$ and only ended with $\boldsymbol{v}$. Any ideas?

Note that $\boldsymbol{x} - \boldsymbol{x}_o$ is the displacement. Does that help? The displacement is related to the average velocity.

$$\overline{\boldsymbol{v}} = \frac{\boldsymbol{x} - \boldsymbol{x}_o}{t}$$

$$\overline{\boldsymbol{v}} = \frac{\text{total displacement}}{\text{total time}} \qquad \text{average velocity} \tag{2-9}$$

It is the average velocity that must be used in Equation (2-8). Let's use the simple trick we employed before. We want to know how the position depends on time, so multiply and divide $\Delta \boldsymbol{x}$ by t:

$$\Delta x = \frac{\Delta x}{t}\,t \qquad (2\text{-}10)$$

where $\Delta x/t$ is the total displacement divided by the time, which is the average velocity. Thus, our formula tells us which velocity we need to use.

The velocity is a linear function of time, which means that the average velocity will be the actual value of the velocity at the midpoint of the time interval $[0,t]$ shown in Figure 2.8. Because the equation of the velocity line can be written as

$$v = \frac{v - v_0}{t}\,t + v_0$$

(slope × t + intercept), at the moment $\frac{1}{2}t$,

$$v = \frac{(v - v_0)}{t}\left(\frac{1}{2}t\right) + v_0 = \frac{1}{2}(v + v_0)$$

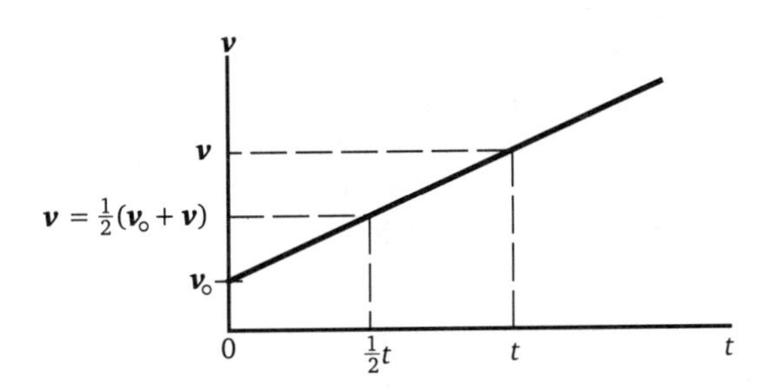

Figure 2.8
For a constant acceleration, the velocity increases linearly, and the average velocity is halfway between v and v_0

From the figure, the average is

$$\overline{v} = \tfrac{1}{2}(v_0 + v) \qquad (2\text{-}11)$$

and occurs at the midpoint in time, $\frac{1}{2}t$. Using this average velocity in Equation (2-10), we find:

$$x - x_0 = \overline{v}\,t$$

$$x = x_0 + \tfrac{1}{2}v_0 t + \tfrac{1}{2}vt\,. \qquad (2\text{-}12)$$

Substituting $v = v_0 + at$ into Equation (2-12):

$$x = x_0 + v_0 t + \tfrac{1}{2}at^2. \qquad \text{position for constant acceleration} \qquad (2\text{-}13)$$

The graph of x vs t is no longer a straight line. The graph of Equation (2-13) is a parabola, as shown in Figure 2.9.

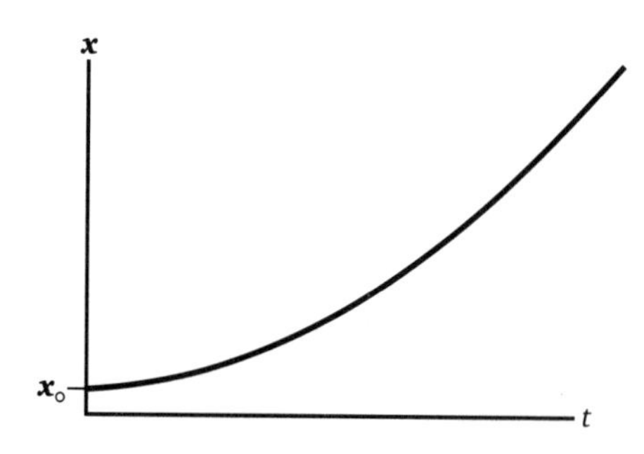

Figure 2.9
Position vs time for an object moving under a constant acceleration.

2.4 ACCELERATION AS THE CURVATURE OF THE POSITION–TIME GRAPH

For positive acceleration, the graph shown in Figure 2.9 has positive curvature. This means that if you draw a tangent line at any point, a curve with positive curvature will curve up from the tangent line, as shown in Figure 2.10. In mathematics what we have called "curvature" is referred to as concavity, but concavity is less descriptive than curvature and is hard to pronounce. Thus, we will use curvature to mean concavity. Which way would the graph curve if the object were slowing down?

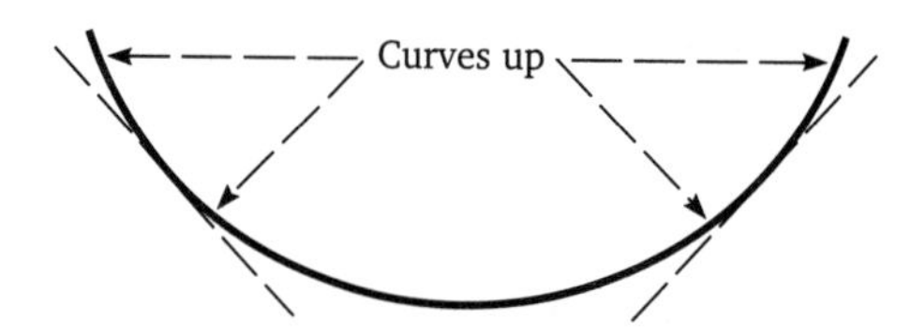

Figure 2.10
A curve with positive curvature always curves up at any point from a tangent line.

What does the curvature of the position-time curve mean? The graph in Figure 2.9 curves up because we have a velocity that is continually changing. The velocity at any given moment is different from that at any other moment. Somehow, we have to associate a velocity with each point in time, but we don't know yet how to make a displacement at a point in time, so how are we to proceed?

Note that in Figure 2.9, we could prove that the object is speeding up by considering time intervals, Δt and looking at the displacement during these time intervals. The average velocity obtained will actually be the instantaneous velocity at some time during the interval. The question is, at which time? From Figure 2.11, we can see that the slope of the dashed line drawn from $\boldsymbol{x_0}$ to $\boldsymbol{x}$ is the average velocity and is the same as the slope of the tangent to the position-time curve at the midpoint of the time interval.

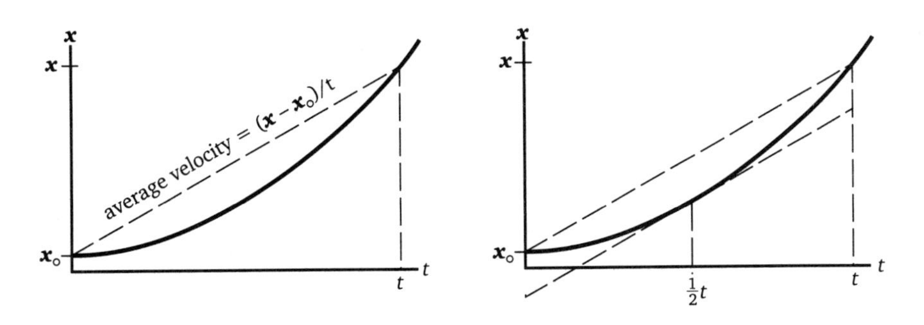

Figure 2.11
Average velocity equals the instantaneous velocity at the midpoint of the time interval where the position – time curve has the slope $(x - x_o)/t$.

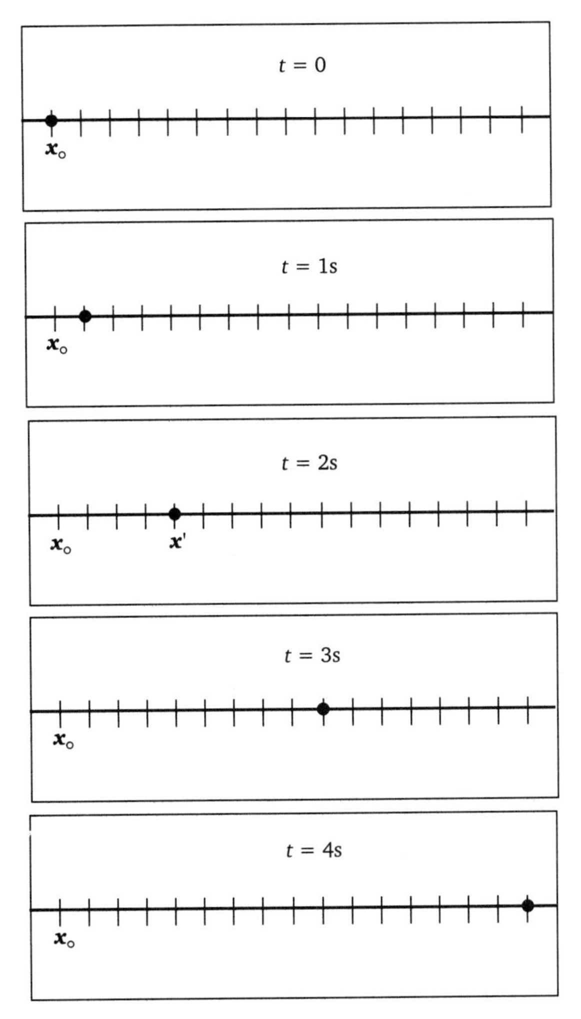

Figure 2.12
Motion of a bead subjected to a constant acceleration.

If we associate the actual *instantaneous* speed of the object with the slope of the tangent line to the position-time curve, then the average velocity is attained at the midpoint of the time interval, as required previously, i.e., the object will actually be moving with the average velocity at the midpoint of the time interval. The essence of the link between the acceleration and the curvature is contained in Figure 2.11 because, although the velocity is evenly split about $\frac{1}{2}t$, the position is not! In the first half of the time interval, the object only travels to $\mathbf{x}'$ and not to the midpoint of the distance interval. Of course, this is a property of any curve that curves up; necessarily, the distance covered in the second half of the time interval must be greater than in the first half if the graph is to curve up. It is the acceleration that causes the graph to curve up, so the acceleration is interpreted as the curvature of the position-time curve, and, necessarily, that implies that an object with positive acceleration will cover more distance with each successive time interval. In Figure 2.12, the bead starts at $\mathbf{x}_o$ at $t = 0$, with zero velocity, and covers one distance interval in 1 second. Where will it be after 2s? 3s? Make a prediction before looking at the figure.

An effective if inane way to remember the convention for curvature is

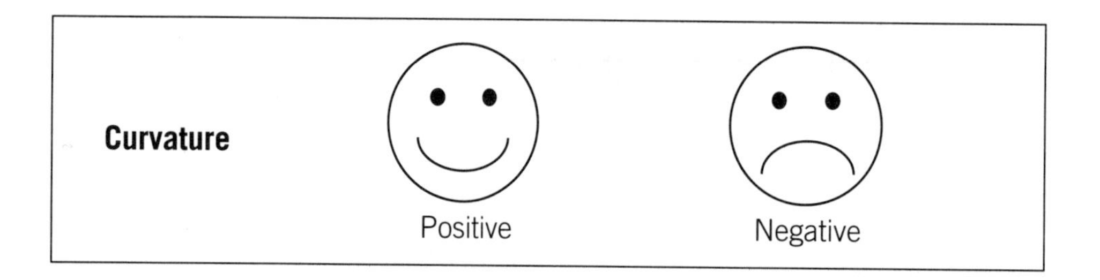

EXAMPLE 2-2

A graph for $\mathbf{x}$ vs t is shown for $\mathbf{x} = \frac{1}{2}\,\mathbf{a}t^2$ in Figure 2.13. Graph **a)** $\mathbf{x} = \mathbf{x}_o + \frac{1}{2}\,\mathbf{a}t^2$ and **b)** $\mathbf{x} = \mathbf{x}_o + \mathbf{v}_o t + \frac{1}{2}\,\mathbf{a}t^2$. Be sure to indicate what $\mathbf{x}_o$ and $\mathbf{v}_o$ are on the graphs.

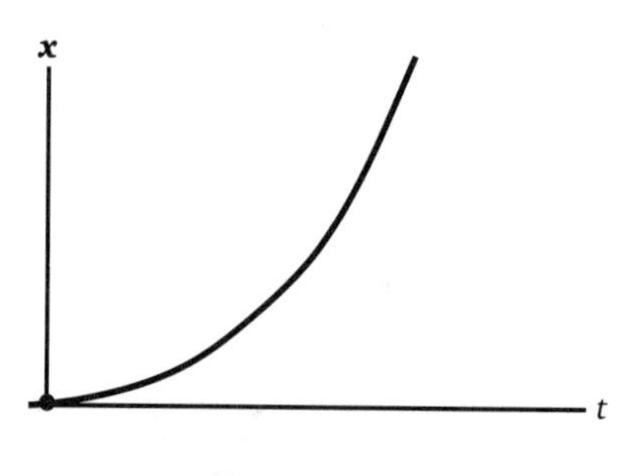

Figure 2.13
The graph of $x = at^2$.

Solution:

a) At $t = 0$, $\mathbf{x}(0) = \mathbf{x}_o$. So $\mathbf{x}_o$ shifts the given graph up the x axis a distance $\mathbf{x}_o$.

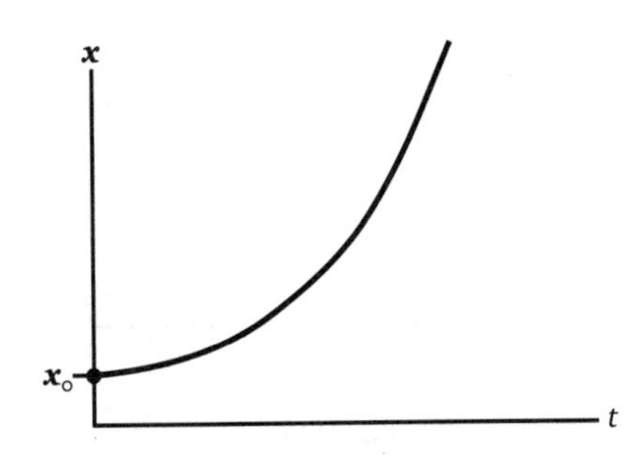

Figure 2.14
The graph of $x = x_o + at^2$.

b) When we add in the $\mathbf{v}_o t$ term, $\mathbf{x}(0) = \mathbf{x}_o$, so it doesn't change in the origin, but it does change the slope at $t = 0$. The slope of the position—time curve is the velocity. At $t = 0$ the velocity by definition is

$$\mathbf{v}\big|_{t=0} = \mathbf{v}_o$$

Thus, the $\mathbf{v}_o$ term gives the graph a positive slope at the origin. The graph has been moved down and to the left by the *($\mathbf{v}_o t$)* term as shown in Figure 2.15.

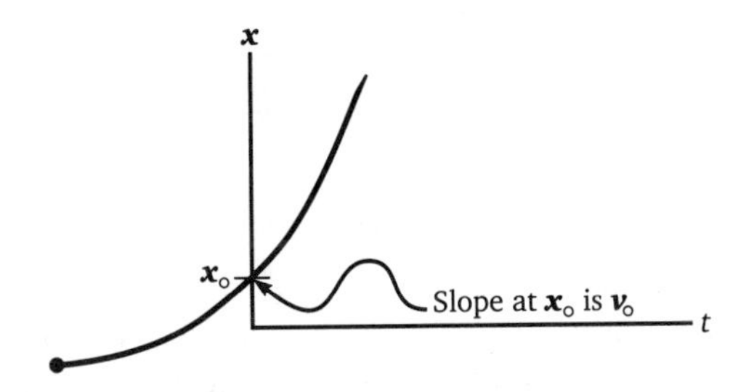

Figure 2.15
The graph of $\mathbf{x} = x_o + v_o t + at^2$.

We have found that the $\mathbf{x}_o$ term moves the curve up, and the $\mathbf{v}_o t$ term moves the graph down and to the left if $\mathbf{v}_o$ is positive. Which way would the graph be moved if $\mathbf{v}_o$ were negative?

EXAMPLE 2-3

Suppose that an object at time $t = 0$ is at the origin, with an initial positive velocity of $\mathbf{v}_o$ and is subjected to an acceleration that varies in time as shown in Figure 2.16. Draw the velocity and the position-vs-time curves that correspond to the acceleration and initial velocity

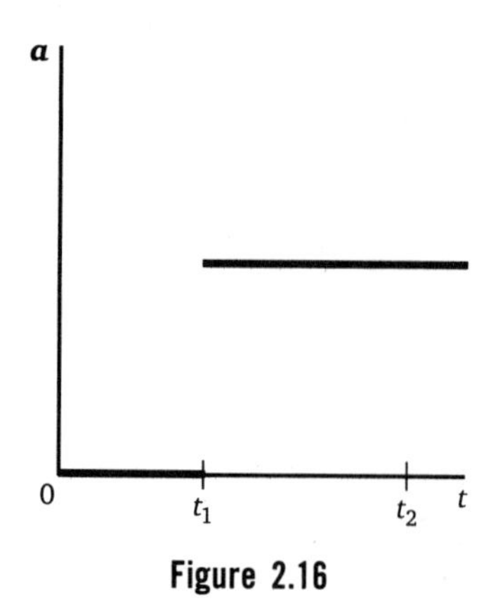

Figure 2.16

Solution:

From $t = 0$ to $t = t_1$, there is no acceleration, so the velocity must have its initial value $\mathbf{v}_o$. The graph will be a constant. From t_1 to t_2 the acceleration is a positive constant, so the velocity must have a constant positive slope.

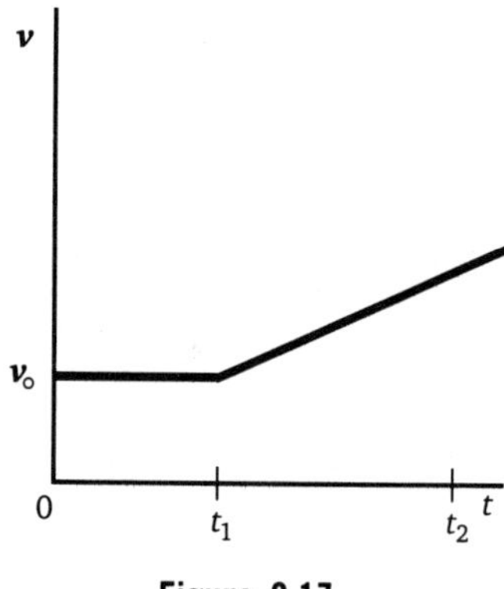

Figure 2.17

From $t = 0$ to $t = t_1$, the velocity has a positive value and it is the slope of the position graph, so the position must have a constant positive slope. Then from t_1 to t_2 the acceleration is the curvature of the position. Thus, the position graph must curve up from t_1 to t_2, producing the curve in Figure 2.18.

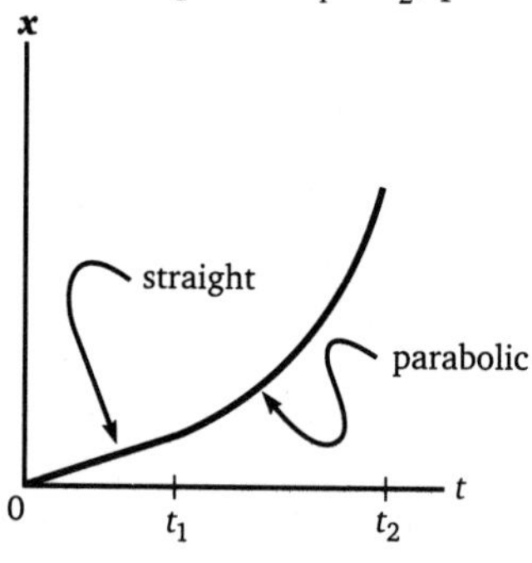

Figure 2.18

Geometrically, we have found that the velocity v_o is the initial slope of the position–vs–time graph, and the acceleration is the curvature.

Visual understanding of what each term does in an equation is a very powerful tool that will allow you to picture the graph of a function without plugging in numbers and calculating points. Often the shape of the curve is far more important and informative than the values of specific points. Whenever you come across new mathematics, you should try to visualize the equations.

2.5 GRAVITY

An important case of constant acceleration is that of an object falling freely near the surface of the earth. At sea level, the acceleration due to gravity is 9.8 m/s^2. There are slight variations in the acceleration with latitude and with elevation as shown in Table 2.1, but for our purposes we consider the acceleration to be a constant, the magnitude of which is denoted by the symbol g.

$$\begin{aligned} g &= 9.8 \text{ m/s}^2 \\ &= 32 \text{ ft/s}^2 \end{aligned} \tag{2-14}$$

TABLE 2.1 Variation in *g* with latitude and elevation

Location	Latitude	Elevation (m)	g(m/s^2)
New York	42° N	0	9.803
Denver	40° N	1650	9.796
Pikes Peak	40° N	4300	9.789
Equator	0°	0	9.780
North Pole (calculated)	90°	0	9.832

If we set up a coordinate system with the y axis pointing up, then the acceleration due to gravity will initially tend to slow down an object thrown straight up. Therefore, it must act opposite to the velocity i.e., in the negative y direction as in Figure 2.19. Using a negative acceleration, the kinematic equations become

$$v = v_o - at$$

$$v = v_o - gt \tag{2-15}$$

$$y = y_o + v_o t - \tfrac{1}{2} gt^2. \tag{2-16}$$

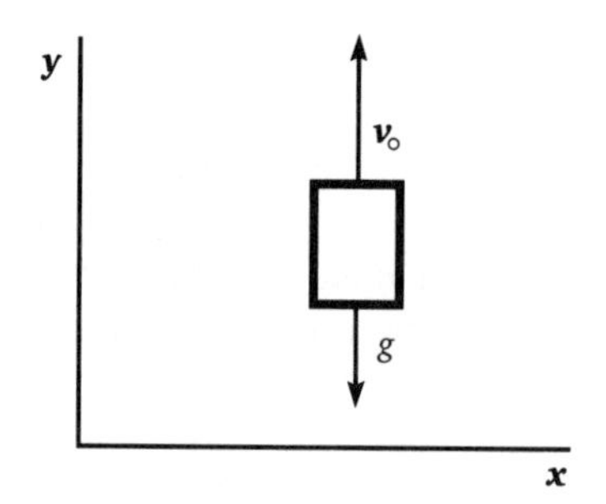

Figure 2.19
Acceleration due to gravity points down, opposite to the velocity of an object thrown straight up with initial velocity v_o.

The symbol g denotes only the magnitude. The downward direction of the acceleration is given by the minus sign in front of the g. The acceleration due to gravity at sea level is a constant in both magnitude and direction, meaning that it is always 9.8 m/s^2, and its direction is down. For a ball thrown straight up in the air, the velocity v is initially straight up, opposite to g, but as t increases, the ball will reach a time when $v_o = gt$ which means the velocity is zero and the ball can go no higher. What happens as time increases when $gt > v_o$? The velocity is negative. When we set up the coordinate system, we choose the positive direction, and once chosen, it cannot be changed. Therefore, if v is negative, it must mean the ball is moving down.

As we watch the time evolution of the ball, it starts moving up with positive velocity. Then at $gt = v_o$, it momentarily stops, and then starts to move down. At the highest point, the velocity of the ball is zero, but this does not mean that the acceleration is zero. The acceleration is still g! Remember that g is a constant. In Figure 2.20 arrows denoting the direction and relative strength of the velocity and acceleration are shown at various times during the ball's flight.

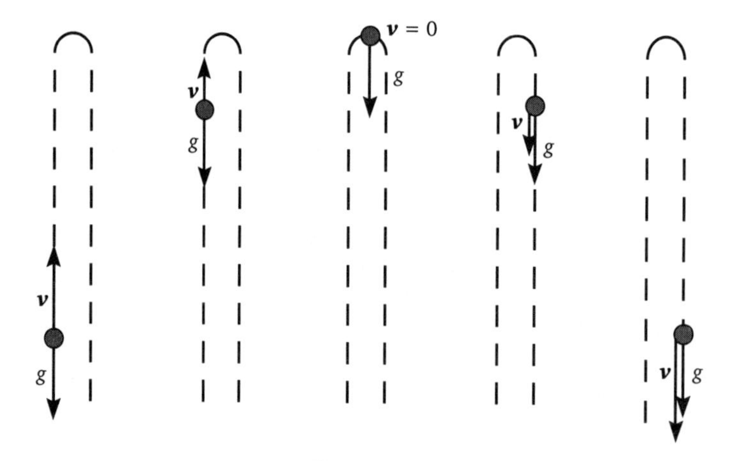

Figure 2.20
Relation between v and g at various moments during the flight of a ball thrown straight up. The up and down paths have been separated for clarity. They should lie on top of one another. The arrow for g is constant throughout the flight.

EXAMPLE 2-4

A ball is thrown straight up, with an initial velocity of 20 m/s. Find out how long it takes the ball to reach its highest point, and find how high the ball goes.

Solution:

If we try to use $\mathbf{y} = \mathbf{y}_o + \mathbf{v}_o t + \frac{1}{2}\,\mathbf{a}t^2$, we run into a roadblock right away. First, we don't know the final height y that the bead attains, and we don't know the time t it takes to attain this height —i.e., we have two unknowns and only one equation. We need another equation to solve this problem. The only other equation we have is

$$\mathbf{v} = \mathbf{v}_o + \mathbf{a}t.$$

If we could find the final velocity, then we could use this equation to solve for the time, and then substitute the time into the first equation to solve for the height y.

At the highest point, the ball is turning around, so its velocity is zero; however, the acceleration is still $-g$, not zero!

Figure 2.21
The velocity at the highest point is zero.

The time at the highest point can be found by setting $\mathbf{v}$ to zero, so that $0 = \mathbf{v}_o - gt$ and $t = \mathbf{v}_o/g$, and substituting for the values,

$$t = \frac{20\ \text{m/s}}{9.8\ \text{m/s}^2} = 2.0\ \text{s}$$

Note how the units cancel, leaving only units of seconds.

> **Units**
>
> Keeping track of the units can often tell you whether you have made a mistake. Always keep track of the units throughout a problem, and above all NEVER, EVER write an answer without units.

Now we can substitute into the first equation to find the height. Assume that the bead starts at zero height. Then $\mathbf{y}_o = 0$, and

$$\mathbf{y} = \mathbf{v}_o t - \tfrac{1}{2}\,gt^2 = (20\ \text{m/s})(2.0\ \text{s}) - \tfrac{1}{2}(9.8\ \text{m/s}^2)(2.0\ \text{s})^2$$

$$\mathbf{y} = 40\ \text{m} - 19.6\ \text{m} = 20.4\ \text{m}$$

EXAMPLE 2-5

During the annual egg-drop contest held from the top of the physics building, one brash young physics student throws a packaged egg up into the air from just over the edge of the building, such that the package, on its way down, just misses the edge of the building and falls to the ground below. If the building is 60 m high and the package is in the air for 4.5 s, then with what initial speed did the person throw the package?

Solution:

First draw a picture showing the major events, as shown in Figure 2.22.

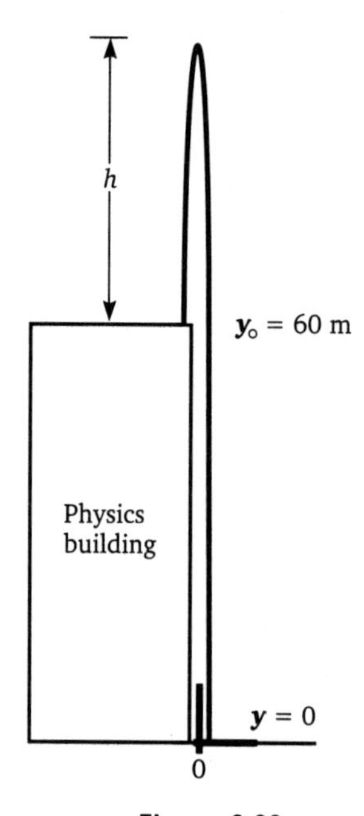

Figure 2.22

Before we even attempt to find any formulas, we need to set an origin and choose the positive y direction. We could set the origin at the top of the building or at the bottom of the building. It is largely a matter of choice. Sometimes, the mathematics works out a little easier one way than another. In Figure 2.22, the origin is set at the bottom of the building, and the given data are marked on the picture. Marking the data on the picture helps determine what formula to use. We can see that we are given information on the position. We know the initial position and the final position. This means we should look for a formula relating the initial velocity to the position.

$$y = y_o + \mathbf{v}_o t - \tfrac{1}{2} g t^2,$$

Since we know the time, we can solve this equation for $\mathbf{v}_o$.

$$\mathbf{v}_o t = y - y_o + \tfrac{1}{2} g t^2$$

$$\mathbf{v}_o = \frac{y - y_o + \tfrac{1}{2} g t^2}{t}$$

$$v_o = \frac{½(9.8\,m/s^2)(4.5\,s)^2 - 60\,m}{4.5\,s} = +8.7\ m/s$$

where the (+) means the ball is thrown up in the positive y direction.

Chapter 2 Problems

[2.1]

An object is moving along a horizontal path, as shown above, with a speed that is uniformly decreasing in magnitude. Which of the following arrows shows the direction of the object's acceleration (if any)?

(a) ↑ (b) ↓ (c) → (d) ← (e) the acceleration is zero because the motion is uniform.

[2.2] Given the following cases of a ball, with acceleration relative to the velocity (assume that the positive direction is to the right) shown in Figure 2.23,

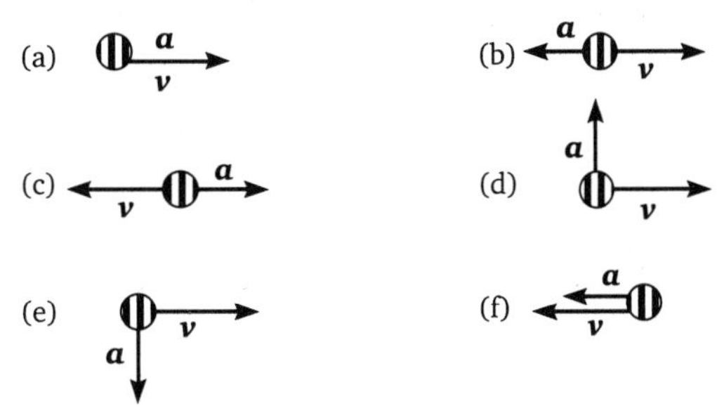

Figure 2.23
Problem 2.2

state in which cases the

i) speed will increase

ii) speed will decrease

iii) speed will not change

iv) next small displacement will be positive

[2.3] Figure 2.24 is a graph of position vs time for a quick little mouse. List all of the lettered points for which the following are true.

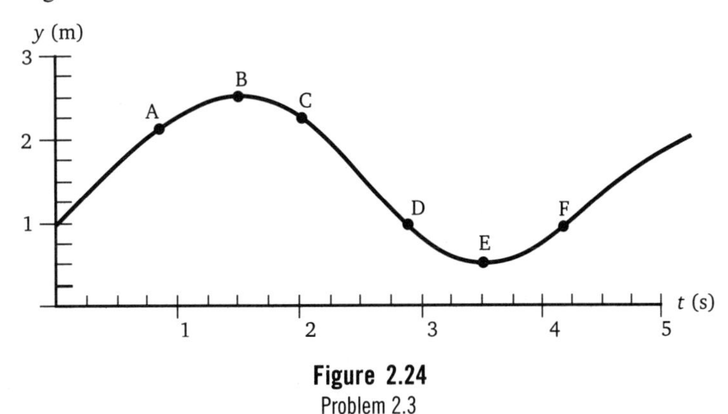

Figure 2.24
Problem 2.3

i) The position is neither increasing nor decreasing.

ii) The velocity is zero.

iii) The position is increasing, and the velocity is increasing.

iv) The position is increasing, and the velocity is decreasing.

v) The position is decreasing, and the velocity is decreasing.

vi) The position is decreasing, and the velocity is increasing.

vii) The position is increasing, more and more slowly.

viii) The position is decreasing, more and more quickly.

ix) The acceleration is positive.

x) The acceleration is negative.

[2.4] Make a graph of the instantaneous speed vs time for the position curve in Figure 2.24.

[2.5] A car on a trip drives 60 mph for the first hour, 30 mph for the next 45 minutes, then stops for gas for 15 minutes, and then travels 60 mph for the last hour. Find the average speed of the car on this trip.

[2.6] A roller coaster starts at rest from point *P* and travels down the track shown in Figure 2.25. At no point does it leave the track. The positive direction is along the track. Label the segments of the track along which the roller coaster will have the following characteristics:

A) speed has an absolute maximum

B) speed has a relative maximum

C) speed has an absolute minimum

D) speed has a relative minimum

E) speed is increasing at the greatest rate

F) speed is decreasing at the greatest rate

G) acceleration is greatest in the positive direction

H) acceleration is greatest in the negative direction

I) speed is constant

J) acceleration is not constant

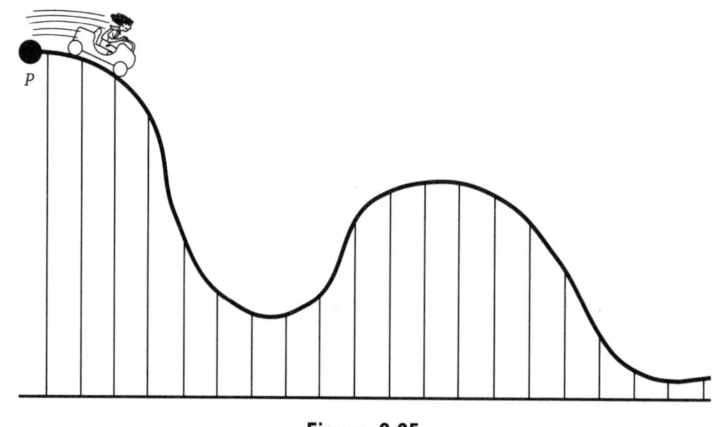

Figure 2.25
Problem 2.6 : A roller coaster track.

[2.7] As in Example 2-2, sketch a graph for $\boldsymbol{x} = \boldsymbol{x}_o - \boldsymbol{v}_o t - \frac{1}{2} \boldsymbol{a} t^2$ where $\boldsymbol{x}_o$, $\boldsymbol{v}_o$, and $\boldsymbol{a}$ are positive.

[2.8] A feather falls freely in a vacuum chamber at sea level. **a)** How far does the feather travel during its first second? **b)** How far does the feather fall during its first 2 seconds? **c)** How far does the feather fall during only its third second of fall?

• **[2.9]** Wyle E. Coyote stands upon the edge of a cliff which breaks. He starts to fall freely. If Wyle E. Coyote falls half his total distance in the last second of fall, **a)** what is the duration of his fall, and **b)** through what height did he fall?

• **[2.10]** Skeletor decides to test the law of gravitation by dropping Man–At–Arms off the top of 800m high Snake Mountain in Eternia and timing how long it takes before he hits the ground. Four seconds after Skeletor lets go, She–Ra arrives at the top of the mountain. With what initial velocity must she dive to save Man–At–Arms just before he reaches the ground? Assume the acceleration due to gravity is 10 m/s^2 in Eternia.

[2.11] A person who lived in southern Utah, downwind from an aboveground nuclear test during 1950, had his urine regularly checked for radioactivity starting on the day of the test. Figure 2.26 is a graph showing the radioactivity of samples of the person's urine over a period of 10 weeks (1 Curie = 3.7×10^{10} decays/s).

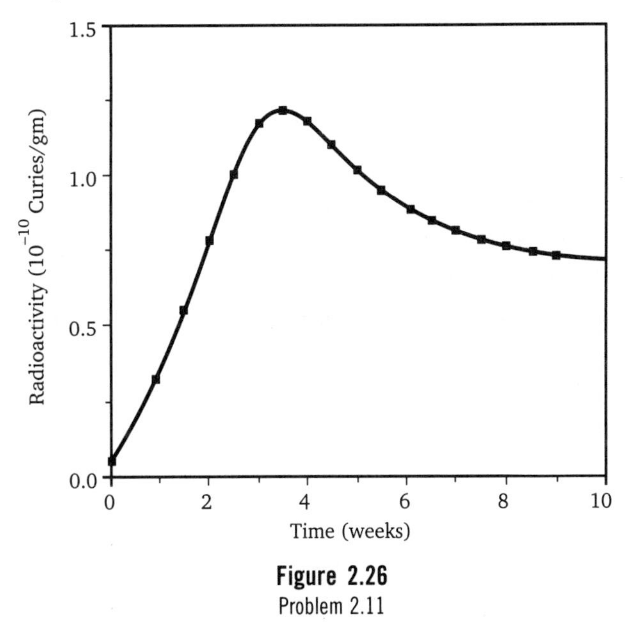

Figure 2.26
Problem 2.11

a) When was the amount of radioactivity greatest? **b)** Between what times was the amount of radioactivity increasing? **c)** Between what times was the amount of radioactivity decreasing? **d)** When was the amount of radioactivity increasing fastest? **e)** When was the amount of radio-activity decreasing fastest? **f)** When was the amount of radioactivity changing the slowest? **g)** What is the greatest amount of radioactivity the person had in his body? **h)** What is the average amount of radioactivity in his body over the 10 week period?

[2.12] Draw approximate graphs of the position, velocity, and acceleration vs time for the motion shown in Figure 2.4.

[2.13] Answer the following questions, referring to graphs A through F in Figure 2.27. Assume that graph F is a parabola.

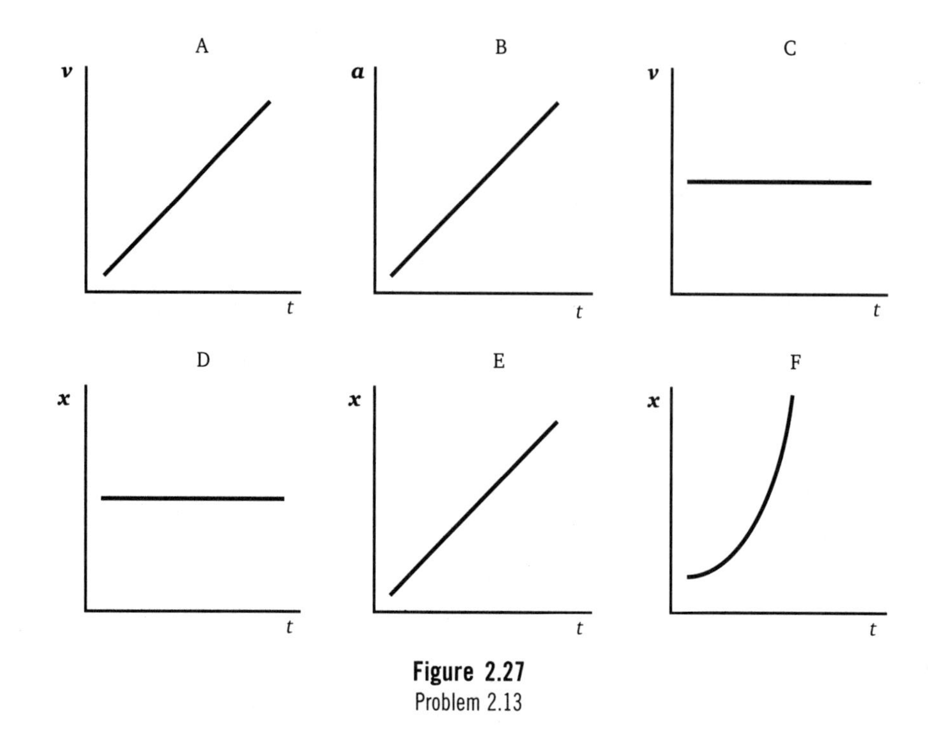

Figure 2.27
Problem 2.13

a) Which graphs represent an object at rest? **b)** Which graphs represent nonzero uniform acceleration? **c)** Which graphs represent the same kind of motion? **d)** For which of these graphs is a graph of position vs time not a line? **e)** Over a long time period, which of these motions will have the largest average velocity?

• **[2.14]** From the graph of position vs time in Figure 2.28, calculate and graph **a)** the average speed between points A and D, **b)** the instantaneous speed at each of the four labeled points, and **c)** the acceleration at each of the four labeled points.

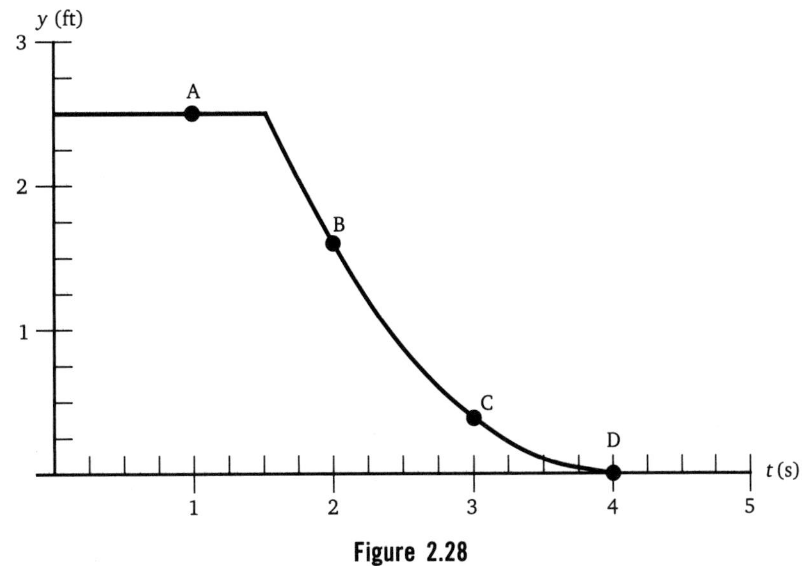

Figure 2.28
Problem 2.14: Position vs time.

[2.15] In Example 2-5, **a)** find the maximum height h above the top of the building that the egg package attains. **b)** Find the velocity of the egg package just before it hits the ground.

[2.16] In the graph of the position vs time in Figure 2.29, for what values of time is the particle speeding up; for what values is the particle slowing down?

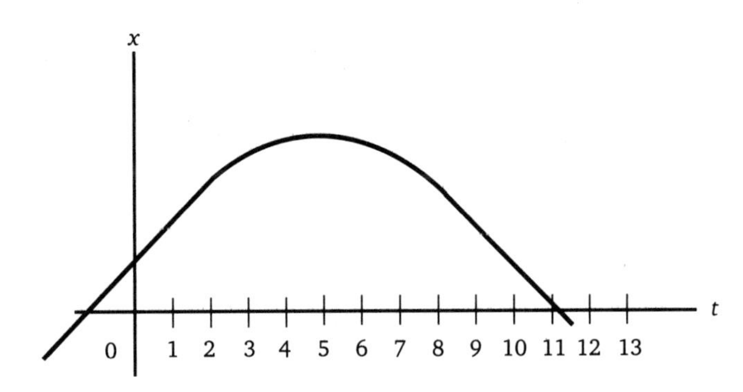

Figure 2.29
Problem 2.16: Position vs time.

CHAPTER 3
Those Special Functions: Logs, Exponentials, and Trig

(If we all split once, then we all split again, how many will we be?)

There exist functions lurking in deep places that intimidate many who pass this way. Purportedly, they exist only to confuse and debilitate, to turn the mind to fodder late on the night before an exam. Functions with long exotic names that are unpronounceable such as logarithm, exponential, or arctangent (to say nothing of the hyperbolic secant). Contrary to rumor, these functions are not the creation of some lunatic mind from the Far Side, but are often used and are important to even a basic understanding of Nature. In this chapter, we will explore the use of some of these functions. They are, in fact, not the bane of your existence, but rather are essential to most physical processes, and without them Nature would be very difficult to understand at all. Interestingly, the functions we describe in this chapter center around two of the most important numbers in mathematics, π and e.

3.1 THE EXPONENTIAL FUNCTION AND THE NUMBER *e*

Consider a function that governs growth. Typically, the growth rate depends on the amount present to create the growth. For instance, the growth of bacteria follow such a process. A given bacterium may divide every hour on average. So, starting with one bacterium, after the first hour, you may have two bacteria, after 2 hours four bacteria, after 3 hours eight. One bacterium was produced in the first hour, two in the second, and four in the third. Thus, the rate at which the colony is growing depends on how many there are. The number at any instant is given by an exponential.

$$N = 2^t \tag{3-1}$$

Obviously, this is a simple example, and we have ignored many important considerations, such as the death rate. Also, other factors influence the answer. Suppose we don't start with one but with N_o bacteria at $t = 0$, and throw in the rate at which the bacteria split. Let N be the number of bacteria after t hours. If x is the rate at which they split (once per hour in the preceding case), then

$$N = N_o + xN_o$$

$$N = (1 + x)\, N_o. \tag{3-2}$$

Suppose, as in our example, all the bacteria are only allowed to split on the hour (a ridiculous proposition), then $x = 1$, and after 1 hour, there would be

$$\begin{aligned} N &= (1 + 1)\, N_o \\ &= 2N_o \end{aligned} \tag{3-3}$$

Bacteria, however, are not prone to doing things all at once. What if half of them reproduce at the half–hour? After 1/2 hour we would have

$$N = (1 + \tfrac{1}{2})\, N_o = \frac{3}{2} N_o \tag{3-4}$$

and then during the next half–hour, half of the new amount replicates again so that after a whole hour

$$\begin{aligned} N &= (1 + \tfrac{1}{2})\,(1 + \tfrac{1}{2})\, N_o \\ &= (2.25)\, N_o \end{aligned} \tag{3-5}$$

Well, you can probably see where this is leading. What if one third of the bacteria split at 1/3 hour? Then, after 1/3 hour,

$$N = (1 + \tfrac{1}{3})\, N_o, \tag{3-6}$$

and after 1 hour, the bacteria will have replicated 3 times.

$$\begin{aligned} N &= (1 + \tfrac{1}{3})(1 + \tfrac{1}{3})(1 + \tfrac{1}{3})\, N_o \\ N &\simeq 2.37\, N_o\, . \end{aligned} \tag{3-7}$$

If $1/n$ of the bacteria are allowed to replicate n times an hour, then after 1 hour there will be

$$N = \left(1 + \frac{1}{n}\right)^n N_o \tag{3-8}$$

There is an important point here. If n is arbitrarily large, as would be the case if the bacteria were reproducing continuously, you might think that as n becomes large, N would also become large. However, $(1 + 1/n)^n$ does not become arbitrarily large but instead approaches a fixed irrational number e, where the limit as n approaches infinity is the fixed number:

$$e = \lim_{n \to \infty} \left(1 + \frac{1}{n}\right)^n \simeq 2.7182818284 \ldots \tag{3-9}$$

The limit is used to express the idea that as n becomes larger and larger the expression in the parenthesis approaches the number e more closely.

For the growth of continuously reproducing bacteria, after 1 hour, we would have

$$N = N_o\, e^1, \tag{3-10}$$

and after t hours

$$N = N_o\, e^t. \tag{3-11}$$

Thus, the number of bacteria start at N_o and grow exponentially. This example points out the utility of the exponential function. Processes in nature that reproduce continuously, or die continuously, do so exponentially, and thus depend on the exponential function

$$exp(t) = e^t \tag{3-12}$$

Any time the rate at which a population changes depends on the number present, there is exponential growth or decay. Examples range from the decay of radioactive nuclei in nuclear physics to the closing of a screen door damped by a piston, to the charging and discharging of a capacitor in an electrical circuit. The more population you have at any given time, the faster the reproduction rate. It is important to understand exponential behavior because this function has so many physical applications in all branches of physics.

Next, we'll need to understand the algebraic properties of this important function. These algebraic properties apply to any number raised to a power so that we can use x rather than the specific number e. Most of these can be ascertained easily by simple manipulations. Consider the product of a variable x raised to two different powers. This can be worked out by using the definition of an exponent.

$$x^2 \bullet x^3 = (x \bullet x)(x \bullet x \bullet x) = x^5 \tag{3-13}$$

Thus, the first important property is that when exponentials are multiplied, the exponents are added.

$$e^a \bullet e^b = e^{a+b} \tag{3-14}$$

Extending the property to division, we find that the exponents are subtracted:

$$\frac{x^3}{x^2} = \frac{x \bullet x \bullet x}{x \bullet x} = x^1 = x^{3-2} \tag{3-15}$$

with the stipulation that if the denominator is raised to a greater power, then the resulting number will be raised to a negative exponent, meaning division, for example,

$$x^{-1} = \frac{x^2}{x^3} = \frac{x \bullet x}{x \bullet x \bullet x} = \frac{1}{x} \tag{3-16}$$

This also has implications for zero exponents. Would e^0 be 0 ? Be careful, we must follow the rules we have set. We could get e^0 by using division of two equal exponents:

$$x^0 = x^{3-3} = \frac{x^3}{x^3} = 1 \tag{3-17}$$

The last algebraic property we need to consider is what happens when a number with an exponent is then raised to some power, such as $(x^3)^2$. Again, follow the basic rules.

$$(x^3)^2 = (x^3)(x^3) = x^6 . \tag{3-18}$$

Thus, an exponent raised to a power becomes multiplied. The general rule is

$$(x^a)^b = x^{ab} . \tag{3-19}$$

Warning! There is always the tendency to confuse the addition-of-exponent rules with the addition of exponentials.

$$x^a + x^b$$

cannot be simplified any further. **It is not x^{a+b}!**

We have seen that a colony of bacteria that continuously reproduces grows exponentially. Let's next see what this means in more depth by trying to find the slope $\Delta N/\Delta t$ of $N_o = N_o e^t$. Suppose that the number of bacteria grows exponentially, as in Equation (3-11). In a time Δt, the number will change by

$$\Delta N = N(t + \Delta t) - N(t) \tag{3-20}$$

$$\Delta N = N_o\,(e^{t+\Delta t} - e^t) \tag{3-21}$$

Given the algebraic properties of exponentials, we can write ΔN as

$$\Delta N = N_o\,(e^t \bullet e^{\Delta t} - e^t) \tag{3-22}$$

$$= N_o\, e^t\,(e^{\Delta t} - 1) \tag{3-23}$$

To interpret $e^{\Delta t}$ in Equation (3-23) we can use an alternate definition of the number e introduced by the great Swiss mathematician–physicist Leonhard Euler (pronounced "oiler"), who defined the number e in terms of a series expansion.

$$e = 1 + \sum_{n=1}^{\infty} \frac{1}{n!} \qquad \text{Euler's definition} \tag{3-24}$$

which when written long-hand is

$$e = 1 + \frac{1}{1} + \frac{1}{2\bullet 1} + \frac{1}{3\bullet 2\bullet 1} + \frac{1}{4\bullet 3\bullet 2\bullet 1} + \cdots \tag{3-25}$$

For $e^{\Delta t}$ this becomes

$$e^{\Delta t} = 1 + \sum_{n=1}^{\infty} \frac{(\Delta t)^n}{n!} \qquad (3\text{-}26)$$

At first, Equation (3-26) might not look like it has introduced any simplification at all. However, the graph of e^t will be a curve, and to find the slope $\Delta N/\Delta t$ of a curve requires using a small Δt so that a small segment of the curve ΔN approximates a straight line rather than a curve. We will treat this topic in more detail in Chapter 4 when we discuss the derivative. If Δt is small, Δt^2 will be very small so the terms raised to a power of 2 or greater can be neglected. Using the first two terms in Equation (3-26), we obtain for $e^{\Delta t}$

$$e^{\Delta t} \simeq 1 + \Delta t. \qquad \text{for small } \Delta t \qquad (3\text{-}27)$$

Using this linear approximation in Equation (3-23)

$$\begin{aligned} \Delta N &= N_o\, e^t\,(1 + \Delta t - 1) \\ &= N_o\, e^t\,(\Delta t) \end{aligned} \qquad (3\text{-}28)$$

Dividing both sides by Δt, we find the slope,

$$\frac{\Delta N}{\Delta t} = N_o e^t. \qquad (3\text{-}29)$$

Because the original function $N = N_o e^t$,

$$\frac{\Delta N}{\Delta t} = N \qquad (3\text{-}30)$$

In the limit, as Δt goes to zero, Equation (3-30) becomes exact. This result is very important. The slope of the exponential function is the exponential function. This means that as the function grows, the slope grows. The more there is, the faster the rate of change. This property is unique to e^t and is so important that it can be thought of as the defining characteristic of *e*.

Often, graphing a function can be easily done by considering how the function behaves under certain limits. This is particularly true for simple increasing functions such as the exponential. Consider the behavior of e^t under the limit as t goes to infinity. Obviously, $\lim_{t \to \infty} e^t = \infty$; that is, it increases without bounds. However, in what manner does it increase? Does e^t increase strongly or very weakly? Strongly or weakly are relative terms. To answer the question, we could compare e^t to a polynomial, such as t^a where a is some integer. From Euler's definition for e^t, we see that e^t contains all powers of t. Hence for large t, e^t will increase faster than any simple polynomial in t, and so, it will curve up dramatically.

$$e^t = 1 + \sum_{n-1}^{\infty} \frac{t^n}{n!} \tag{3-31}$$

At the origin $e^t = e^0 = 1$. For negative t, $\lim_{t \to -\infty} e^t = \frac{1}{\infty} = 0$. By exploring what happens at the origin and the limits of large t and large negative t, we have a pretty good idea of what the function looks like. At the origin, it has a value of 1 and curves up strongly in the positive t direction. It dies to zero in the negative t direction, as shown in Figure 3.1.

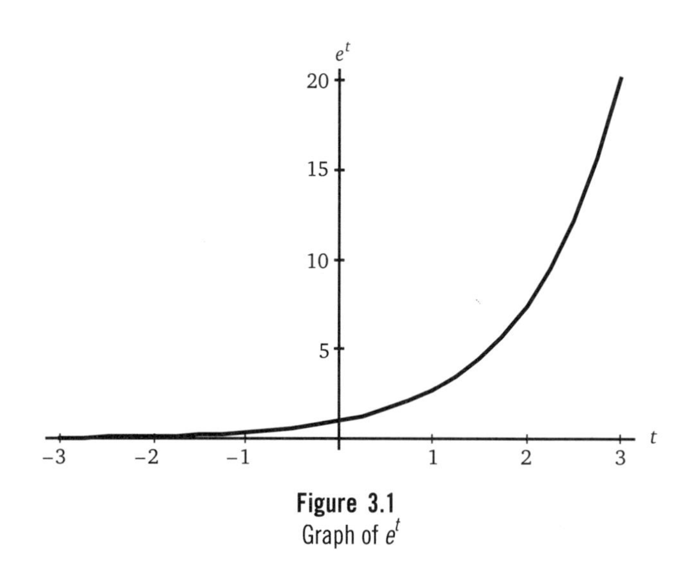

Figure 3.1
Graph of e^t

From the graph in Figure 3.1, we see that e^t is a monotonically increasing function that describes growth processes where the amount of growth depends on the amount present. It is a strongly increasing function. As you may have guessed, e^t also applies to human growth. However, we would expect that growth to be tempered by the death rate caused by diseases and other limiting factors, so it would not increase exponentially. Unfortunately, the human birth rate is so much higher than the death rate that the world population is still increasing exponentially! To add to the problem, the world population curve is no longer on the flat beginning part of the exponential growth. In 1845, it is estimated that the world population reached 1 billion people. It took about 1 million years (give or take a few hundred thousand) for the world population to reach 1 billion people. In the past 140 years, the population increased to 5 billion (1986). Compare the growth rate. In 1650, the world population was about 0.5 billion, in 1750 about 0.7 billion, whereas in 1980, the population was 4.5 billion, and yet by only 1986 we reached 5 billion.

As with any natural growth process, nature always finds a way of limiting the growth from a true exponential. In the birth-rate problem, the limiting factor may well prove to be AIDS (Acquired Immune Deficiency Syndrome). The number of AIDS cases is also on an exponentially increasing curve, and unfortunately it shows no signs of abating yet! *

3.2 EXPONENTIAL DECAY AND TERMINAL SPEED

Populations can not only grow, but also decay. Mathematically, we can describe the extinction process by making the slope of the exponential negative rather than positive.

$$\frac{\Delta N}{\Delta t} = -N \tag{3-32}$$

The solution to this equation is

$$N = N_o e^{-t} \tag{3-33}$$

which we can see from the graph of e^{-t} shown in Figure 3.2.

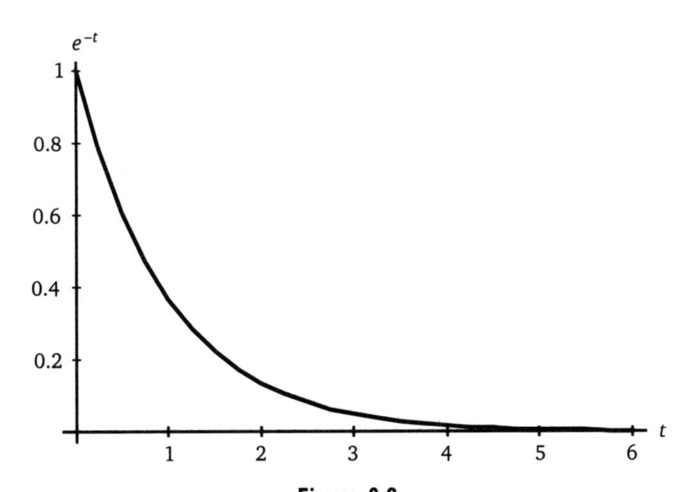

Figure 3.2
A decaying exponential e^{-t}.

Note that the function starts at 1 and decays to 0. Theoretically, it takes an infinite amount of t for the exponential to actually reach zero,

$$e^{-\infty} = \frac{1}{e^{\infty}} = \frac{1}{\infty} = 0 \tag{3-34}$$

but practically, it is almost zero quite a bit sooner. A measure of how much is "sooner" is the next topic of discussion.

We can affect the time scale of the decay by dividing the time by a number. The larger the number by which we divide the time, the longer it will take for the function to decay. For instance, what if we divide the time by 8, as in Figure 3.3?

*See Sherman Stein, and Anthony Barcellos, *Calculus and Analytic Geometry*, 5th ed., p. 374, McGraw-Hill (1992).

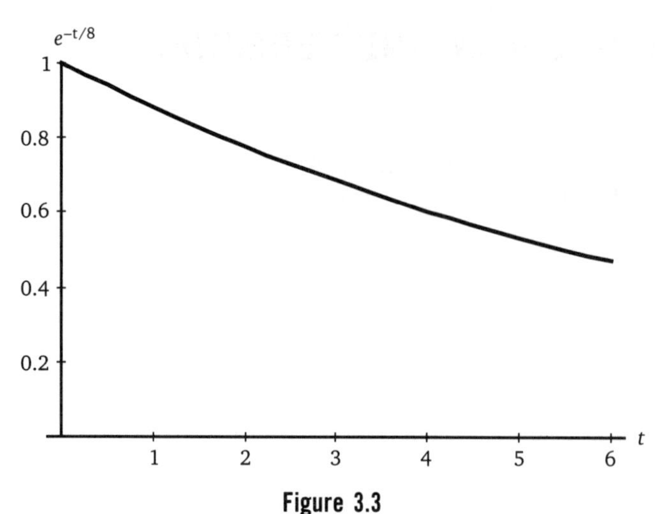

Figure 3.3
Extending the scale of the decay by dividing the *t* by 8.

Relatively, the function plotted in Figure 3.3 takes much longer to decay than the function in Figure 3.2. Both functions require $t \rightarrow \infty$ to reach 0, so what do we mean by "longer to decay"? Typically, this is answered by defining a decay constant, τ, for the function. The decay constant is the number by which we divide t in e^{-t}, as in Equation (3-35)

$$e^{-t/\tau} \tag{3-35}$$

In the Figure 3.3, the decay constant is

$$\tau = 8 \tag{3-36}$$

Note that if t has units of time, then the decay constant has units of time so that the units cancel. You cannot take the exponential of a unit, be it a second or a meter! In problems involving a decay in time, the decay constant is also termed the *time constant.*

Because a larger time constant causes the graph to decay more slowly, the decay constant sets the overall scale for the problem. The graph of e^{-t} shown in Figure 3.2 can be thought of as having a decay constant of 1. Note that by about 6 decay constants, the function has, for all practical purposes, died away to 0. The function shown in Figure 3.3, $e^{-t/8}$=, has a decay constant of 8. After 6 decay constants, $t = 48$. Will it also have died away to approximately 0? The answer is shown in Figure 3.4.

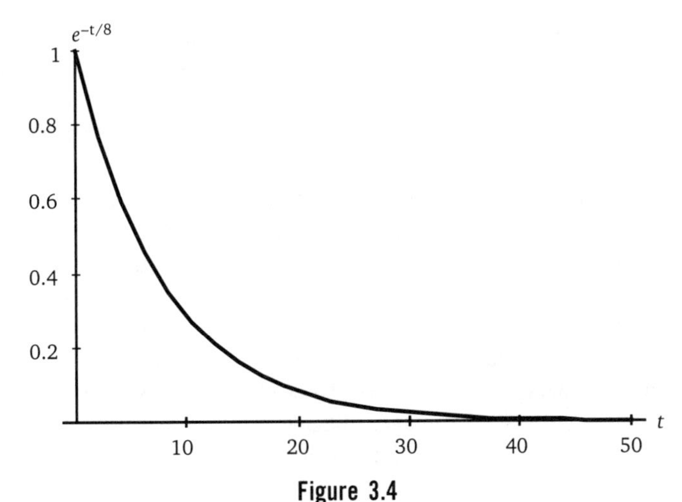

Figure 3.4
Decaying function with a decay constant of 8.

Compare Figures 3.2 and 3.4. With the exception of the labeling on the t axis, they are essentially identical. Thus, the decay constant represents a very useful quantity for plotting the function. To obtain a pleasing graph of a decaying exponential, make the horizontal axis, the t axis, about 6 decay constants long. Note that the graph in Figure 3.3 was not even 1 decay constant long and therefore does not give a good sensory impression of the function that is being plotted.

EXAMPLE 3-1

During the testing of an antibiotic drug, the number of bacteria are seen to decrease exponentially in time, with a decay constant of 6 hours. If there are N_o bacteria present when the drug is introduced, find the fraction of the bacteria left after 1 decay constant has elapsed.

Solution:
The formula for the decay is $N = N_o e^{-t/6}$. We need only to plug in for the time to determine the number remaining. At 1 decay constant, $t = 6$.

$$N_6 = N_o e^{-6/6} = \frac{N_o}{e}$$

$$N_6 = 0.368\ N_o$$

The fraction remaining is

$$\frac{N_6}{N_o} = 0.368$$

that is, after 1 decay constant, approximately 37% of the bacteria are remaining. This is a general property of exponential decay. For any exponential decay process, there will be approximately 37% left after 1 decay constant has elapsed.

The exponential function also occurs in the treatment of terminal speed. As an object falls through the air, it accelerates, gaining speed; however, the speed does not increase unabatedly. The faster the object moves, the greater becomes the air resistance. If the object falls for a long enough time, the air resistance becomes large enough to completely negate the effect of gravity. When this point is reached, the object has attained a terminal speed and no longer accelerates. Raindrops falling from clouds reach a terminal speed, typically of 5–20 m/s, which is a good effect because otherwise, they would be traveling fast enough to do quite a bit of damage. Imagine if there were no air. Then the speed of a raindrop falling about ½ km from a cloud would be

$$v = gt, \tag{3-37}$$

where t is found from

$$y = \tfrac{1}{2} gt^2 \Rightarrow t = \sqrt{\frac{2y}{g}} \tag{3-38}$$

$$v = \sqrt{2gy} \tag{3-39}$$

For the ½ km distance,

$$v = \sqrt{2(9.8\,\text{m/s}^2)(0.5\times 10^3\,\text{m})} = 100\ \text{m/s}.$$

Ouch!

Terminal speed is not reached immediately but is approached exponentially. We want a function that builds from 0 to 1, which we can construct by taking 1 and subtracting $e^{-t/\tau}$ from it. In this way, the exponential starts at 1, canceling the first 1, and then dies to 0, leaving a result of 1 at $t \rightarrow \infty$. The velocity given by

$$v = v_o(1 - e^{-t/\tau}) \tag{3-40}$$

will be 0 at $t = 0$ and then will build to v_o as the exponential dies away. The terminal speed is then v_o. The function $(1 - e^{-t})$ is plotted in Figure 3.5. You can see it approach 1 as t becomes large.

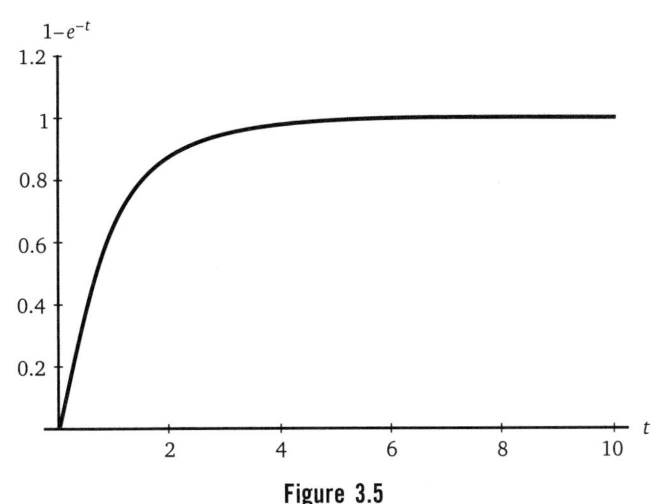

Figure 3.5
An exponential function used in terminal speed. As t increases, the $1 - e^{-t}$ approaches 1 exponentially.

3.3 LOGARITHMS

While it is useful to be able to use exponentials and to know what they look like, you also must know how to handle them. This means, given an equation involving e^x, you should be able to solve for the x. How do you get variables out of the exponent? This is really a question of defining the inverse of exponentiation. Most processes have an inverse, except of course multiplication by zero, the inverse of which—division by zero—is undefined. The logarithm is the inverse of exponentiation. The logarithm is usually abbreviated *log* except in the case of e, when its logarithm is termed the *natural log* and has a special abbreviation of *ln.* There can be different types of logarithms, depending on the number being exponentiated. For instance, if we wish to solve the equation

$$2^x = 8 \tag{3-41}$$

we need to undo 2^x, so we would take the log of both sides of the equation, giving

$$\log_2 (2^x) = \log_2 (8) \tag{3-42}$$

By definition, the log is the inverse of exponentiation, so $\log_2 (2^x) = x$, just as $2x$ can be undone by dividing by 2, $\frac{2x}{2} = x$. Because $2^3 = 8$, this is a particularly easy equation to solve:

$$x = \log_2 8 = 3 \tag{3-43}$$

The subscript 2 in $\log_2$ is termed the *base* of the log. It says that this log undoes 2 raised to a power. If the base is left off the log, it usually means that the base is 10. Also, *ln* has a base of *e*.

EXAMPLE 3-2

Find $\log_3 81$.

Solution:

Knowing that the log undoes the exponent operation, we need only to write 81 as 3 raised to a power.

$$\log_3 81 = \log_3 3^4 = 4$$

The different bases complicate things somewhat. The preceding example was very easy because 81 is an integral power of 3, but suppose we had asked for $\log_{10} 81$? 81 is not an integral power of 10, but we can simplify it because by definition, the log undoes exponentiation, whatever the base.

$$\log_3 3 = \log_3 3^1 = 1 \tag{3-44}$$

In Example 3-2, we could have written

$$\log_3 81 = 4 \log_3 3$$

and so too, we can simplify $\log_{10} 81$ as

$$\log_{10} 81 = \log_{10} 3^4 = 4 \log_{10} 3 \tag{3-45}$$

If we know what $\log_{10} 3$ is, then we know $\log_{10} 81$.

$$\log_{10} 3 = 0.477$$

$$\log_{10} 81 = 4(0.477) = 1.9 \tag{3-46}$$

The defining property of logarithms is then

$$\log a^x = x \log a \tag{3-47}$$

There are several useful properties of logs that follow directly from the definition and the exponent laws. For instance, given

$$a^x a^y = a^{x+y}, \tag{3-48}$$

taking the log of both sides gives

$$\log(a^x \bullet a^y) = (x+y) \log a = x \log a + y \log a,$$

so that

$$\log(a^x a^y) = x \log a + y \log a. \tag{3-49}$$

The properties of logarithms and exponents are listed in Table 3.1.

TABLE 3.1

Logarithms	Exponents
$\log a^x = x \log a$	$(a^x)^y = a^{x \cdot y}$
$\log_a a = 1$	$a^1 = a$
$\log (a \cdot b) = \log a + \log b$	$a^x a^y = a^{x+y}$
$\log \frac{a}{b} = \log a - \log b$	$\frac{a^x}{a^y} = a^{x-y}$
$\log_a 1 = \log_a a^0 = 0$	$a^0 = 1$
$\log \frac{1}{a} = -\log a$	$a^x = \frac{1}{a^x}$
$\log \sqrt{a} = \frac{1}{2} \log a$	$\sqrt{a} = a^{\frac{1}{2}}$

Because the log undoes exponentiation, it is custom made for solving equations where the variable exists in the exponent. Suppose we have an exponential decay function given by

$$N = N_o\, e^{-t/\tau} \tag{3-50}$$

and we wish to find the time when only half the sample is left. This time is termed the *half-life*, $t_{1/2}$. After one half-life, $N = \frac{1}{2} N_o$.

$$\tfrac{1}{2} N_o = N_o\, e^{-t/\tau} \tag{3-51}$$

The N_o cancels out, but we still have to solve for *t*. We can do this by taking the natural log (*ln*) of both sides of the equation.

$$\tfrac{1}{2} = e^{-t/\tau}$$

$$\ln \tfrac{1}{2} = \ln(e^{-t/\tau})$$

$$-0.693 = -t/\tau \tag{3-52}$$

Thus, the half-life is a simple multiple of the decay constant.

$$t_{1/2} = 0.693 \cdot \tau \tag{3-53}$$

EXAMPLE 3-3

While laying a new road, a work crew uncovers the remains of an ancient boat. The archeologist employed to investigate brings a wood sample to you for dating. It weighs 50 grams and shows a carbon-14 activity of 200 disintegrations per minute. If the half-life of C^{14} is 5730 years and the C^{14} activity of living plants is 12 disintegrations per minute per gram, estimate the age of the wood.

Solution:

The only thing you need to know to determine the age is that the activity obeys an exponential law

$$A = A_o\, e^{-t/\tau}$$

where initial activity is that of living plants

$$A_o = (12 \text{ decays/min} \bullet \text{g}) \times 50\ g = 600 \text{ decays/min}$$

and the decay constant is

$$\tau = \frac{t_{½}}{0.693} = 8270 \text{ years}$$

A is the current activity, which is $A = 280$ decays/min. Solving the exponential equation for t:

$$ln\left(\frac{A}{A_o}\right) = -\frac{t}{\tau}$$

$$t = \tau\, ln\left(\frac{A_o}{A}\right) = (8270 \text{ years})\, ln\left(\frac{600}{200}\right) = 9000 \text{ years.}$$

The trees from which the wood came were cut down 9000 years ago.

• 3.4 LOG, SEMI-LOG AND LOG-LOG GRAPHS

The log is a very slowly growing function of x. $y = ln\ (x)$ and $x = e^y$ are one and the same statement because the logarithm and exponential functions are inverse functions. Thus, $y = ln\ (x)$ is the mirror reflection about the line $y = x$ of $y = e^x$, as shown in Figure 3.6. Because e^x grows at such a large rate, the log grows slowly.

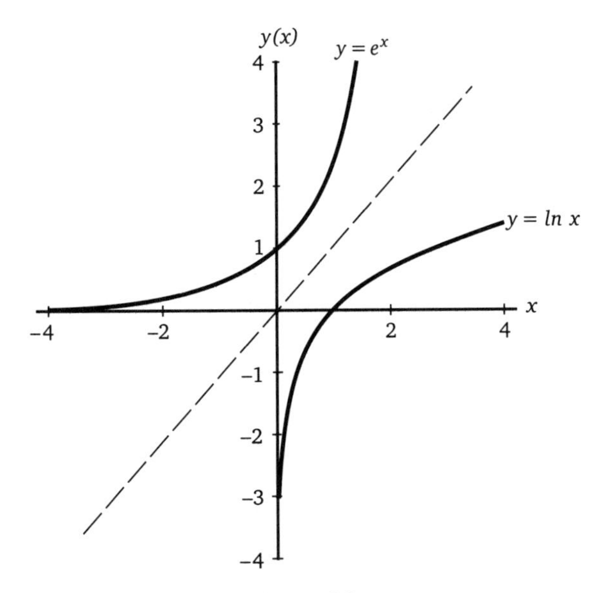

Figure 3.6
The natural log, *ln(x)*, is a slowly growing function of *x*. It is the reflection of e^X about the line $y = x$.

In Figure 3.7 *ln (x)* is graphed. Note that it goes through 0 at 1 and is negative for $0 < x < 1$. Although the logarithm will produce a negative number, there is no value of y that will produce a negative value of x in $x = e^y$. This means that you cannot take the log of a negative number. The negative numbers are not part of the domain of the logarithm.

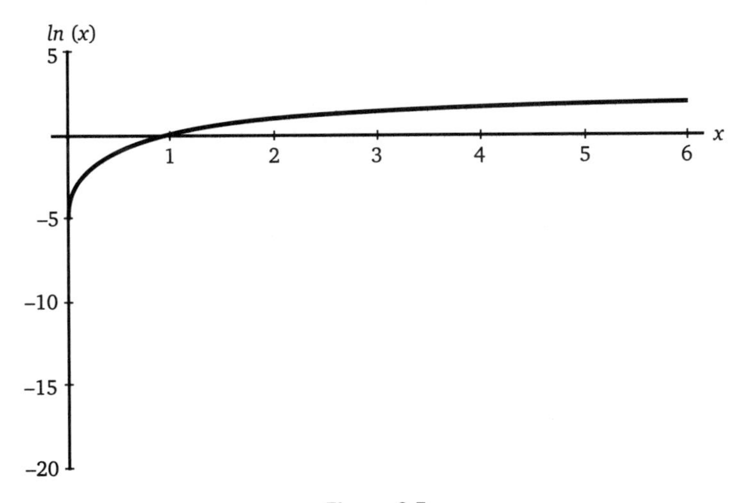

Figure 3.7
The logarithm is a slowly increasing function of *x*. The domain is all positive real numbers.

It is the slowing increasing property of the logarithm that makes it very useful for graphing processes that span many orders of magnitude. Two examples are the range of sound intensities we can hear and the frequency spectrum of electromagnetic waves. The softest sound we can hear, just at the threshold of hearing, has an intensity of 1×10^{-12} watt/m^2. In contrast, a rock concert can produce a sound of 1 watt/m^2 which is at the pain threshold. Our ears are sensitive to sounds for an intensity range that can span 12 orders of magnitude! Hearing is a most remarkable sense. By contrast, our eyes are only sensitive to visible light, which represents a small part of the electromagnetic spectrum. Light is an electromagnetic wave. One way to graphically depict the full range of a phenomenon is to use a logarithmic scale. Thus, log 1 = 0, log 10 = 1, log $10^2 = 2$, and log $10^3 = 3$, so that many orders of magnitude can be spanned in a short space. Graphic representations of the sound intensity scale and the electromagnetic spectrum are shown in Figure 3.8. Note that each tick on the graph represents a power of ten. This means that the scale is *not* linear. Also note that with a logarithmic scale, there is no zero because log(0) is undefined.

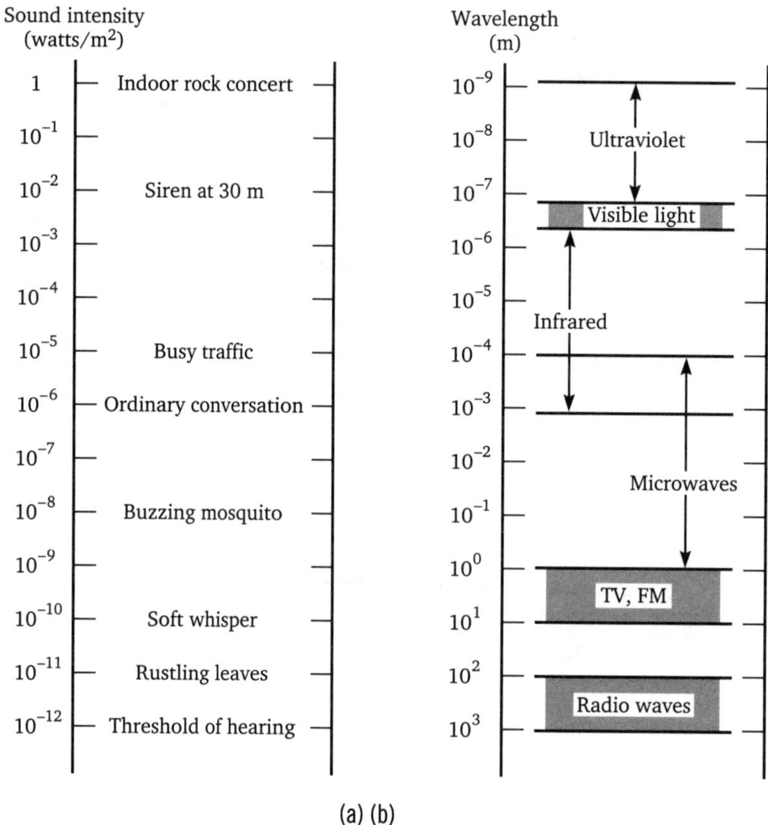

(a) (b)
Sound intensity levels Electromagnetic wave spectrum

Figure 3.8

Often, when plotting a function to see its behavior over a large range of values, or if $f(x)$ changes over many orders of magnitude for a relatively small change in x, then a *semi--log* graph can be used. This is made by plotting the vertical axis, using a log scale, and using a linear scale for the horizontal axis. For instance, you may have noticed in the graph of e^x shown in Figure 3.1 that by the time x reached only 3, e^x had exceeded 20; e^x increases very quickly. If we had wanted to plot e^x to $x = 1000$, we would have a very large graph, indeed. On the other hand, if we plot the natural log of e^x, we can make the plot quite easily, as shown in Figure 3.9. Of course, $ln(e^x) = x$, so we obtain a straight line.

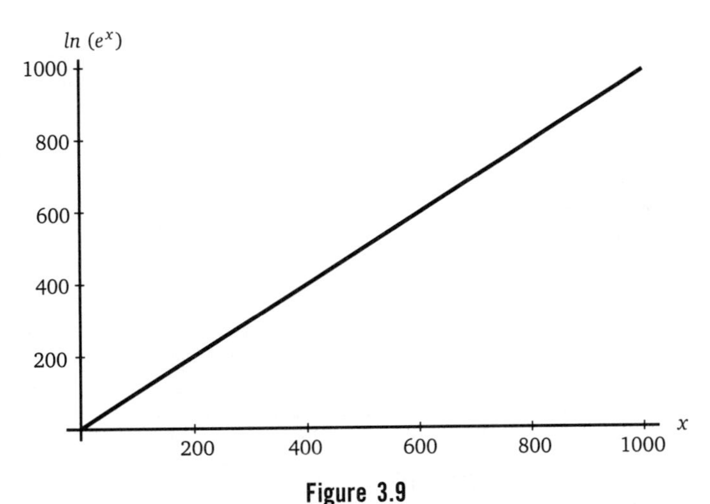

Figure 3.9
A semi-log graph of e^x. The ordinate is a natural log scale.

Often, in laboratory work, you will be trying to establish a functional relationship, and you might not know the relationship. In fact, there may be no theory giving you any hint as to what that relationship might be. How do we determine what the relationship is? Let's start with an example we know and see how it works.

EXAMPLE 3-4

The period T of a pendulum is the time it takes to swing from its farthest extent from vertical out to its other farthest extent and return to its original position. This period depends on the length L of the pendulum, according to the relation

$$T = 2\pi \sqrt{\frac{L}{g}} \tag{3-54}$$

where g is the acceleration due to gravity. If we were to measure T for various Ls, we would get the graph shown in Figure 3.10.

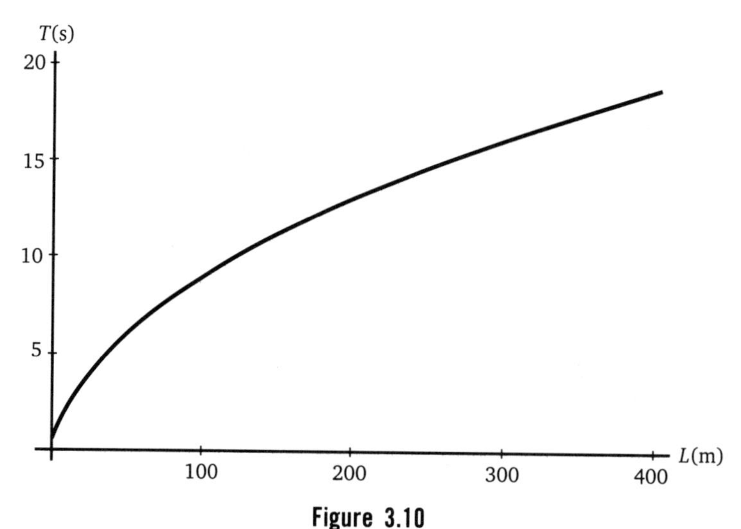

Figure 3.10
A graph of period vs length for a very long pendulum.

If you are lucky, you might guess that the functional relationship is $T^2 \propto L$, and a plot of T^2 vs L would be a straight line. For a simple pendulum, we are lucky to have a well-established theory to guide us. However, what if we didn't know the relation between T and L? If it is a power-law relation, we can use logs to determine it from the data. Taking the log of Equation (3-54) gives

$$\log T = \tfrac{1}{2} \log L + [\log 2\pi - \tfrac{1}{2} \log g] \tag{3-55}$$

where everything in the square brackets is constant. If we plot log(T) vs log(L), Equation (3-55) gives the equation of a straight line with slope ½. Thus, by making a log–log graph of T vs L, we can easily determine the power by measuring the slope, as in Figure 3.11. The y–intercept gives the constant term in the square brackets.

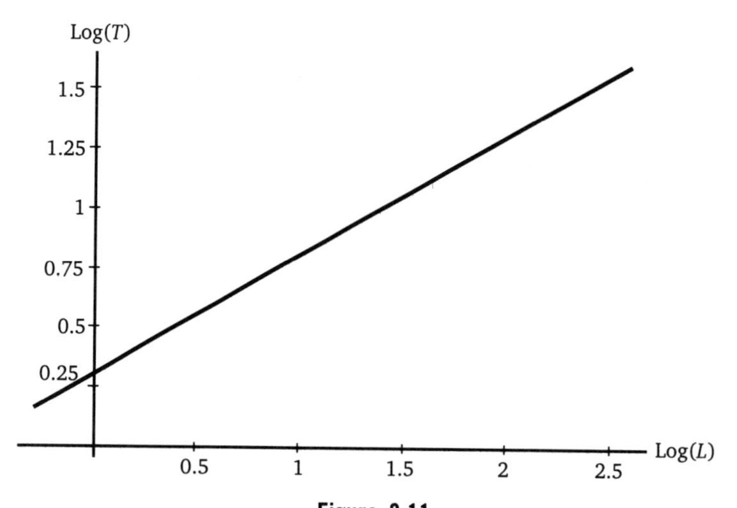

Figure 3.11
The slope of the log-log graph of T vs L gives the power of L in the relation between T and L.

This is a very powerful technique for experimentalists. Another bonus for experimentalists comes from the smoothing of the data produced by the compression of the orders of magnitude spanned by a log-log graph. As stated by Douglas McColm, who is an atomic physicist, "The secret that students discover is that lousy data looks great in a log-log graph."

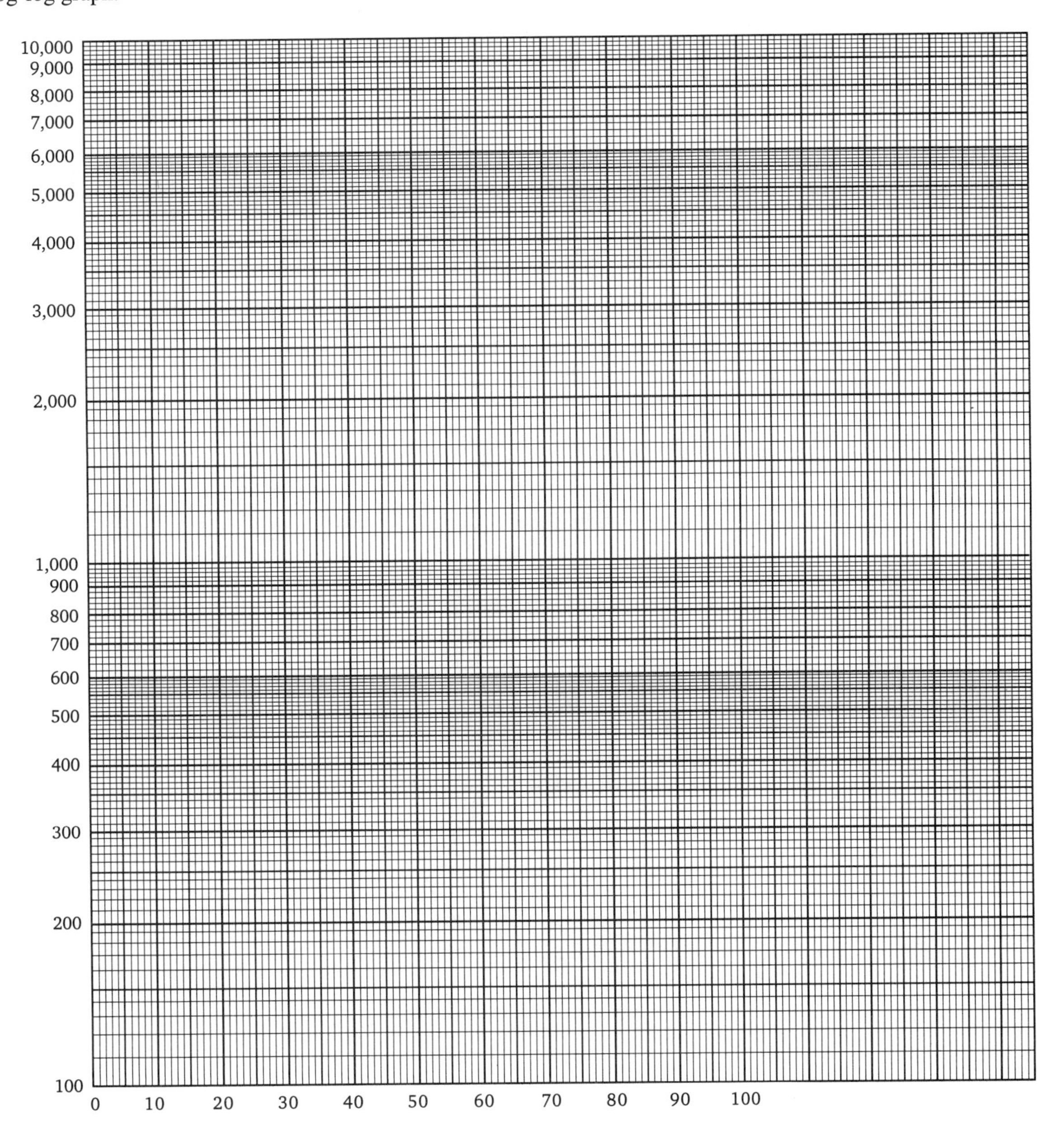

Figure 3.12 (a)
Semi-log paper with sample axes.

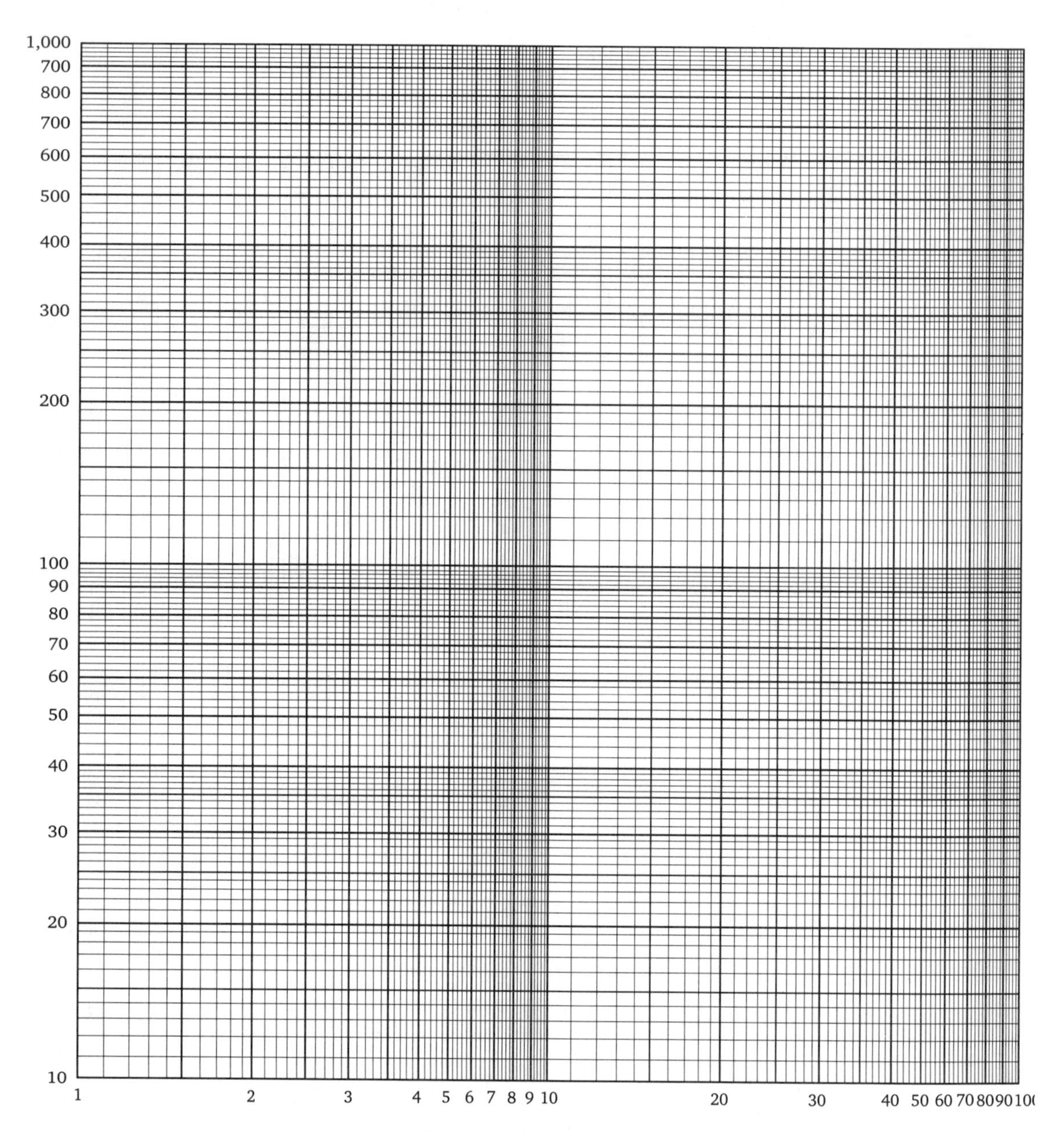

Figure 3.12 (b)
Log-log paper with sample axes.

Because the log scale is nonlinear, handmade semi-log graphs and log-log graphs are made on special paper that is "ruled" with a logarithmic scale. Samples of semi-log paper and log-log paper are shown in Figure 3.12. The axes have been labeled with sample data to show how you might use such paper. When you use semi-log paper and log-log paper, you do not have to compute the values of the logarithms before plotting but can label the graph with the values of the functions directly. Presently, virtually all scientific graphing software packages that run on a personal computer (PC) will make semi-log and log-log graphs, so the use of these special papers is decreasing.

3.5 CIRCULAR SYMMETRY AND ARC LENGTH

Besides logarithms, another topic that can be confounding is the geometry of circles, not so much because it is hard, but rather because we are not accustomed to using polar geometries regularly. A rectangular Cartesian coordinate system is perhaps the simplest type of coordinate system to visualize, but there are times when it is not the easiest to use. Try to describe the position of an object on a circle, and you will see what I mean. Using a grid, you need two coordinates (x, y); however, the two are not independent of one another. They are related by the formula for a circle:

$$r = \sqrt{x^2+y^2} \tag{3-56}$$

where r is the radius of the circle. Thus, given x, we can find y at the expense of some algebra. There is a simpler way. If there is only one independent coordinate for an object traveling in a circle, we should be able to formulate the problem in terms of only one coordinate. The natural coordinates for a circle are the radius and the angle measured from the x axis, as shown in Figure 3.13.

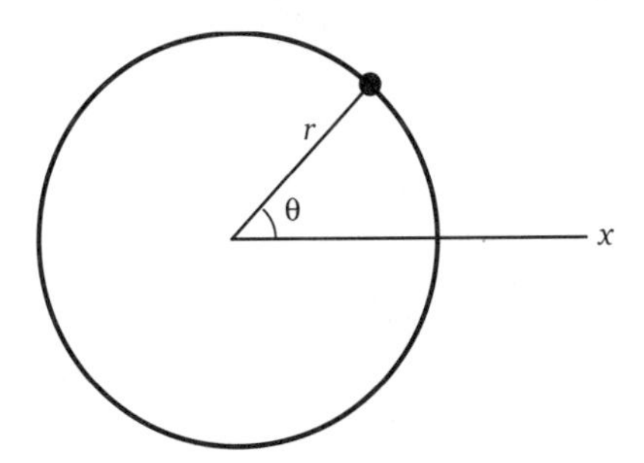

Figure 3.13
The position of an object moving on a circle is given in polar coordinates (r,θ).

If the object continues to move along the circle, then r is constant, so we only need to specify how θ changes, i.e., we have reduced the dimension to one changing coordinate, θ. This brings up the next important point. What units should we use to measure θ? Normally units are a matter of convenience, so we use the units that make problem solving the easiest. In polar coordinates, the units that are the easiest to use need some clarification.

Everyone is familiar with degrees. There are 360° in a circle. Why? Why an off-the-wall-number like 360? Why not a nice even number such as 100?

The answer dates back to ancient Babylon. Then, only priests could do the higher arithmetic associated with fractions. This becomes important when you want to set up a monetary system for trade. If no one can handle fractions, then all the different portions of your monetary scheme better divide into one another and give integers, so that the populace only has to deal with integer arithmetic. Thus, we are looking for an integer that—when divided by another integer—gives an integer result. Well, 100 might suit the bill. It is integer divisible by 2, 4, 5, 10, 20, 25, and 50. However, there is a smaller number that even beats 100: 60, which is an integer divisible by 2, 3, 4, 5, 6, 10, 12, 15, 20, and 30! So using a multiple of 60 allows you to avoid fractions to as large an extent as possible.

The Babylonian priests were becoming fairly skilled astronomers, as well as successful traders. When they were able to determine that the time it took for the earth to go around the sun is about 360 days, the number 360 (a multiple of 60) took on a mystical significance in their practice of numerology. Of course, we now know that the earth takes 365¼ days to make the journey, and so the number 360 loses its mystique. Still though, dividing a circle into 360° makes a lot of sense if you wish to avoid adding fractions, so 360° in a circle is important to us for that reason. (See Figure 3.14.)

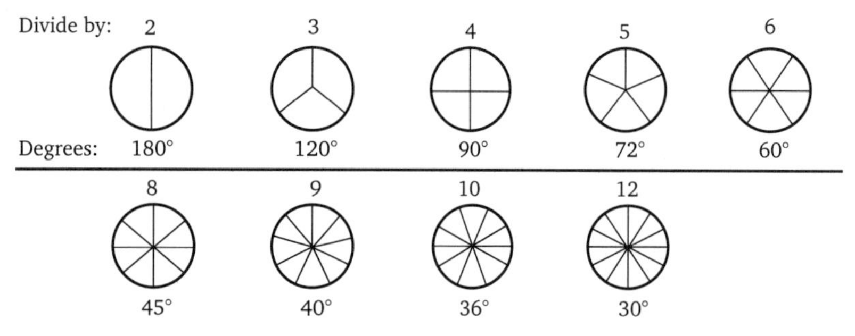

Figure 3.14
A partial pictorial list of the integer divisions of 360°.

If you are not limited to integer math, then there is another important unit, more important even than degrees, for measuring angles. This unit, the *radian,* is important when you try to relate the arc length to the angle it subtends. Next, we derive this relationship.

If we travel around the whole circle, we will have covered a distance of $2\pi r$, which is just the circumference of the circle. If we only cover a fraction of the circle, then we only cover that fraction of the circumference. For instance, if we went one quarter of the way around the circle, then we would have gone through a distance of $\frac{1}{2}\pi r$. Note also that we would have covered one quarter of the 360° around the circle. So the angle we travel through must be directly proportional to the distance along the circle we travel. This distance along the circle is called the *arc length* and is usually represented by the variable s. Can you determine the relationship between arc length and angle?

Of course $s = r\theta$, provided the angle is measured as a fraction of 2π—i.e., if we go all the way around the circle, we must travel the circumference $2\pi r$, and so θ must be 2π. The radian unit is a fraction of 2π. Thus, to use the formula for arc length, θ must be measured in radians.

Arc Length Formula

$$s = r\theta \tag{3-57}$$

θ must be measured in radians

Note that s and r both have units of distance, and this implies that radians are dimensionless; that is, you never need write down "radians" after an angle. If there are no units on an angle, then you are to assume that the angle is in radians. The conversion between radians and degrees is

$$\pi = 180°$$

Because the radian measure is interpreted as the fraction of 2π that an angle subtends, we can also use the table in Figure 3.14 to convert common degree measures to radian measure. For instance, 45° is $\frac{1}{8}$ of a circle, so $45° = 2\pi/8 = \pi/4$. Viewed this way, radians are not mysterious at all; however, the price we pay is that we must use fractions.

The arc-length formula is important whenever you need to measure distance along a section of a circle. In general, both degrees and radians have their usefulness. You must be able to convert easily between the two, which means you must practice converting between degrees and radians until it comes naturally.

3.6 ANGULAR KINEMATICS

In addition to angular position, we can also define angular speed. Suppose we are traveling with a constant speed around a circle, as shown at one particular moment in Figure 3.15.

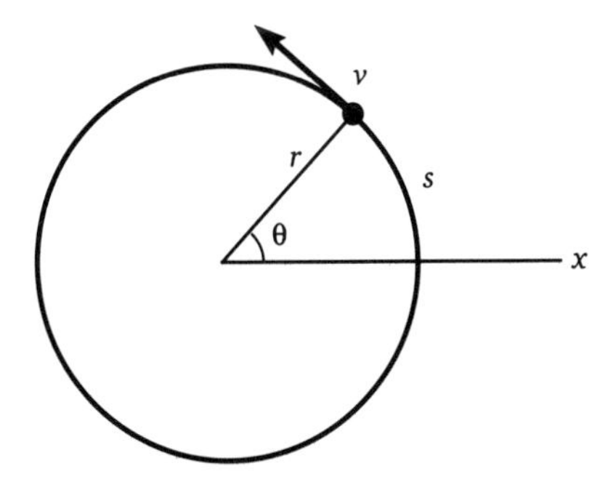

Figure 3.15
For an object traveling around a circle the velocity is always tangent to the circle.

The linear velocity will always be changing its direction because it must always be tangent to the circle. The velocity's magnitude—the speed—will be constant. The linear speed is the distance traveled per unit of time. The distance traveled is again an arc length s with constant r.

$$v = \frac{s}{t} \tag{3-58}$$

Because r is constant,

$$v = r\frac{\theta}{t} \tag{3-59}$$

where θ is changing in time. The rate at which the object is sweeping out an angle is θ/t and is called the *angular speed* ω (the Greek lowercase letter omega).

$$v = r\omega \qquad \text{linear to angular velocity} \tag{3-60}$$

The angular speed can at first appear confusing, but for an object moving with constant speed, we can always easily find ω. Let's call the time it takes to travel once around the circle the *period* T. To find ω , we only need to find how much angle is swept out in one period. Of course! It is 2π! Thus,

$$\omega = \frac{\Delta\theta}{\Delta t} = \frac{2\pi}{T} \qquad \text{angular velocity} \tag{3-61}$$

and remember ω **must** have units of radians/second. Angular speed is important when considering the motion of a rigid body, such as a spinning disk or ball. In order to remain intact, all points on the body must have the same angular speed.

EXAMPLE 3-5

A person travels with a constant linear speed of 7 m/s around a circle of radius 6 m. Find how long it takes to complete one cycle.

Solution:

We can blindly use the formulas we have derived.

$$T = \frac{2\pi}{\omega} \quad \text{where} \quad v = r\omega$$

$$\Rightarrow T = \frac{2\pi}{\frac{v}{r}} = \frac{2\pi r}{v}$$

$$\Rightarrow T = \frac{2\pi(6\text{m})}{7\text{m/s}} = 5.4s.$$

However, it is more instructive to take the distance traveled in one cycle = $2\pi r$ and to divide that by the rate. We get

$$T = \frac{2\pi r}{v},$$

which is the same thing we derived using the formulas. Sometimes, the derived formulas get in the way and obscure a relatively simple process. It is always better to think the problems through than to rely on formulas.

EXAMPLE 3-6

In Figure 3.16, both particles shown start together at time t_o, but Particle 1 is on a smaller circular track than Particle 2. The particles are then shown (in gray) at a later time t. In which case do the particles have the same linear speed, and in which case do the particles have the same angular speed?

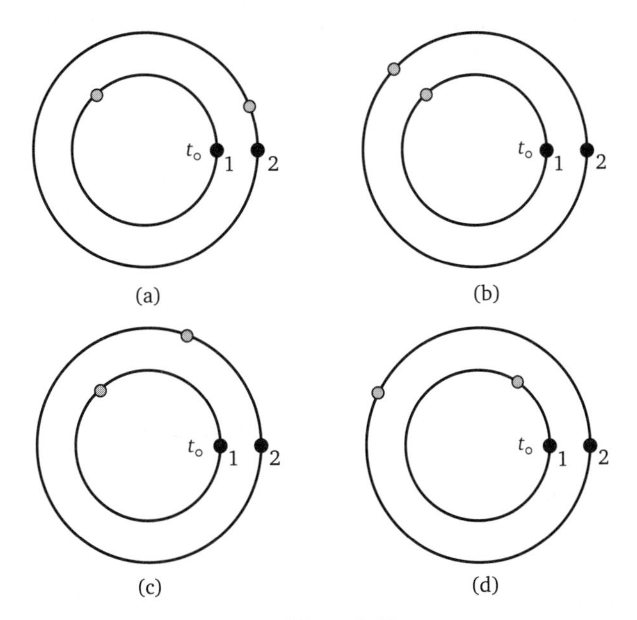

Figure 3.16
The particles in gray are the particles at a later time.

Solution:
In (a), Particle 2 is obviously moving much more slowly than Particle 1 in every sense. In (b), both particles have carved out the same angle in the same time, so they have the same angular speed, but Particle 2 has moved through a larger arc length, so 2 has the greater linear speed. In (c), Particle 1 has traveled through the greatest angle, so it has the greater angular speed, but both particles have roughly traveled the same distance, so the linear speeds will be about the same. In (d), Particle 2 is traveling much faster than 1 in every sense.

3.7 CIRCULAR FUNCTIONS

If you think of the period as the number of seconds per cycle, then the inverse of the period is the number of cycles per second—i.e., the frequency (f). Thus,

$$f = \frac{1}{T} \qquad \text{linear frequency} \qquad (3\text{-}62)$$

Since the period is related to the angular speed by

$$\omega = \frac{2\pi}{T}$$

$$\omega = 2\pi f \qquad \text{angular to linear frequency} \qquad (3\text{-}63)$$

So we have a linear relationship between the angular speed and the linear frequency. Now what in the world does the angular speed have to do with the frequency? It turns out that there is a very fundamental relationship between the two, and understanding it starts by considering the simple trigonometric functions sine, cosine, and tangent. We can gain quite a bit of insight into the properties of these functions by using circular motion. The position of an object at one moment in time is shown in Figure 3.17. Note that if the angle θ is changed, x and y also change. It seems plausible that they must be connected to θ.

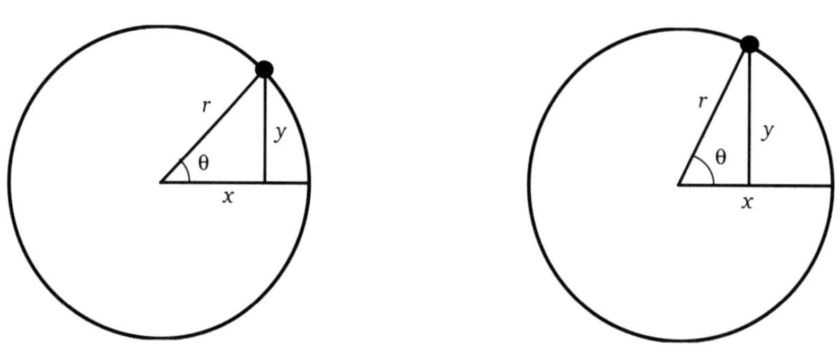

Figure 3.17
If θ is changed, x and y also change.

The angle θ is obviously related to the slope of r. This relationship defines the tangent. Similarly, the relation between θ and y defines the sine, and the relation between x and θ defines the cosine.

Trigonometric Functions

$$\tan\theta \equiv \frac{y}{x} = \text{slope} \qquad (3\text{-}64)$$

$$\sin\theta \equiv \frac{y}{r} \qquad (3\text{-}65)$$

$$\cos\theta \equiv \frac{x}{r} \qquad (3\text{-}66)$$

These relations between the trigonometric functions and the circle also allow us to easily estimate the values of the trig functions without a calculator. For example, as illustrated in Figure 3.18 (a), at $\theta = 0$, we find $y = 0$. Because the y-leg of the triangle is zero, the $\sin\theta$ also must be zero.

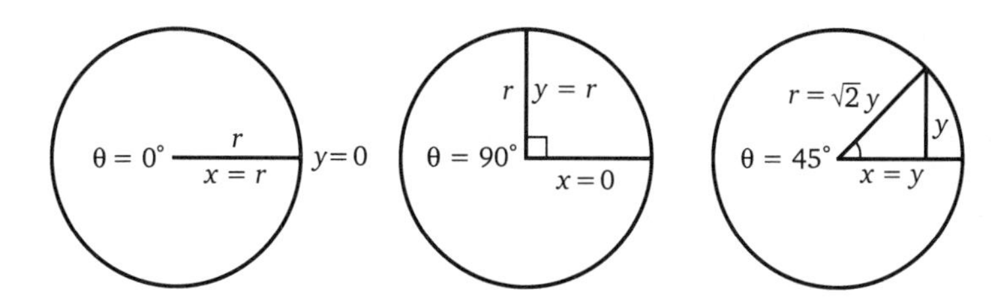

Figure 3.18
a) for $\theta = 0°$, the y-leg is zero. b) For $\theta = 90°$, the y-leg equals the r-hypotenuse. (c) For $\theta = 45°$, the x-leg and the y-leg are equal.

What happens to y as θ approaches 90°? As shown in Figure 3.18 (b), the y-leg of the triangle is now equal to r so the $\sin 90° = r/r = 1$. At 45°, the two legs are equal, so $r = \sqrt{2}y = \sqrt{2}x$, and thus, $\sin 45° = \frac{y}{\sqrt{2}y} = \frac{\sqrt{2}}{2}$. Can you estimate the value of the cos 45°? From Figure 3.18 (c), the x and y-legs of the triangle have the same length so $\cos 45° = \sin 45° = \frac{\sqrt{2}}{2}$. In fact, by looking at the circle, we can trace out all the properties of the trig functions.

EXAMPLE 3-7

Using the behavior of the legs of the triangle shown in the circle for Figure 3.17, estimate the graph of $\sin\theta$ from 0 to 2π.

Solution:
In this exercise, we are not trying to guess values so much as to determine the properties of the sine. For instance, as we allow θ to increase by swinging the radius around the circle, the largest value y can have is r. Because $\sin\theta = y/r$, the largest value the sine can have is $r/r = 1$. Next, what happens to the sine for small θ? From Figure 3.19 (a), the y-leg of the triangle is approximately equal to the arc of the circle s. This approximation holds from $\theta = 0°$ up through $\theta = 20°$.

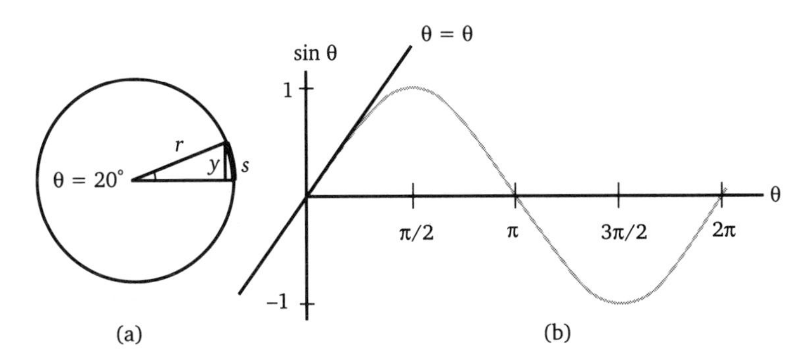

Figure 3.19
The sine behaves linearly for small θ because the y-leg on the triangle approximates the arc of a circle.

Because the arc of a circle increases linearly with θ, i.e., $s = r\theta$, $y \simeq r\theta$, and the $\sin\theta \simeq \theta$. Thus, near the origin the sine behaves as a straight line of slope 1, as shown in Figure 3.19 (b).

As we increase θ further and approach $\theta = \pi/2$, we are reaching the maximum value for y; hence, the curve must bend over from straight line behavior and reach a maximum value of 1.

As θ is increased past $\pi/2$, the sine will now decrease, because the y-leg on the triangle becomes compressed by the downward curve of the circle. Figure 3.20 compares the y-leg of the triangle for the two angles, $\theta = 2\pi/3$ and $5\pi/6$, and shows the corresponding behavior of the sine (highlighted in black) in this region.

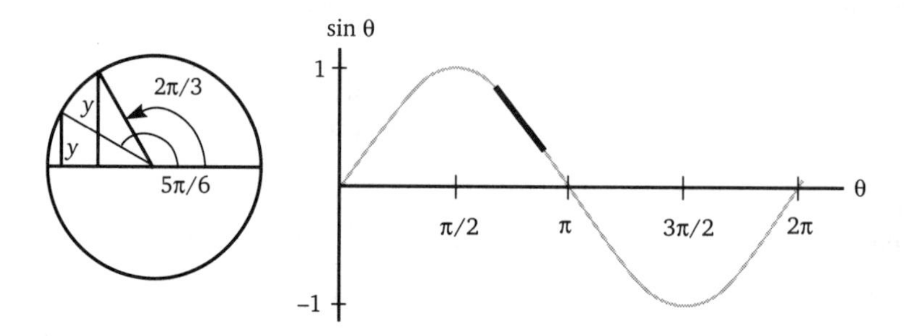

Figure 3.20
As θ increases from $2\pi/3$ to $5\pi/6$, the sine decreases because the y-leg of the triangle decreases under the curve of the circle.

As θ increases past π, the triangle turns upside down, so we can think of the y-leg as being negative because it is below the x axis, as shown in Figure 3.21.

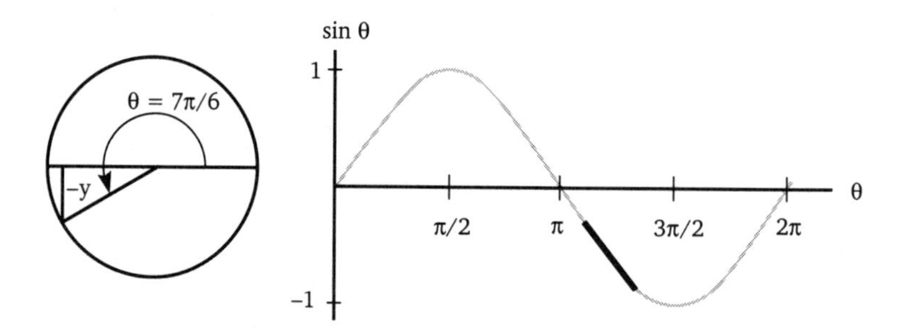

Figure 3.21
As θ increases beyond π, the y-leg becomes negative, so the sine becomes negative.

The rest of the sine curve mirrors the positive behavior. Thus, by letting θ increase, and watching what happens to the vertical leg of the triangle, we can deduce the behavior of the sine for all values of θ. You should also make graphs of the cosine and tangent based on the behavior of the legs of the triangle in the circle.

If we solve for x and y from equations (3-65) and (3-66), then we obtain the transformations between the Cartesian and polar coordinates.

$$\begin{aligned} x &= r\cos\theta \\ y &= r\sin\theta \end{aligned} \tag{3-67}$$

Can you find the inverse transformations?

$$r =$$

$$\theta =$$

For r, use the Pythagorean theorem, and for θ, use the inverse tangent.

$$r = \sqrt{x^2 + y^2}$$

$$\theta = \tan^{-1} \frac{y}{x} \tag{3-68}$$

Just as an inverse function, the logarithm, exists for the exponential function, so too, an inverse function must exist for the sine, cosine, and tangent. These inverse trigonometric functions are referred to either by placing a $^{-1}$ superscript right behind the function name, or by prefixing the name with *arc.* Thus, the inverse cosine could be denoted either as $\cos^{-1} x/r$, or as $\arccos x/r$. Note that the inverse trig functions return an angle. There are two points that need to be mentioned. First, the notation $\cos^{-1} x/r$, means, "Take the inverse cosine of x/r," as in Example 3-7 and does *not* mean

$$\frac{1}{\cos\frac{x}{r}} \quad \text{This is not the inverse cosine!}$$

Second, the domain and range of these inverse functions require special attention. First note that for the inverse tangent, when $x = 0$, the argument y/x is undefined, but from Figure 3.17 you can see that if $x = 0$, then ω must be $\pi/2$ giving $\tan^{-1} \infty = \pi/2$. The tan, however, is undefined at all odd multiples of $\pi/2$, such as $3\pi/2$ or $-3\pi/2$; therefore, you must be careful to draw a picture of the geometry in the problem you are working to successfully determine the angle. Next, note that in Figure 3.17 x can never be larger than r so that x/r, must never be larger than 1. Another way to see this is that the sine and cosine are never larger than 1 so you must never try to take the inverse sine or inverse cosine of any number larger than 1.

EXAMPLE 3-8

Using a calculator find: **a)** $\cos^{-1}(0.6)$, $\sin^{-1}(0.6)$, $\tan^{-1}(0.6)$, **b)** $\tan^{-1}(100)$, $\sin^{-1}(100)$.

Solution:

First locate on your calculator the sin, cos, and tan buttons. Some calculators used in business do not have these buttons, but all scientific calculators do. Next locate the "INV' button or the "2nd F" key; it is the one that allows you to use the functions typed above the function keys. We'll call this key the second function key. This key may have a different label on your calculator so you may have to consult your instruction manual. You should see $\sin^{-1}$ or *arcsin* printed above the sin key, so to take the $\sin^{-1}$ you will have to push the second function key before pushing the sin key. **a)** To find $\cos^{-1}(0.6)$, enter ".6." Next, push the second function key. Finally push the *sin* key. Most calculators will return the result either in radians or degrees. If your calculator is in degree mode, you should obtain the result

$$\cos^{-1}(0.6) = 53.1°$$

Next try the sine and tangent:

$$\sin^{-1}(0.6) = 36.8°, \quad \tan^{-1}(0.6) = 30.9°$$

In radian mode,

$$\cos^{-1}(0.6) = 0.9272952, \quad \sin^{-1}(0.6) = 0.6435011, \quad \tan^{-1}(0.6) = 0.5404195$$

For **b)** note that the large value of 100 means that the angle for the tangent must be approaching 90° where the tangent approaches infinity. And this is confirmed by the calculator

$$\tan^{-1}(100) = 89.4°$$

The sine, however, cannot produce a number larger than 1, so $\sin^{-1}(100)$ is undefined. Try taking the inverse sine of 100 on your calculator to see what it does. Most will produce an error message, either a capital *E* or *err.*

Now we are ready to consider the connection between frequency and angular speed. If we look at the shadow of the object along the x axis, its position is given by

$$x = r\cos\theta$$

as shown in Figure 3.22.

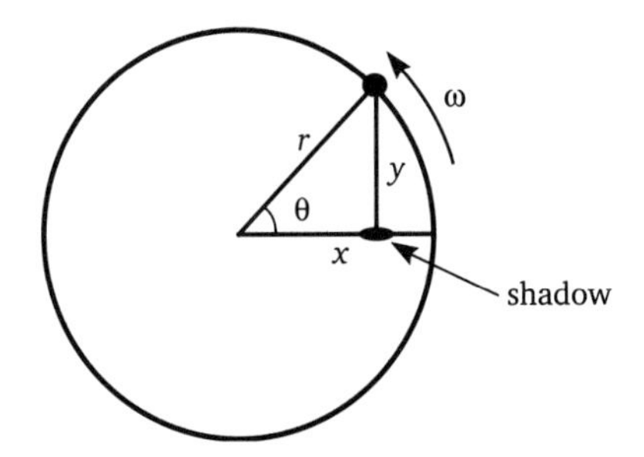

Figure 3.22
As the ball rotates to the left with angular speed ω, the shadow moves to the left along the *x* axis.

The position of the object changes in time and is determined by the angular speed. Assume that the initial angle is $\theta_o = 0$. Then

$$\omega = \frac{\Delta\theta}{\Delta t} = \frac{\theta - 0}{t - 0} \tag{3-69}$$

$$\Rightarrow \theta = \omega t. \tag{3-70}$$

The position of the shadow is

$$x = r \cos \omega t. \tag{3-71}$$

In Equation (3-70) lies the connection between the angular frequency and the angular speed. The shadow will bounce back and forth as the object rotates around the circle. As the cosine goes from 1 to −1 and back to 1, the time goes from 0 to one period T, which means at $t = T$, $\omega T = 2\pi$, or

$$\omega = \frac{2\pi}{T} = 2\pi f$$

Thus, the ω used in $x = r \cos \omega t$ is the angular frequency at which the shadow bounces back and forth. What we have shown is that this angular frequency is the same as the angular speed of the ball. It is because of this relationship between frequency, angular speed, and the transformations between Cartesian and polar coordinates that the trigonometric functions are termed *circular functions.*

To succeed in physics, you don't need to know a tremendous amount of trigonometry, but the trig you need to know must be mastered extremely well. You will find that you will use sine, cosine, and tangent all the time. The other trig functions you will probably use infrequently. Here I have presented pretty pictures for the definitions of the trig functions; however, in practice, it is much like working on a car. In the manual, there is always a pretty, upright picture of some car part, but when you go to find the part on a real car, it is covered in layers of crud and twisted up in some impossible location. The triangles used in physics are rarely pointing in the right direction. Usually, they are part of a complicated drawing, and the triangle is twisted up with many other triangles. You must be practiced enough to quickly identify the hypotenuse and which trig functions are the ratios of which sides. In Figure 3.23, you must be able to quickly determine that side x is $r \sin \theta$! You must, therefore, practice your geometry and trigonometry.

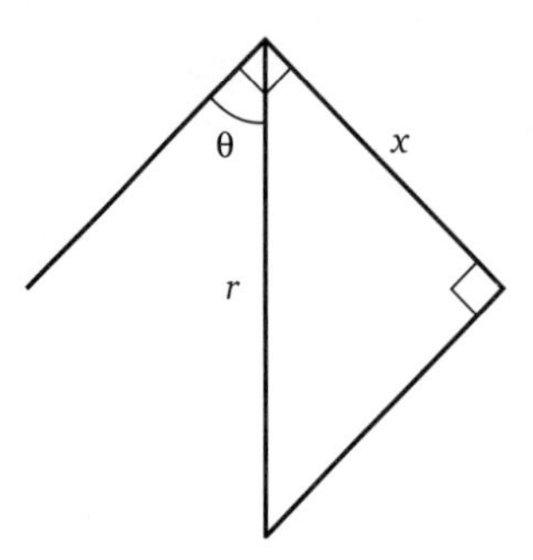

Figure 3.23
Quickly relate *r, θ, and x!*

You should also know the trig functions, angles, and sides for some of the commonly used triangles. These common triangles are the 45° right triangle, the 60°–30° right triangle, and the 3–4–5 right triangle. The sides and angles are shown in Figure 3.24. You should have these triangles memorized. For instance, if you are working a problem and see a 36.9° angle in one of the right triangles, you should know immediately that it is a 3–4–5 triangle. Or if you have a 60°–30° right triangle, you should know that the side opposite the 60° is $\sqrt{3}/2$. You should have these triangles "down cold."

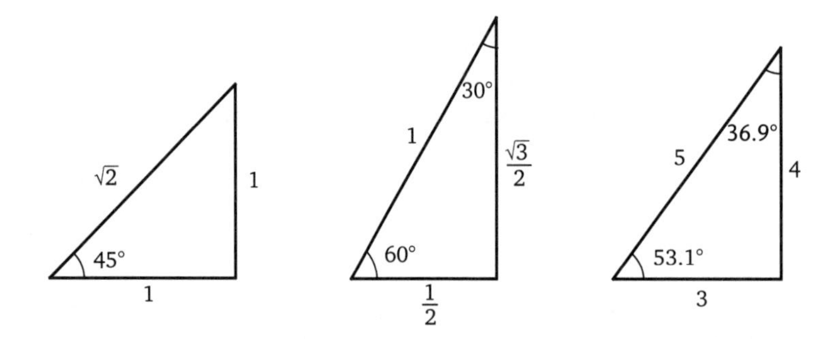

Figure 3.24
Side ratios and angles for some commonly used triangles.

Chapter 3 Problems

[3.1] Suppose that \$5000 is invested at 6½% interest, compounded annually. **a)** Show that the formula for the amount of money, M, available after t years is

$$M = \$5000\ (1.065)^t$$

b) How long will the money take to double, and then to triple?

[3.2] The number of VCRs that an assembly-line worker can assemble daily after t days of training is given by $N = 60(1 - e^{-0.04\,t})$. After how many days does the worker assemble 38 VCRs daily?

[3.3] G raph the following on the same set of coordinate axes: **a)** $y = e^x$, **b)** $y = e^{3x}$, **c)** $y = e^{x+3}$, **d)** $y = e^x + 3$.

[3.4] Simplify $\left(\frac{-y^{-6}z^3}{8x^4}\right)^{-2/3}$.

[3.5] Solve for x: **a)** $10^x = 0.001$, **b)** $5^{x-2} = 625$, **c)** $8^x = 16$, **d)** $\log_6 x = 3$, **e)** $\log_7\left(\frac{1}{343}\right) = x$, **f)** $\log_x 64 = \frac{3}{2}$, **g)** $\log_4 x - \log_4 (4x + 6) = -2$.

[3.6] If possible, write the following as the sum or difference of simpler logs: **a)** $\log_5\left(\frac{5x}{y^2 z}\right)$ **b)** $\log_6 (6 + x)$, **c)** $\log(xy)$ **d)** $(\log x)^2$ **e)** $\log(x^2)$.

• **[3.7]** The voltage reading V across a capacitor in an electrical circuit that has just had the voltage source turned off obeys the equation $V = 10\,e^{-t/0.003}$, where V is in volts and t is in seconds. **a)** Determine the decay constant. **b)** Make a graph of V vs t. What values of t should you plot along the abscissa? **c)** What is the voltage at the initial moment? **d)** How long will it take the voltage to reach 9 volts?

[3.8] Wyle E. Coyote steps out of a plane, realizing only afterward that his parachute is still in the plane. His velocity is given by $v = 120(1 - e^{-t/5.5})$ where v is in mi/hr and t is in seconds. **a)** What is the coyote's initial speed? **b)** Make a graph of v vs t. **c)** Realistically, how long does it take the coyote to reach his terminal speed?

[3.9] The decibel scale used for the apparent loudness of sound is a logarithmic scale. **a)** Make a semi-log plot of the decibel sound level β vs the sound intensity for the data in Table 3.2. Along the horizontal axis plot $\log {}^{I}/_{I_o}$) and along the vertical axis plot β. **b)** From the graph, derive a formula for the relation between the sound intensity I and the decibel level β of the form $\beta = Af\left(\frac{I}{I_o}\right)$ where $I_o = 10^{-12}\,W/m^2$, and A and $f.\ (I/I_o)$ are to be determined.

TABLE 3.2

Intensity (W/m^2)	Decibel β (db)
10^{-12}	0
10^{-10}	20
10^{-8}	40
10^{-6}	60
10^{-4}	80
10^{-2}	100
1	120
10^{2}	140

• **[3.10]** In an experiment, you have made the following measurements of x and y shown in Table 3.3. **a)** Make a log–log graph of y vs x. **b)** Determine the relation between y and x in the form $y = f(x)$.

TABLE 3.3

y	x
1	1
343	7
512	8
1728	12
4096	16
5832	18
9261	21
15625	25

• **[3.11]** In an experiment, you have made the following measurements of F and z shown in Table 3.4. **a)** Make a log–log graph of F vs z. **b)** Determine the relation between F and z in the form $F = f(z)$.

TABLE 3.4

F	z
6.3	39.7
8.3	68.9
10.4	108.2
11.1	123.2
12.8	163.8
13.5	182.3
15.7	246.5
16.3	265.7

[3.12] A swimmer swims in a circular arc of radius 560 m and swims a distance of 90 m. Through what angle (in degrees) has she swum?

[3.13] If a stuntwoman travels with a speed of 44 m/s around a circle with radius of 7m, how many times per second will she complete a circle?

[3.14] Refer to Figure 3.25. **a)** Find θ. **b)** Find y.

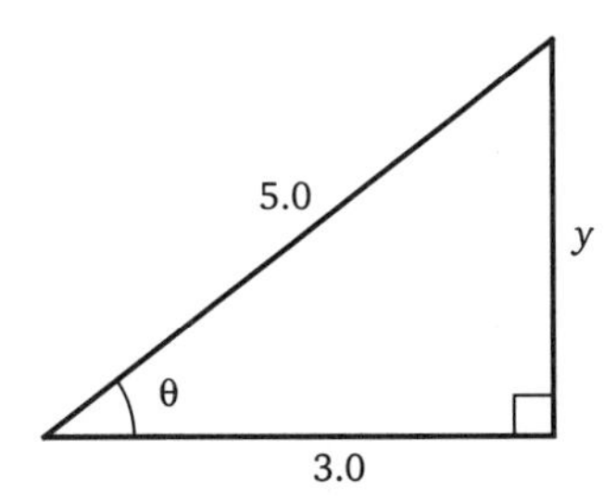

Figure 3.25
Problem 3.14.

[3.15] For the arrangement of triangles shown in Figure 3.26, find θ. The vertical lines are parallel.

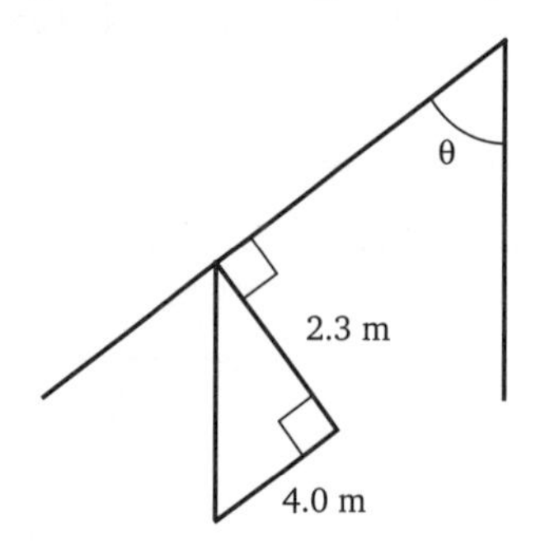

Figure 3.26
Problem 3.15.

[3.16] For the triangle shown in Figure 3.27, if $a = 6.0$ m, and $b = 8.0$ m, then find r. This is not a right triangle.

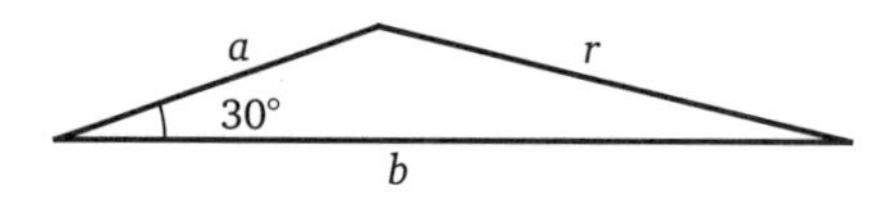

Figure 3.27
Problem 3.16.

[3.17] Evaluate the following: **a)** $\cos^{-1}\left(\frac{\sqrt{2}}{2}\right)$ **b)** $\sin^{-1}$ **c)** $\tan\left(\sin^{-1}\frac{13}{7}\right)$.

[3.18] A water tower stands on a cliff with the base of the tower 600 ft above a level plain. The angles of elevation of the top and bottom of the tower with the plain are 45° and 30°, respectively. How tall is the tower?

[3.19] On a hot summer day, a buzzard is flying in a circle of radius 20 m. He completes one whole circle every 17.9 s. Find **a)** his period, **b)** his linear frequency, **c)** his angular frequency, **d)** his angular speed, and **e)** his linear speed.

• **[3.20]** An object moves in a circle of radius r with a constant angular speed ω. Assuming the object starts at the initial angle θ_o, derive a formula for θ as a function of time. It should be the angular equivalent of the linear formula $x = x_o + v\,t$.

• **[3.21]** When viewed from the earth, the moon perfectly eclipses the sun. The moon has a diameter of 3.5×10^6 m and is 3.8×10^8 m from the earth. If the sun has a diameter of 1.4×10^9m, how far is the sun from the earth?

[3.22] Using the behavior of the legs of the triangle shown in the circle for Figure 3.17, sketch the graph of $\cos\theta$ from 0 to 2π.

• **[3.23]** Using the behavior of the legs of the triangle shown in the circle for Figure 3.17, estimate the graph of $\tan\theta$ from $-\pi/2$ to $+\pi/2$.

CHAPTER 4
Elements of Approximation and Graphing

(Pretty Steep Business!)

In this chapter we consider how to describe processes. In a process quantities change. Change one quantity and it causes changes in other quantities. Here we will use the ideas of algebra and slope to describe functional relationships. In the first two sections of the chapter we will expand on the ideas introduced in Chapter 2. You will review the geometrical basis of functional relationships but most of the discussion will be directed towards how slope is used in physics. And as we investigate functions that depend on one variable, you will discover that slope is an invaluable tool for exploring nature.

In the last three sections of this chapter we consider an often neglected, but very important topic, that of approximation. In attacking a problem a student will often strive to learn the exact solution with the mistaken idea that the exact solution will tell him all there is to know about the phenomenon. Complete solutions, however, are available in only a handful of problems, and even then, they are often complicated. The details and the complicated result of an exact solution often mask the underlying principles and behavior of a system. Often you will learn far more by determining how the system behaves under certain limiting cases. A useful limiting case will have a simple approximation that you can readily understand. These limiting cases can be much easier and quicker to calculate than the full solution, and can give you insight into the physics of the phenomenon. Approximation techniques are tools for exploration. There is nothing wrong with finding an approximate solution. Do not become academically "macho" and assume that approximation is only the recourse of mental weaklings. An experienced scientist knows not only how to approximate, but also when.

4.1 THE CHAIN RULE

The mathematics of physics is the mathematics of change. At the base of this mathematics is the relation that describes how a function responds to a change in a variable. Often several variables may depend on one another and the changes must be strung together or *chained.* The chain rule is one of the most powerful and important concepts in mathematics, and it is also one of the most useful in science. Often one must calculate the response of a system to a change in the range of the variables. For example, one could ask how the pressure in the containment vessel of a nuclear reactor would change given a change in temperature. The result is important not only for your future employment, but also because thousands of lives may depend on the calculation being correct. We will first consider a function of one variable $f(x)$. Given a change in x, Δx, what is the corresponding change in the function Δf? We can begin by looking at the slope of a function, as in Figure 4.1.

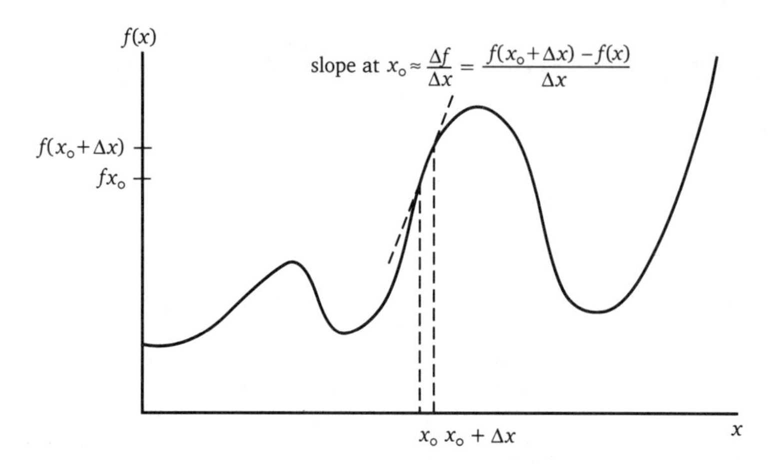

Figure 4.1
Provided the intervals are small, the slope at x_o is approximately $\Delta f/\Delta x$.

Provided the intervals are small enough, the slope, m, of the function at x_o will be approximately the same as the change in the function divided by the change in x.

$$m \approx \frac{\Delta f}{\Delta x} = \frac{f(x_o + \Delta x) - f(x)}{\Delta x} \tag{4-1}$$

$$\Delta f = \left(\frac{\Delta f}{\Delta x}\right) \Delta x. \tag{4-2}$$

Equation (4-2) may have the appearance of algebraic double-talk; however, it is a very powerful technique. The slope $\Delta f/\Delta x$ is often a quantity of physical importance. We have used this technique in kinematics to construct the relationship between the velocity and the displacement.

$$\Delta \mathbf{x} = \frac{\Delta \mathbf{x}}{t} t \tag{4-3}$$

where the $\Delta \mathbf{x}/t$ is the average velocity, $\mathbf{v}$ so that we obtain

$$\mathbf{x} - \mathbf{x}_o = \mathbf{v}t. \tag{4-4}$$

We can extend the technique to nested functions of functions by *chaining* together the various slopes of the functions involved. For example, suppose f is a function of x, and x, in turn, depends on time. Then to find how f will change if time changes, we easily obtain the correct rule by chaining together slopes:

$$\frac{\Delta f}{\Delta t} = \frac{\Delta f}{\Delta x}\frac{\Delta x}{\Delta t} \qquad \text{chain rule} \tag{4-5}$$

or

$$\frac{\Delta f}{\Delta t} = v\frac{\Delta f}{\Delta x}$$

where $\Delta x/\Delta t$ might represent the velocity of some part of the system. Equation (4-5) is termed the *chain rule* in calculus, but it is often used algebraically in physics as well.

EXAMPLE 4-1

Consider the function f to represent the height of a traveling wave pulse. You are a stationary observer watching the pulse come towards you. Given that from the chain rule

$$\frac{\Delta f}{\Delta t} = v\frac{\Delta f}{\Delta x} \tag{4-6}$$

why would the rate of change of the height depend on the velocity and the slope?

Solution:

The more quickly the pulse comes towards you, the more quickly you will see the height change, i.e., the duration of the rise and then fall of the height will be shorter. Also, if f is coming towards you with a steep profile to the wave, it will cause f to change quickly, as illustrated in Figure 4.2.

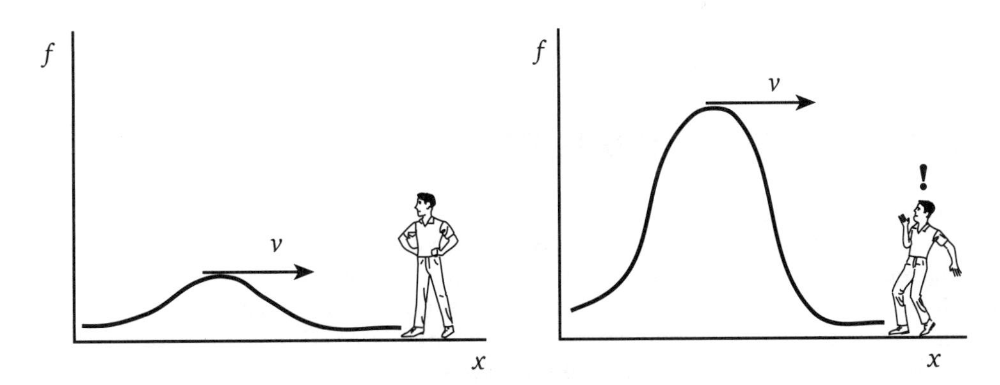

Figure 4.2
The first wave is never very steep so df/dt is always small.
The second wave is steep so you see a large $\Delta f/\Delta t$.

EXAMPLE 4-2

A storage tank is being filled with gasoline of specific gravity 0.9 by a pipe of radius 9.0 cm. If the speed of the flow in the pipe is 12 m/s, find the rate at which gasoline is pouring into the tank in kg/s.

Solution:

First ask yourself what are we being asked for. The units, kg/s, indicate that we need to calculate the $\Delta m/\Delta t$ pouring into the tank. Next ask what are the things that can affect that rate. They are:

1) how fast the fluid is flowing, $v = \Delta x/\Delta t$
2) how large the pipe is, cross-sectional area, A,
3) how dense the fluid is, $\rho = \Delta m/(A\Delta x)$.

Applying the chain rule to include these effects gives

$$\frac{\Delta m}{\Delta t} = \frac{\Delta m}{A\Delta x}\frac{A\Delta x}{\Delta t} \tag{4-7}$$

Δx is the distance the flow moves in a time Δt, so $A\Delta x$ is the volume the flow will sweep out in the pipe in a time Δt. The first term on the right-hand side is the mass per volume which is the density, ρ, and the $\Delta x/\Delta t$, is the speed, v.

$$\frac{\Delta m}{\Delta t} = \rho v A. \tag{4-8}$$

A specific gravity of 0.9 means that the gasoline is 0.9 times as dense as water.

$$\frac{\Delta m}{\Delta t} = (0.9 \times 1000 \text{ kg/m}^3)(12 \text{ m/s})\, \pi\, (0.09 \text{ m})^2 = 275 \text{ kg/s}$$

This Equation, (4-8) is quite important in fluid flow. The amount of mass passing per unit time through a surface is called the flux, and represents one of the fundamental quantities for describing a flowing fluid.

The chain rule is a very powerful tool for solving many difficult problems in physics. You will use it often.

4.2 SLOPE AND CURVATURE

The slope of a function is used for more than just finding rates of change. Slope forms the foundation for describing any process in which a change in one quantity causes a change in another. For instance, suppose we are trying to describe the function shown in Figure 4.3. Note that it is not a simple linear function, but contains many curves and at first glance appears rather formidable. One might start by trying to describe where the maxima and minima occur.

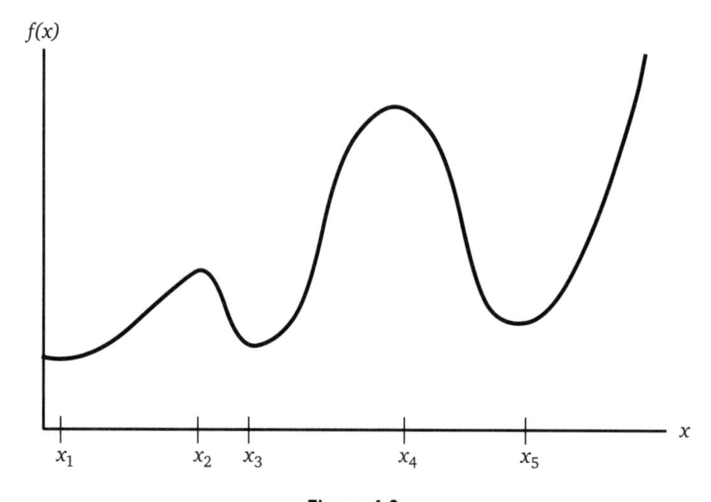

Figure 4.3
A complicated function of one variable.

The lowest point in the function occurs at the minimum, x_1. Because this is the lowest point on the function, it is called the absolute minimum. There are also minima at x_3 and x_5 which are called relative minima. There can be many relative minima but there can be only one absolute minimum for a given function. Because none of the maxima at x_2 and x_4 are the largest value of the function, they are both relative maxima. There is one thing that all the minima and maxima have in common. Can you find it? If you are having trouble, look at Figure 4.4.

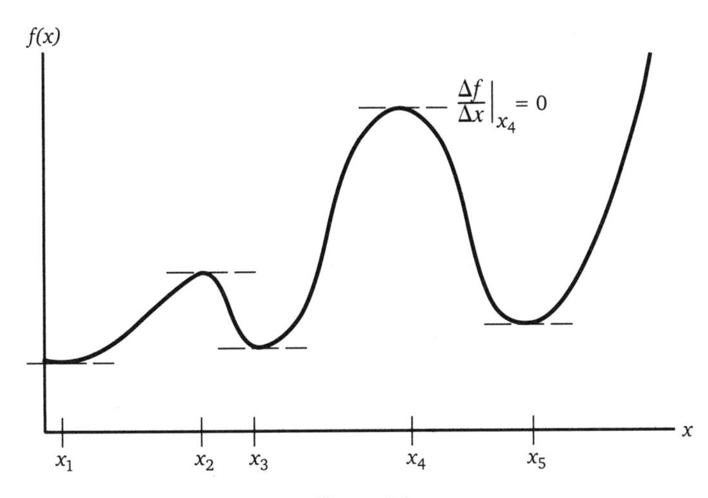

Figure 4.4
The slope of the function is zero at a minimum or a maximum.

Of course, the slope of the function is zero both at the maxima and at the minima. Extrema refers to both maxima and minima. Thus to find where the extrema occur, we only need to find where the slope of the function is zero!

EXAMPLE 4-3

Find the location of the extrema of the function

$$f(x) = 4x^2 + 3x + 7.$$

Solution:

$f(x)$ is a parabola, so there will be one extremum, a minimum, at the vertex. For a parabola of the form

$$y = ax^2 + bx + c,$$

the vertex is at

$$x = \frac{-b}{2a}$$

$$x = -\frac{3}{8}.$$

$f(x)$ is a parabola in which the $3x$ term has moved the minimum of the parabola to the left of the origin. If instead $f(x) = 4x^2 + 7$, then $b = 0$ and the minimum is at $x = 0$.

In the previous example it was easy to tell whether the extremum was a maximum or a minimum because we used a familiar function, but what if the example were complicated? How can we tell whether an extremum is a maximum or a minimum?

The only way to tell if an extremum is a maximum or a minimum is to look at the curvature. If you stand on the top of a hill, the land will curve down no matter which direction you look. So too, if you stand at the lowest point in a valley, the land will curve up in every direction. What we need is a way to quantify this. One measure is the change in the slope of the function or the "slope of the slope". At a minimum, the slope must increase as x increases, because the function curves upwards as at x_5 in Figure 4.5.

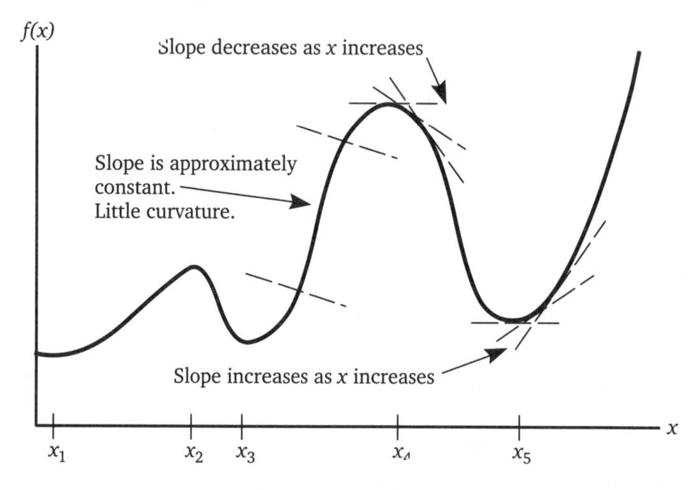

Figure 4.5
At a minimum the slope increases as x increases, and at a maximum the slope decreases as x increases.

Thus, at a minimum the change in the slope must be positive. Similarly, at a maximum the slope decreases, becoming more negative as x increases, so the change in the slope must be negative, i.e., the function curves down and has negative curvature.

The funny thing about all this is that the curvature tends to have its greatest magnitude in the region where the slope goes to zero—that is, at the maximum or minimum. And furthermore, where the slope has its largest absolute values, that is, where the function is steepest—the curvature tends to be near zero. This can be seen from Figure 4.5. It's always important to remember that just because a function has zero slope at a specifed point does not mean that the curvature is zero at that point.

EXAMPLE 4-4

For the function $f(x) = x^3 - 4x^2 - 16x + 100$

a) find the positions of the extrema, b) find the regions in which the curvature is positive, c) find the regions in which the curvature is negative, d) find the regions in which the curvature is approximately zero.

Solution:
One way to find the extrema is to graph the function as in Figure 4.6.

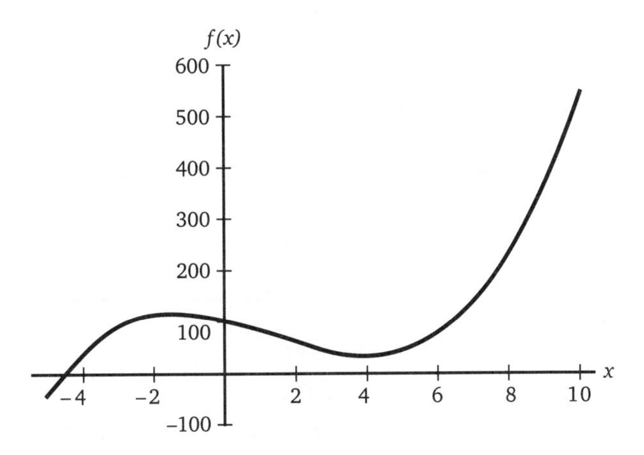

Figure 4.6
A graph of $f(x) = x^3 - 4x^2 - 16x + 100$.

a) From the graph in Figure 4.6 you can see that there is a relative maximum at $x = -4/3$, and a relative minimum at $x = +4$. As x goes to infinity, the function increases without bounds so there are no absolute extrema. b) The function has positive curvature where it forms a smiley face which is in the range $2 < x < 6$. c) The function has negative curvature where it forms a frowney face which is in the range $-4 < x < 0$. From $0 < x < 2$, the function is a straight line so the curvature is approximately zero. Also in the range $8 < x < 10$, the curvature is very small and the function may be approximated with a straight line.

4.3 EXPANSION PARAMETERS AND THE BINOMIAL THEOREM

The basis behind all approximation techniques is that when something small is added to a large quantity, the small thing does not significantly change the large quantity. For instance, consider the addition of $7.5 \times 10^4 + 0.003 = 75000.003$. Obviously, the 0.003 is insignificant compared to the 7.5×10^4. In fact, the notation on the 7.5×10^4 leads one to believe that it is not known to enough significant places to bother with the 0.003. So if we have a function $f(x) = x \pm d$, and $d \ll x$, then $f(x) \simeq x$. The d is insignificant compared to x.

You will see approximations denoted in various ways. Some use $\simeq$, $\cong$, or $\approx$. All of these symbols mean an approximation, but they are used differently, depending on the individual. For example, some people set up a hierarchy where $\cong$ denotes a very good approximation, $\simeq$ a fair approximation, and $\approx$ a rough approximation. These conventions tend to be rather loose, so take the usage with a grain of salt.

When we multiply two numbers together, it becomes more confusing, but there are some general rules to follow: d is small relative to x, so $x + d \simeq x$, but if x is large and d is small, then $(x \cdot d)$ may be a medium number that we really should not ignore.

$$x + x \cdot d + d \simeq x + x \cdot d$$

A general rule of thumb is that small numbers may be ignored when added or subtracted from a large number, but you cannot ignore the product of a small number times a large number.

Consistency is another important rule. When dealing with powers, you must consistently retain those at the same level of approximation. For example, if $x + xd + d \simeq x + xd$, then $x + xd^2 + xd + d \simeq x + xd$. The xd^2 is a large number multiplied by two small numbers $(d \cdot d)$ and therefore is small. We can make this discussion more formal by introducing an expansion parameter ε *(epsilon)* where $\varepsilon \ll 1$.

Expansion Parameter

The expansion parameter must be a small quantity $\varepsilon \ll 1$. You can form an expansion parameter if there are two widely different numbers. For instance, if $x \gg d$, then

$$\varepsilon = \frac{d}{x} \qquad (x \gg d)$$

or if $d \gg x$, then

$$\varepsilon = \frac{x}{d} \qquad (d \gg x).$$

EXAMPLE 4-5

As a first example, we approximate $x^2 + xd + d^2$ for $x \gg d$ by constructing a small expansion parameter ε.

Solution:

First, pull out the highest power of x, such that,

$$x^2 + xd + d^2 = x^2\left(1 + \frac{d}{x} + \frac{d^2}{x^2}\right) \tag{4-9}$$

Because d is very small relative to x, $\frac{d}{x}$ is a very small number compared to 1. Thus, our expansion parameter is $\varepsilon = \frac{d}{x}$, so that,

$$x^2 + xd + d^2 = x^2(1 + \varepsilon + \varepsilon^2). \tag{4-10}$$

Because $\varepsilon \ll 1$,

$$x^2 + xd + d^2 \simeq x^2. \tag{4-11}$$

This technique of using a small expansion parameter has the advantage of clearly showing the order of the approximation being made. For example, if we want a better approximation, we could keep all terms of first order in ε; i.e.,

$$x^2 + xd + d^2 \simeq x^2(1 + \varepsilon), \tag{4-12}$$

where, to be consistent we need to keep all terms that depend on the first power of ε. Of course, in this case, there is only one term; however, using a small expansion parameter gives us a consistent means of approximating functions. *Order* refers to the power of the expansion. "First order in x" signifies all terms raised to the first power of x, and "second order in x" refers to all terms raised to the second power of x.

We can also extend these ideas to ratios of polynomials. Remember, you can drop d when adding or subtracting from a relatively larger number x but not when multiplying or dividing, as Example (4-8) indicates. Using this property, we can sometimes approximate a polynomial quickly without doing a formal expansion.

EXAMPLE 4-6

Find the first order expansion for $\frac{x^2d+x}{x+d}$ when $x \gg d$.

Solution:

The $x + d$ in the denominator may be approximated by x so that

$$\frac{x^2d+x}{x+d} \simeq \frac{x^2d+x}{x} = xd + 1. \tag{4-13}$$

Note that the d in the numerator cannot be dropped because it multiplies x. We do not know whether x^2d is large or small compared to 1, so we cannot justifiably approximate this further. Also, we were told to find the approximation through first order, so we must include all terms involving the first power of x.

There are tricks that allow us to deal with special cases, such as those involving square roots in the denominator. In these cases, you must rationalize the denominator before approximating.

EXAMPLE 4-7

Approximate to first order in x the function $\frac{1}{x-\sqrt{x^2+d^2}}$ .f or $x >> d$.

Solution:

First, rationalize the denominator by multiplying the numerator and denominator by the conjugate,

$$x + \sqrt{x^2+d^2}.$$

$$\frac{1}{x-\sqrt{x^2+d^2}} = \frac{1}{x-\sqrt{x^2+d^2}} \bullet \frac{x+\sqrt{x^2+d^2}}{x+\sqrt{x^2+d^2}}$$

$$\frac{1}{x-\sqrt{x^2+d^2}} = \frac{x+\sqrt{x^2+d^2}}{x^2-(x^2+d^2)} \tag{4-14}$$

Note that so far we have not yet made any approximations. The d in the denominator cannot be dropped because the x^2 would cancel, leaving zero in the denominator. We can, however, drop the d in the numerator because it is added to x^2

$$\frac{1}{x-\sqrt{x^2+d^2}} \simeq -\frac{2x}{d^2}. \tag{4-15}$$

Often a function is represented in the form of a series of terms, each a higher power than the previous term. Such a series is termed a *power series.* One of the most useful power series expansions is the *binomial theorem* because it represents sums raised to a power, and many laws in physics turn out to be power laws. To apply the binomial theorem, we start with $(x + d)^n$, where n is an integer, and construct an expansion parameter ε where $\varepsilon < 1$.

$$(x + d)^n = x^n \left(1+\frac{d}{x}\right)^n = x^n (1 + \varepsilon)^n \tag{4-16}$$

The binomial theorem is

The Binomial Theorem

$$(1 \pm \varepsilon)^n = 1 \pm n\varepsilon + \frac{n(n-1)}{2!}\varepsilon^2 + \ldots + (\pm 1)^n \frac{n!}{(n-r)!\,r!}\varepsilon^r + \ldots + (\pm 1)^n \varepsilon^n$$

where $0 < \varepsilon < 1$, and n is an integer. If n is not an integer, the series still holds but becomes an infinite series.

Although the binomial theorem holds for any $0 < \varepsilon < 1$, it would pretty much be a mathematical curiosity were it not that for small ε, $|\varepsilon| \ll 1$, we can neglect terms of higher order than ε, such that,

$$(1 \pm \varepsilon)^n \simeq 1 \pm n\varepsilon \quad \text{for } \varepsilon \ll 1. \tag{4-17}$$

EXAMPLE 4-8

A commonly occurring case in physics is when $n = -½$. Find the linear approximation for

$$f(x) = \frac{1}{(x-3)^{½}} \quad \text{if } x \gg 3.$$

Solution:

First we need to create a very small expansion parameter. Let $\varepsilon = \frac{3}{x}$. Then, applying the binomial theorem:

$$f(x) = (x - 3)^{-½}$$

$$f(x) = x^{-½}(1 - \varepsilon)^{-½} \simeq x^{-½}(1 - (-½)\varepsilon)$$

$$(x-3)^{-½} \simeq x^{-½}\left(1 + \frac{\varepsilon}{2}\right)$$

$$f(x) = \frac{1}{x^{½}} + ½\frac{3}{x^{3/2}} \tag{4-18}$$

EXAMPLE 4-9

A case that can really throw you happens when a very small number d is combined under a square root with a large number x and then subtracted from x, as in the formula

$$y = x - \sqrt{x^2 - d^2},$$

where completely dropping the d term gives zero. Plugging numbers into a calculator will also often yield zero. Evaluate y using $x = 100$ and $d = 0.000003$ with **a)** your calculator, **b)** the binomial theorem.

Solution:

a) Using a calculator gives $d^2 = 9 \times 10^{-12.}$ Plugging in the formula produces

$$y = 100 - 100 = 0.$$

b) We need to create a very small expansion parameter. Let $\varepsilon = \frac{d}{x}$. Then, applying the binomial theorem to the square root and keeping only the first two terms:

$$(x^2 - d^2)^{1/2} = x(1 - \varepsilon^2)^{1/2} \simeq x\left(1 - \frac{1}{2}\varepsilon^2\right).$$

Approximate y:

$$y \simeq x - x + \tfrac{1}{2}\frac{d^2}{x^2}x$$

$$y \simeq \tfrac{1}{2}\frac{d^2}{x} = \frac{1}{2}\frac{0.000003^2}{100} = 4.5 \times 10^{-14}.$$

Obviously, the result is small, but it is not zero. Without an approximation scheme, you would have had to wait for the next generation of calculators to get the result.

4.4 APPROXIMATIONS FOR TRIGONOMETRIC FUNCTIONS

The sine, cosine, and tangent are termed *circular functions* because they give the projections for a unit circle. It is fairly easy to obtain approximations for the circular functions by relating them to the unit circle. In terms of the unit circle shown in Figure 4.7, the circular functions are

$$\sin\theta = \frac{y}{1} = y$$

$$\cos\theta = \frac{x}{1} = x$$

$$\tan\theta = \frac{y}{x} \qquad (4\text{-}19)$$

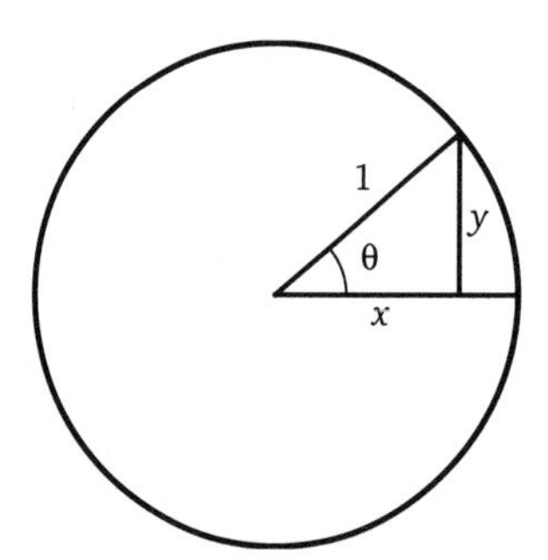

Figure 4.7
The trig functions from a unit circle

Suppose that θ is small, so that $\theta \ll 1$, where θ is in radians. Note that 1 rad = 57.3°, so that "small" compared to 1 radian can be as large as 20°. When θ is small, y will also necessarily be small, as shown in Figure 4.8.

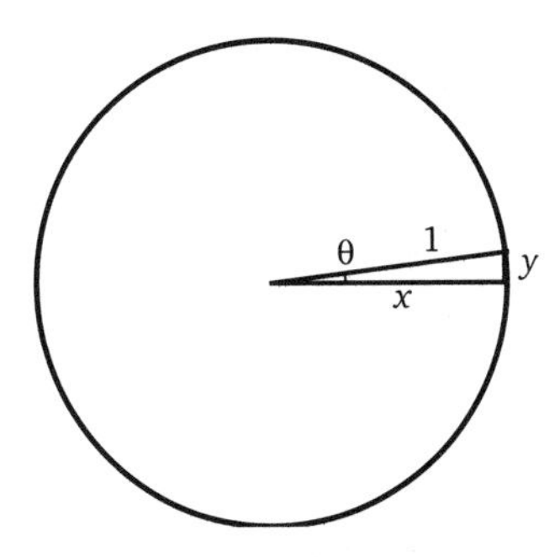

Figure 4.8
When θ is small, y is also small compared to 1.

Furthermore, y is approximately equal to the arc length of the circle $y \simeq s$ where $s = r\theta = \theta$. We have thus arrived at the approximation

$$\sin\theta \simeq \theta \,. \tag{4-20}$$

We can also approximate the tangent because there is very little difference between the radius of the unit sphere and x, $x \simeq 1$:

$$\tan\theta \simeq \frac{\theta}{1} = \theta \,. \tag{4-21}$$

We have found that when θ is small, there is little difference between the tangent and the sine. This approximation is used very often in optics and in simple harmonic motion. It is termed the *small-angle approximation.*

The Small Angle Approximation

For $\theta \ll 1$ (in radians):

$$\sin\theta \simeq \tan\theta \simeq \theta \,. \tag{4-22}$$

The small-angle approximation is a very good approximation up to about $\theta = 20°$.

For small angles, the cosine is at its maximum, so it changes very little compared to the others. From Figure 4.9,

$$\cos\theta = \frac{x}{1}\,, \tag{4-23}$$

but since $x \simeq 1 \Rightarrow$

$$\cos\theta \simeq 1 \tag{4-24}$$

The first variation in the cosine at the origin is in the θ^2 term of the power series expansion because the cosine at $x = 0$ is at a maximum, and so it varies little with θ. So, to the same order as the sine and tangent, the cosine is a constant in the small-angle approximation.

The approximations change if we choose a different point to approximate the functions. For example, the cosine changes most rapidly at $\theta = \pi/2$ so we could expand the cosine about $\pi/2$. This turns out to be a very good point to use for the cosine because x is small, as shown in Figure 4.9.

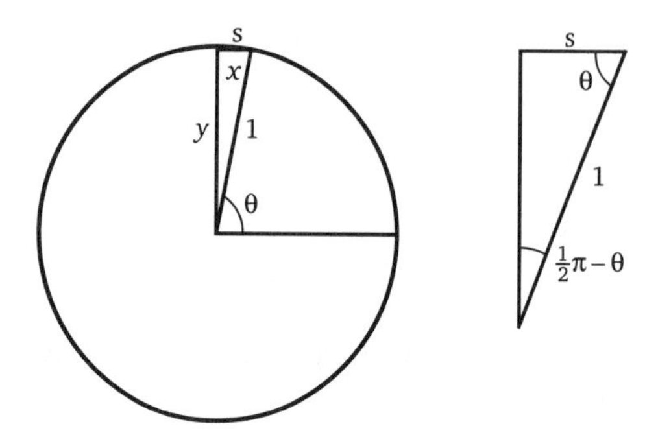

Figure 4.9
Near $\pi/2$, $x \simeq s$ and $y \simeq r$.

Also, x is very nearly s, so that

$$\cos\theta = x \simeq s$$

$$s = 1 \cdot \left(\frac{\pi}{2} - \theta\right)$$

$$\cos\theta \simeq \frac{\pi}{2} - \theta\,. \qquad (4\text{-}25)$$

Remember that expansion Equation (4-25) only holds near $\theta = \pi/2$, where the cosine is almost zero.

4.5 SLOPE AND NUMERICAL APPROXIMATION

Heretofore we have discussed techniques for approximating specific functions; however, we will need general techniques that work in a wide variety of cases. In this section we will discuss a fairly simple yet powerful approximation technique called linear approximation that will work with a wide variety of functions.

Suppose we have a complicated function of a variable x like that shown in Figure 4.10. The idea behind an approximation scheme is not to approximately reproduce the whole function, but to simply reproduce a segment of the function. We'll start by trying to reproduce the function in the neighborhood of the point x_o where the function has the value $f_o \equiv f(x_o)$.

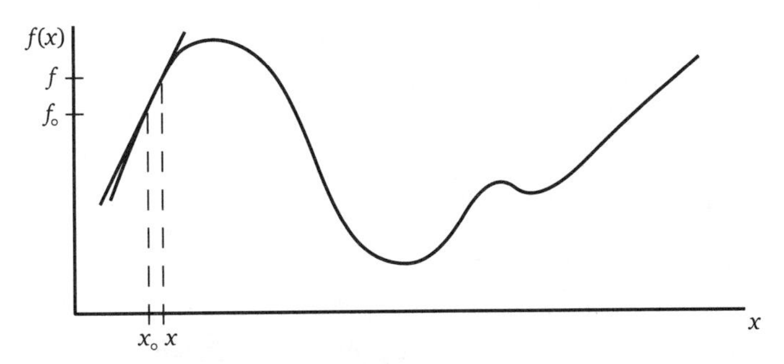

Figure 4.10
A complicated function $f(x)$ can be approximated in a small region about x_o by a straight line.

In a small region about x_o, the function is almost a straight line, so we could approximate $f(x)$ with the equation of a straight line. We can again use the slope, m, to write

$$\Delta f = \frac{\Delta f}{\Delta x}\Delta x \approx m\,\Delta x. \tag{4-26}$$

The change in f can be taken between x_o and an arbitrary point x to the right of x_o. If m_o is the slope of the function at x_o,

$$f - f_o \approx m_o(x - x_o)$$

Solving for f:

$$f(x) \approx f_o + m_o(x - x_o) \qquad \text{linear approximation} \tag{4-27}$$

All we have done is approximate the function $f(x)$ with a straight line of slope m_o. This will be accurate if the curvature in f is small or we keep $x - x_o$ small. We can use the linear approximation to approximate the trigonometric function we studied in the last section.

EXAMPLE 4-10

Find the linear approximation for sin θ about the origin.

Solution:
We always must specify the point about which we are expanding the function. In this case we wish to know the behavior of sin θ for small θ; i.e., those values of θ near θ = 0. The variable is now θ, so $x = \theta$ and $x_o = 0$. Substituting sin θ for $f(x) \Rightarrow$

$$\sin\theta \approx \sin(0) + m_o(\theta - 0) \tag{4-28}$$

We need the slope of the sin θ near the origin which we can calculate from using a small interval say $0 \le \theta \le 1°$.

$$m_o \approx \frac{\Delta(\sin\theta)}{\Delta\theta} = \frac{\sin 1° - \sin 0}{1° \cdot \frac{\pi}{180°}}$$

where $\Delta\theta$ is in radians.

$$m_o \approx \frac{0.01745}{0.01745} = 1 \qquad (4\text{-}29)$$

Combining Equations (4-33) and (4-34),

$$\sin\theta \approx \theta \qquad (4\text{-}30)$$

which is the result we found in the last section. In fact, back in Chapter 3, we argued that the sine must be linear near the origin. We have just found that the slope of the sine is approximately 1 at the origin.

In Example 4-13, we found a result that goes past the small-angle approximation found in Equation (4-24). There are two points worth mentioning. The first point is that the linear approximation works fine for odd functions, but it does not work for even functions. If we look at the Taylor series for the other circular functions, we find that the expansions for the sine and tangent use only odd-powered terms, although the expansion for the cosine involves only even-powered terms.

Series Expansions for the Circular Trig Functions

$$\sin\theta = \theta - \frac{\theta^3}{3!} + \frac{\theta^5}{5!} - \frac{\theta^7}{7!} + \ldots$$

$$\cos\theta = 1 - \frac{\theta^2}{2!} + \frac{\theta^4}{4!} - \frac{\theta^6}{6!} + \ldots$$

$$\tan\theta = \theta + \frac{\theta^3}{3} + \frac{2}{15}\theta^5 + \frac{17}{315}x^7 + \ldots$$

From the graphs of the trig function in Figure 4.11, we see that the cosine is symmetric about the y axis; i.e., the positive side curves down from a maximum, and so does the negative side. On the other hand, the sine and tangent are termed *antisymmetric* about the y axis; i.e., the behavior on the positive side is exactly opposite to the negative side.

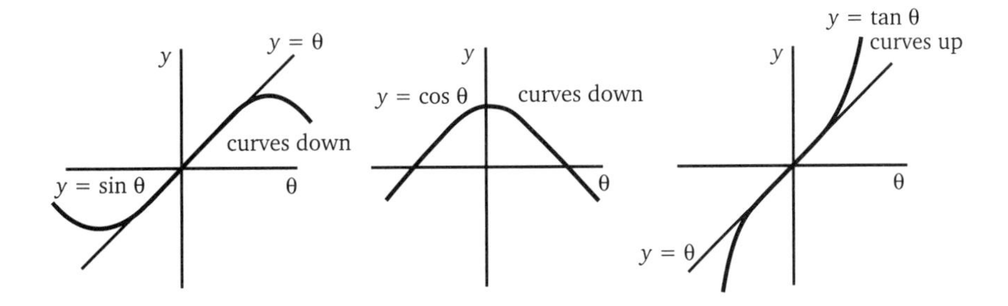

Figure 4.11
1) The cosine is symmetric about the y *axis*, but the sine and tangent are antisymmetric.
2) The sine and cosine curve down from straight–line behavior ,while the tangent curves up for positive θ .

For symmetric behavior, $f(\theta) = f(-\theta)$, which can only be reproduced by even-powered terms. Likewise, for antisymmetric behavior $f(\theta) = -f(-\theta)$, which is a property of the odd-powered terms.

The second point is that the sign of the next highest nonzero-powered term after the linear term (the one that's $\propto \theta^3$) represents the curvature and will be either positive if the function curves up from straight-line behavior, as does the tangent, or negative if the function curves down from straight line behavior as do both the sine and the cosine.

Chapter 4 Problems

[4.1] A wave with a velocity of 12 m/s is passing by a buoy. Find the slope of the wave profile at the moment the buoy is increasing in height at 1.3 m/s. Is the slope positive or negative?

[4.2] A balloon is being filled with air that flows through a small filling tube with a velocity of 4 m/s. If the mass of the air per length in the tube is 3 g/m, then find the rate at which the mass of the air in the balloon is increasing in g/s.

[4.3] Captain Nemo's submarine has just received a damaging hit and is sinking at a rate of 4 m/s. If the pressure (force per unit area) is increasing by 10,000 N/m^2 with each meter of depth, then determine the rate at which the pressure is increasing.

• **[4.4]** A storage tank is being filled with gasoline of specific gravity 0.9 by a pipe of radius 7.0 cm. If the speed of the flow in the pipe is 12 m/s, **a)** find the rate of volume at which gasoline is pouring into the tank in m^3/s. **b)** If the tank has a volume of 12,000 m^3, how long will it take to fill the tank?

• **[4.5]** **a)** Find the speed of electrons in a beam exiting a cyclotron accelerator if there are 5×10^{14} electrons/s coming out of the accelerator, and the density of the electrons in the beam is 2.5×10^6 electrons/m. **b)** Current is measured in amperes (A), and is the amount of charge per unit time coming out of the accelerator. Charge is measured in Coulombs (C), so that, 1 A = 1 C/s. If the charge on an electron is 1.6×10^{-19} C, then find the beam current.

[4.6] **a)** Make a graph of $v = (2.0)\ t$ where the velocity v has SI units. Allow t to go from 0 to 4 seconds **b)** Make a graph of the slope of v from zero to 4 seconds.

[4.7] Find whether the function $f(x) = x^3 - 9x^2 + 3x + 14$ has positive or negative curvature at $x = 2$.

[4.8] The velocity as a function of time is given as:

$$v = 8t - t^2$$

in SI units. Consider the time interval from $t = 0$ to $t = 10$, find **a)** the range of time when the speed is increasing and the range when it is decreasing, **b)** the time at which the velocity is a maximum and its maximum value, and **c)** the time at which the speed is a maximum and its maximum value.

[4.9] List all the regions where the function f shown in Figure 4.12 has **a)** positive curvature, **b)** negative curvature, and **c)** zero curvature.

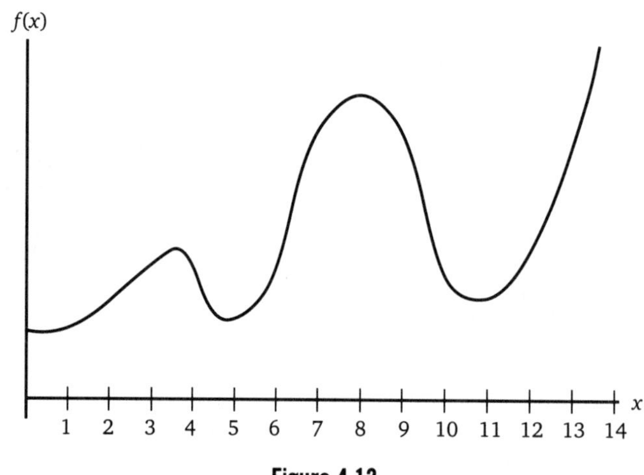

Figure 4.12
Problem 4.9.

[4.10] **a)** Make a sketch of $f(t) = Ae^{-t/\alpha}$. **b)** Is the curvature of $f(t) = Ae^{-t/\alpha}$ positive or negative? **c)** Make a sketch of $f(t) = A(1 - e^{-t/\alpha})$. **d)** Is the curvature of $f(t) = A(1 - e^{-t/\alpha})$ positive or negative?

[4.11] An arrow with an initial velocity v_o starting from an initial position y_o is shot straight up into the air. **a)** Write a formula for the height y as a function of time. **b)** Find all the values of t and y where the extrema occur. **c)** Prove whether each point is a maximum or a minimum by finding if the curvature of y is positive or negative. Be sure to thoroughly explain conclusions. **d)** Make a sketch of y vs t.

[4.12] On a clear day if you look southwest from Davis, California towards San Francisco, you can see Mount Diablo. With your arm fully extended, Diablo subtends the same angular width as three finger nails of your little finger. If Mount Diablo is about 3 miles wide, how far away is Mount Diablo?

[4.13] **a)** Compare the value of $\tan\theta$ to that for $\sin\theta$ for $\theta = 10°$, and for $\theta = 30°$. What do you conclude? **b)** Explain your conclusion to part **(a)** by using a triangle in which θ is small but whose hypotenuse and side adjacent to θ are very large compared to the side opposite θ.

[4.14] Find an approximation for

$$f(x) = \frac{ax - a^2}{x^2 - 2ax + a^2}$$

when $x \gg a$.

[4.15] **a)** Use your knowledge about the power series behavior of sin θ to estimate sin 20°. Do not use your calculator. This is purely a mental exercise. **b)** Compare your estimate in part **(a)** to the value obtained from your calculator. Give the percentage error in your estimate.

[4.16] Derive the linear approximation for tan θ at $\theta = 0$.

[4.17] Find the first two non-zero terms in the approximation for

$$f(x) = \frac{1}{(x - d)^5}$$

where $x \gg d$.

[4.18] Derive the linear approximation for $e^{x/d}$ where $x \ll d$.

• **[4.19]** Derive the linear approximation for $\dfrac{1}{\sin\left(\dfrac{x}{d}\right) + 1}$ where $x \ll d$.

• **[4.20]** Calculate the slope of e^t at $t = 0$ by using successively smaller intervals in Equation 4-1. Let $t_1 = 0$ and find the slope if **a)** $t_2 = 1$, **b)** $t_2 = 0.5$, **c)** $t_2 = 0.1$, and **d)** $t_2 = 0.01$. What do you conclude about the slope of e^t at the origin? Which result do you feel is the most accurate?

CHAPTER 5
Probability and Error

(The odds must be crazy)

This chapter teaches you how to handle errors and inaccuracies in experiments. These are necessary skills that will increase your appreciation and understanding of science. In the television series *Star Trek,* Spock plays a parody of the quintessential scientist. When queried, he always responds with the complete formula, accurate to the nth degree. You will find that making an accurate measurement is the result of painstakingly careful work, and that even in the best measurements, there will always exist an uncertainty, either from random errors produced by the experiment itself, or at best, by uncertainties introduced by the structure of the universe. Quantum mechanics describes the ultimate limit on any measurement. Heisenberg's Uncertainty Principle places a theoretical limit that no experiment, no matter how clever, can avoid. Uncertainty is a fundamental part of the world we live in. It is built into the very fabric of the universe. You must know how to handle uncertainty and be able to faithfully communicate the uncertainty in your results to others.

5.1 PROBABILITY

Very often, in physics, the problems become very complex because there are many measurements or many components to a system. Consider for a moment the problem of determining the thermodynamic properties of a gas from the molecular motions. Temperature and pressure are properties of a mole of a substance, not of a molecule. You know that a mole contains Avagadro's number of molecules, 6.023×10^{23} molecules/mole. The human mind can deal with 3 or 4 variables at a time, but not 10^{23}! There must be some way to reduce the variables in the problem. Such systems with many measurements or variables are often treated statistically to reduce the number of variables. For instance, rather than specifying a speed for each of the 10^{23} particles, we might only talk of the average speed. If we were trying to determine the grades on a test given to a class of 50 students, we would find the mean score and some measure of how wide each grade category should be. The success of the exam can then be judged by looking at the mean and the width of the distribution, rather than by looking at the 50 individual scores. We begin with a brief introduction to probability.

The Probability Distribution

Any discussion of probability begins with a discussion of the probability distribution. The easiest case is that of a perfect die. A die has six sides so each side has an equal probability of appearing on top. The probability of the ith side showing one sixth where i runs from one to six. Probability is typically reported such that the sum of the probabilites of all possibilites is one

$$\sum_{i=1}^{6} p_i = 1 \qquad \text{normalization condition} \tag{5-1}$$

If all the probabilities sum to 1, the distribution is said to be normalized to 1. Here, each of the sides of the die is equally likely to occur so all the $p_i = 1/6$. The 1/6 means that out of 600 throws, a given side will occur about 100 times. You must be careful here because probability does not apply to a given throw of the die, but only to many throws. A given side can even occur several times in a row. On any given throw, all sides are equally likely to occur. There is no relation between a given throw and another. For example, if you have thrown a four 2 times in a row, a four is just as likely to occur on the next throw as any of the other possibilities. However, for 3 throws, there are many more possibilities than for 1 throw. You could have thrown one, three, and five for example, or a four, two and three. There are many more combinations, so the total probability of throwing a four, 3 times in a row is much lower than that of throwing a four once.

What if the die is not perfect, or what about an exam, where all the possible scores are not equally likely to occur? How would we arrive at the probability for a given side or a given score? One way would be to do it experimentally. Suppose that there are m outcomes (for a die $m = 6$; for an exam, m might be 100), and the ith outcome occurs n_i times, and there are N total trials. The probability of ith outcome occurring is then

$$p_i = \frac{n^i}{N} \,. \tag{5-2}$$

If we throw a die 100 times, and one particular side comes up 20 times, then the probability of that side appearing is 20/100 = 1/5. Because there are N total throws, if we sum all the possible outcomes, we must get N.

$$\sum_{i=1}^{m} n_i = N \tag{5-3}$$

Summing all the p_i will give 1.

$$\sum_{i=1}^{m} p_i = \frac{\sum_{i=1}^{m} n_i}{N} = 1 \tag{5-4}$$

This probability is thus normalized to 1.

The Mean and RMS Values

Certain values help to characterize a probability distribution. Some of the most common are the mean or average value, the most probable value, the median value, and the root-mean-square or rms value. Each has a particular advantage in certain situations. The *median value* is just the middle value. The *most probable value* is the one with the largest p_i, which means the one that occurs most often. The mean—as anyone who has taken an exam graded on a normal curve can testify—is very important and is given by summing all the test scores and dividing by the number of students.

$$\overline{x} = \frac{\sum_{i=1}^{N} x_i}{N} = \frac{x_1+x_2+\ldots+x_i+\ldots+x_N}{N} \tag{5-5}$$

The rms value is the square *root* of the average (*mean*) of the sum of the *squares.*

$$x_{rms} = \sqrt{\frac{\sum_{i=1}^{N} x_i^2}{N}} \qquad \text{root-mean-square} \tag{5-6}$$

Use of these values is illustrated in the next example.

EXAMPLE 5-1

A quiz worth 10 points is given to 17 students. The following scores were recorded:

1, 3, 3, 4, 4, 4, 5, 5, 6, 6, 6, 6, 7, 8, 8, 9, 10.

Find the median, the mean, the most probable, and the rms values.

Solution:

This problem is just using the definitions. Make sure you understand them. The median value is the middle score, which in this case is the ninth score and is 6. The most probable is the one that occurs the most often, which is also 6. The mean score s is given by

$$\overline{s} = \frac{1+3+3+4+4+4+5+5+6+6+6+6+7+8+8+9+10}{17} = 5.6$$

and the rms value by

$$s_{rms} = \sqrt{\frac{1^2+3^2+3^2+4^2+4^2+4^2+5^2+5^2+6^2+6^2+6^2+6^2+7^2+8^2+8^2+9^2+10^2}{17}}$$

$$s_{rms} = 6.0$$

Note that the rms value is higher than the mean because when the numbers are squared, the higher numbers contribute more strongly than the lower numbers.

Looking at the previous example, it would have been more efficient to have just counted how many times each exam score occurred. Then, the mean would be

$$\overline{s} = \frac{1+2(3)+3(4)+2(5)+4(6)+7+2(8)+9+10}{17} = 5.6$$

In this example, 2/17 is the probability of a score of 3 occurring, and 4/17 is the probability of a score of 6 occurring. Thus, the average score is found by multiplying the ith score by its probability $p_i = n_i/N$ and summing from 1 to the number of possible scores, m.

$$\overline{s} = \sum_{i=1}^{m} s_i \frac{n_i}{N} \tag{5-7}$$

For the quiz in the preceding example, m = 10. This technique for finding the average value of a quantity is standard. To find the average value of any quantity, sum over the value times its probability. For instance, to find the average value of the speed, let the ith particle have a speed v_i with a probability of $p_i = n_i/N$ where N is the total number of particles. The average is given by

$$v = \sum_{i=1}^{m} v_i p_i \tag{5-8}$$

At this point, you might be a little perplexed. Don't the mean, the most probable, and the median have the same value? Also, what's all this rms stuff? The answer to my first question is "Sometimes, but not always." In fact, rarely is the mean the most probable outcome. In general, the mean and the most probable are the same **only** for the normal probability distribution of a continuous variable. A normal probability distribution is also called a "bell curve" because it has a bell shape. It has the distinct property of being symmetric around the mean. For a discrete, normal distribution of integer scores, the mean can have a fractional value, but the most probable, and the median, having integer values, have the same value which you can see from Figure 5.1. In Figure 5.1, the mean, the most probable, and the median have a value of 50.

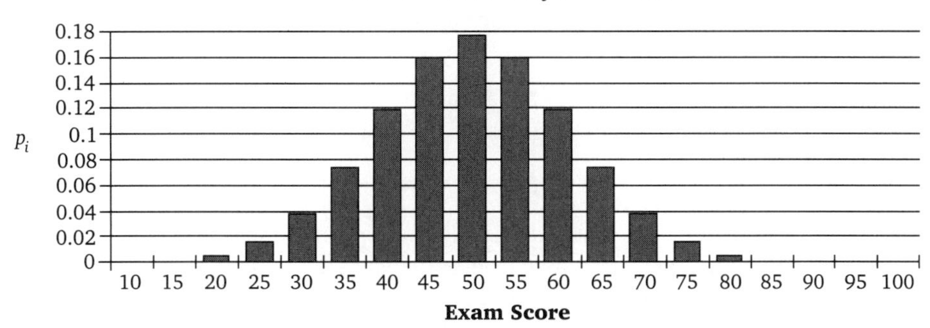

Figure 5.1
A normal distribution with a mean of 50 and a standard deviation of 16.

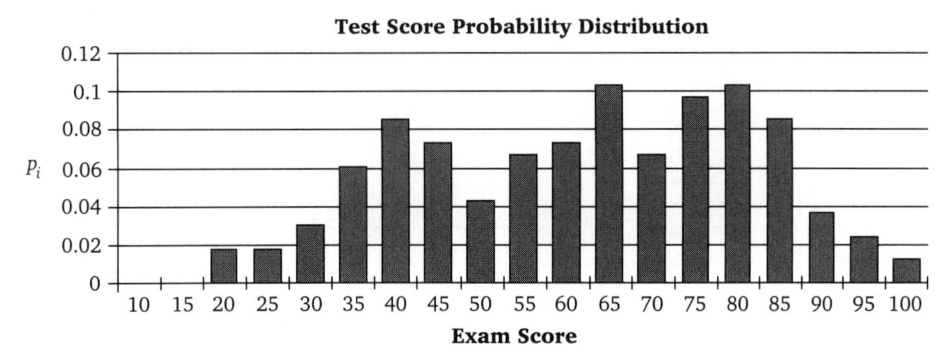

Figure 5.2
Histogram of a real exam. The mean is 60, and the standard deviation is 20.

Unfortunately, Mother Nature rarely gives us a normal distribution. Usually, things are quite skewed, as shown in the histogram of a real exam in Figure 5.2.

In the case shown in Figure 5.2, the mean is neither the most probable (80), nor the median (65), nor the rms (64). If you had taken this exam, you most certainly would not want a grade of C to be set by the most probable score! This example points out the problem of using the most probable as an indicator of average performance. The most probable is susceptible to runaway problems. A statistical fluke can cause a particular value to run away and have many more occurrences than it would normally. Can you determine from Figure 5.2 why the mean is less than the median?

The other question, "What's all this rms stuff?" is also related to the type of distribution used. The rms value is never equal to or less than the mean, because squaring the scores tends to make the high scores more important. In probability parlance, the high scores carry more weight than the low scores. What makes the rms value so useful is that in science, you are very often concerned about kinetic energy (KE), which is given by the formula

$$KE = \tfrac{1}{2}mv^2 \tag{5-9}$$

where m is the mass of a particle and v is the speed. The average kinetic energy depends on the average of the square of the speed. Hence, it depends on the rms speed and not the average speed.

$$\overline{KE} = \tfrac{1}{2} m (v_{rms})^2 \tag{5-10}$$

There might also exist a case where the average value is zero but the root mean square is not. For instance, you might have a gas of molecules in random motion. If the molecules are truly in random motion, there should be as many moving to the right as to the left. More precisely, the distribution for the x component of the velocity should show two peaks: one positive with half the particles in it, and the other negative with the other half in it, as shown in Figure 5.3.

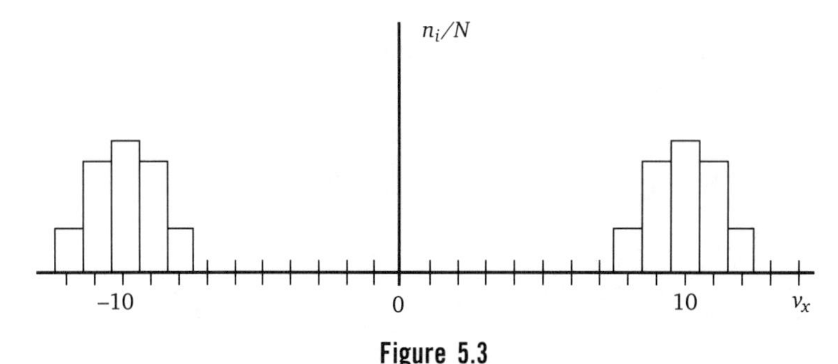

Figure 5.3
Velocity distribution for a gas in random motion. The average velocity is zero.

The average speed of the particles in each peak should be the same. In this case, the average x−velocity will be zero. Thus, the average velocity, being zero, is not a good predictor of how fast the particles are moving. To get out of this dilemma, we could square all the velocities, making them positive, so they couldn't cancel one another. Then, we could average the squares and then take the square root, thereby obtaining the rms speed.

5.2 ERROR IN EXPERIMENT: TYPES OF ERROR

Physics is an *experimental science*, which means that the theories we develop about a system must ultimately be compared to actual measurements made upon that system. Physics is not a discipline of pure logic. The logic, no matter how cunning, must always face the ultimate test: comparison with a real system. Unfortunately, obtaining accurate measurements of a real system is a very demanding and time−consuming task. All sorts of things can affect your measurements. For instance, one of the first laboratory experiments in a typical physics class involves a a small piece of aluminum, called a "glider," which floats upon an air track. The air track has a triangular cross section and is filled with compressed air. There are holes drilled in the surfaces that face the aluminum glider, and the air flowing through the holes forms an air bearing, causing the aluminum glider to float above the track on a thin cushion of air, as shown in Figure 5.4. This airbearing reduces friction to an almost negligible value.

Suppose we are to measure the time it takes for a glider start from rest and glide a given distance down an inclined track.

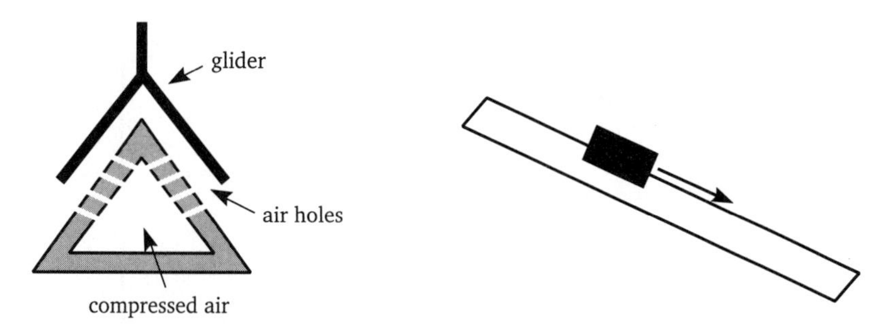

Figure 5.4
(a) Cross section of an air track and glider. (b) Side view of an inclined track with glider.

Because friction has been greatly reduced, you might think that it can be ignored; however, that turns out not to be the case. Although friction is indeed almost zero, it will still introduce an annoying error into your experiment. The gliders are slightly unstable and tend to rock slightly as they glide. When a glider rocks, the air will tend push the glider either more quickly down the track or more slowly down the track, depending on whether the glider rocks forward or backward. The effects average out over a large number of runs; however, any two runs can differ measurably from one another. The result is that our measurements will be unsteady, and there is not too much we can do to change this.

Effects over which we have little control always will affect any physical measurement. One objective is to reduce these effects by designing a good experiment, but no matter how well designed an experiment is, there will always be limits to the accuracy. We next discuss how to handle errors in your experiment. First, we define *error* as the difference between the observed value and the true value.

Error

Error = observed value - true value

The true value may not be known; i.e., you may be doing the experiment to find the true value which makes the error analysis be very difficult indeed. Whenever you do an experiment, you try to publish an honest estimate of the error. Because you typically don't know the true value, it inevitably involves guesswork.

There are three types of error, and you must learn how to deal with each.

Types of Error

Illegitimate Error—Mistakes and blunders—such as misreading a ruler or copying down a wrong number.

Systematic Error—A reproducible error that biases the data in a given direction—such as that a timer you used was slow, so all the times measured were slow by the same factor.

Random Error—Fluctuation in the results of a measurement when the measurement is repeated—such as that when you measured the distance by which you moved a mirror three times, obtaining the values 2.145 cm, 2.143 cm, 2.142 cm. Random error is not the fault of you or of your equipment.

Dealing with error basically calls for integrity. If you have done an experiment, then you are the only person who knows what happened during that experiment. It is your responsibility to report it accurately, first so that you do not confuse yourself, and second so that you do not mislead others. For instance, suppose you have been working for years building an apparatus that will find a certain type of particle—let's call it a "Zeetron." Nobody has ever observed a Zeetron before, and further, nobody knows whether a Zeetron even exists. You have been taking data for several weeks and have seen nothing; however, late that night you are inspecting the apparatus while drinking some coffee. You slip and spill some coffee into the electronics, which immediately goes haywire. You note later though that just at the moment you spilled the coffee into the machine, the detectors recorded a pulse that could be a Zeetron. What is more, the pulse has all of the right characteristics of a Zeetron. What do you do? Rush to the newspapers with the discovery of the Zeetron? No.

You never ever publish or use a measurement you know or even suspect is an illegitimate error!

In a catastrophic error, such as happened in the discovery of the Zeetron, you would most certainly suspect your data and not publish it. However, just because error has occurred does not always mean that the data are no good or must be thrown out. In cases of systematic error where, typically, a faulty piece of equipment produced the same error each time it was used, the data can be corrected for the error. In order to do this honestly, the error must be reproducible. This does not give you free license to use fudge factors. (Gee, all my data are off from what I should be getting—I'll just correct it here.) You must find the faulty piece of equipment and measure accurately the error it introduced. For instance, suppose we have used a faulty timer to measure the period of a pendulum. The timer runs slow, so every measurement of the period will be too short by the same factor. We can measure the factor by comparing our faulty timer to an accurate timer, and we adjust our data accordingly.

When all the other types of error have been eliminated, there will still be fluctuations in our data. We assume that these errors are random, in that they are not correlated to one another, any faulty piece of equipment, or a faulty operator. The fluctuations arise from the precision of our equipment and our experimental design. The more closely the fluctuations resemble random events, the more accurately our analysis will be able to deal with them. We need to quantify our concept of observed value and error. Let Δx be the *uncertainty*, which is a measure of the error in an observed value x. There are three standard ways of reporting the uncertainty: as an absolute uncertainty, a relative uncertainty, or a percent uncertainty.

Uncertainty	
$x \pm \Delta x$	absolute
$\frac{\Delta x}{x}$	relative
$\frac{\Delta x}{x}(100\%)$	percent

The *absolute uncertainty* places bounds telling you that the actual value is expected to fall between $x - \Delta x$ and $x + \Delta x$. The absolute uncertainty is a fairly easy way to report the error. It also makes it easy to check that you aren't reporting too many significant figures. For example, suppose we had made a measurement of 10.02 ± 0.3m. Because the reported value is uncertain in the tenths place, it is wrong to report it to the hundreds place. It should be rounded to agree with the uncertainty—i.e., 10.0 ± 0.3m. Another point is that both the reported value x and the uncertainty Δx must have the same units.

Although it is hard to beat the simplicity of absolute error, it doesn't give a sense of how large the error is. For example an absolute error of 0.5m would be outrageous uncertainty in the measurement of a 2m bookcase, but that same 0.5m error, in measuring the distance to the moon, might be quite good. Relative error and percent error are both ways to report a sense of how good the measurement is. For the bookcase, the 0.5 absolute uncertainty gives

$$\frac{0.5\text{ m}}{2\text{ m}} = 0.25 \qquad \text{relative error}$$

$$(0.25)(100\%) = 25\% \qquad \text{percent error}$$

and for the moon

$$\frac{0.5\text{ m}}{3.8\times10^{8}\text{ m}} = 1.3 \times 10^{-9} \qquad \text{relative error}$$

$$(1.3 \times 10^{-9})(100\%) = 1.3 \times 10^{-7}\% \qquad \text{percent error}$$

The ways of estimating the random error are more involved, so we'll leave that to a section of its own.

5.3 RANDOM ERROR AND BEST ESTIMATORS

In this next section, we explore some of the standard estimators of error. We will cover only a brief introduction to a very intricate field. Should you need more information on these ideas please consult a book on statistics. As stated, if the fluctuations in the measured value of a quantity are random, we can use statistics to provide estimators of the actual value and the uncertainty. All of what we discuss herein assumes a *bell curve* or a *normal distribution* for the observed values. Assume we make a large number of measurements of a variable x with the ith measurement returning a value x_i, and next, construct a histogram for the fraction of times a given value is observed. If the data are truly random, and we are not, for instance, near a threshold, we'd expect the histogram to be close to a normal distribution, which is symmetric around the mean as shown in Figure 5.5.

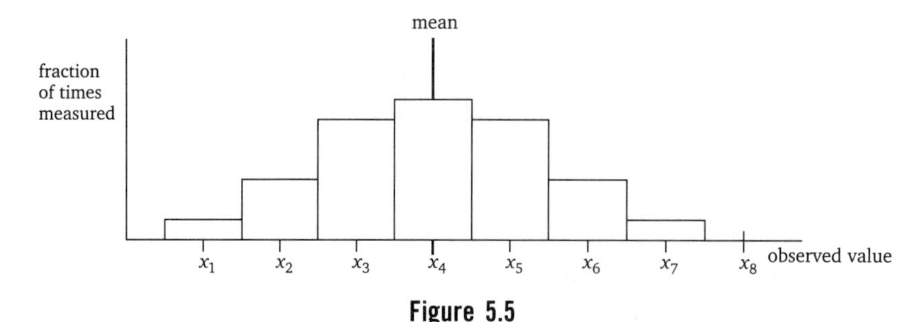

Figure 5.5
A normal distribution is bell-shaped with the maximum value occurring at the mean.

We can see from the distribution shown in Figure 5.5 that the mean is the best estimator of the actual value. If the data are not normally distributed; however, the mean may not be the best estimator. When handling just a few measurements, the mean is sensitive to extreme measurements or outcomes. If you include an illegitimate data point that is far from the mean, it will tend to skew the mean to the high or low side of the mean for the legitimate data. Of course the more data points you have, the less an extreme value will affect the mean.

Mean—Best Estimator of the Actual Value

$$\overline{x} = \frac{\sum_{i=1}^{N} x_i}{N} = \frac{x_1 + x_2 + \ldots + x_i + \ldots + x_N}{N} \tag{5-11}$$

Note that the mean is not necessarily the best estimator if the data are distributed differently than on a normal curve. There are other predictors for the best estimator of the actual value, such as the median value, the most probable value, and the rms value. Each predictor has its strengths and weaknesses, depending on the type of distribution and the quantity being measured. Although we do not go any further into the different types of estimators in this volume, you should be aware that others exist, and you may need to use other estimators in the future.

5.4 STANDARD DEVIATION

Now that we have the best estimator of the actual value from our data, how do we estimate the error? We certainly would not want to use the largest deviation from the mean to estimate the error, because as we record more data, the chances of measuring a larger deviation would increase. We would then be in the unhappy position of seeing our uncertainty increase as we obtain more data. Roughly, the error should be a measure of how wide the distribution is. For the two distributions shown in Figure 5.6, the flat distribution (b) has a greater uncertainty in the data.

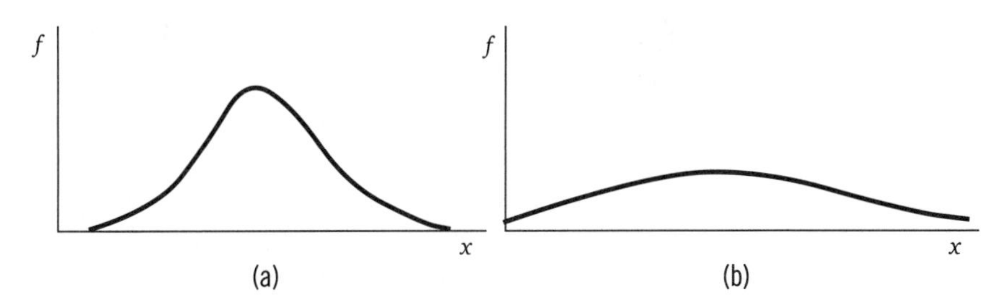

Figure 5.6
The distribution in (b) has a greater uncertainty.

One of the measures we could use for the uncertainty is the deviation δ from the mean. What we want is something like the average deviation from the mean. However, there is a problem with the average deviation. The data is symmetrically distributed around the mean, so there will be just as many deviations below the mean as above it. Defining the ith deviation to be

$$\delta_i \equiv \overline{x} - x_i, \qquad \text{deviation} \qquad (5\text{-}12)$$

half the deviations will be positive (those below the mean), and half will be negative. Thus, if we try to sum the deviations we get zero

$$\sum_{i=1}^{N} \delta_i = 0$$

$$\overline{\delta} = \frac{\sum_{i=1}^{N} \delta_i}{N} = 0 \qquad (5\text{-}13)$$

and so, too, the average deviation is zero. Now we have encountered this type of problem before. Remember that the average value of the molecular velocity in a gas is zero. What did we do then? Any ideas as to what we should do now?

The problem is that half the deviations are negative and half are positive, so they cancel one another. There are two possible ways to proceed. First, we could take the absolute value of the deviations and find the mean of them. This is termed the *mean absolute deviation.* The mean absolute deviation provides us with a good measure of the error, but it is not mathematically tractable because the absolute values are difficult to combine algebraically. The biggest problem with the mean absolute deviation, however, is that it has fallen out of favor with statisticians and is not implemented in any statistic packages that run on hand-held calculators. Do not use this scheme to estimate errors.

A more acceptable solution to the problem of the positive and negative deviations canceling is to square the deviations. Then, they would no longer cancel each other, and we could just average the squared deviations. In this spirit, we define the square root of the average of the squares of the deviations to be the *standard deviation.*

Standard Deviation - best estimator of uncertainty

$$s = \sqrt{\frac{\sum_{i=1}^{N} \delta_i^2}{N-1}} \qquad (5\text{-}14)$$

When we average the square of the deviations, we divide by $N - 1$ rather than by N because there are only $N - 1$ independent deviations. Although there are N deviations, one of them is determined by Equation (5-13) and so isn't independent. Like the absolute mean deviation, the standard deviation provides a good estimate of the error; however, it turns out to be more tractable algebraically. The standard deviation is also implemented in all statistics packages, so you can use a calculator or standard statistics package on a computer to calculate it from your data. In general, the standard deviation is used as the best unbiased estimator of the uncertainty or error.

There is one other advantage to using the standard deviation to estimate the error. Note that the standard deviation does not give us the largest possible error. Some of our data points will lie outside the range $\overline{x} - s$ to $\overline{x} + s$, as shown in Figure 5.7. In fact, only 68% of our data will lie within $\overline{x} - s$ to $\overline{x} + s$. Using $\overline{x} - s$ is called "one sigma" uncertainty because σ is often used for the standard deviation. In addition, 95% of the data points will lie in the range $\overline{x} - 2s$ to $\overline{x} + 2s$. When a very conservative estimate is needed, you can use this "two sigma" uncertainty.

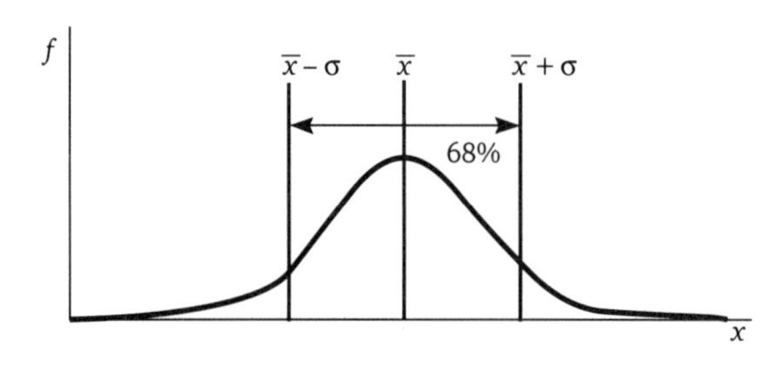

Figure 5.7
Within one standard deviation of the mean lie 68% of the data.

Because you must use the mean and standard deviation often, it is important to be able to calculate both from a set of data points. Although the definitions of the mean and standard deviation, Equations (5-11) and (5-14), respectively, can be used, it is very laborious to do so. Most scientific calculators will find the mean and standard deviation, and you are strongly encouraged to purchase one and know how to use it. All calculators that take the mean have a statistical mode that you must first activate. This is usually located on a key marked "stat" as shown in Figure 5.8. For the calculator shown, the stat mode is activated by pushing the "2nd F" key and then pushing the "on/C" key which is also marked "STAT".

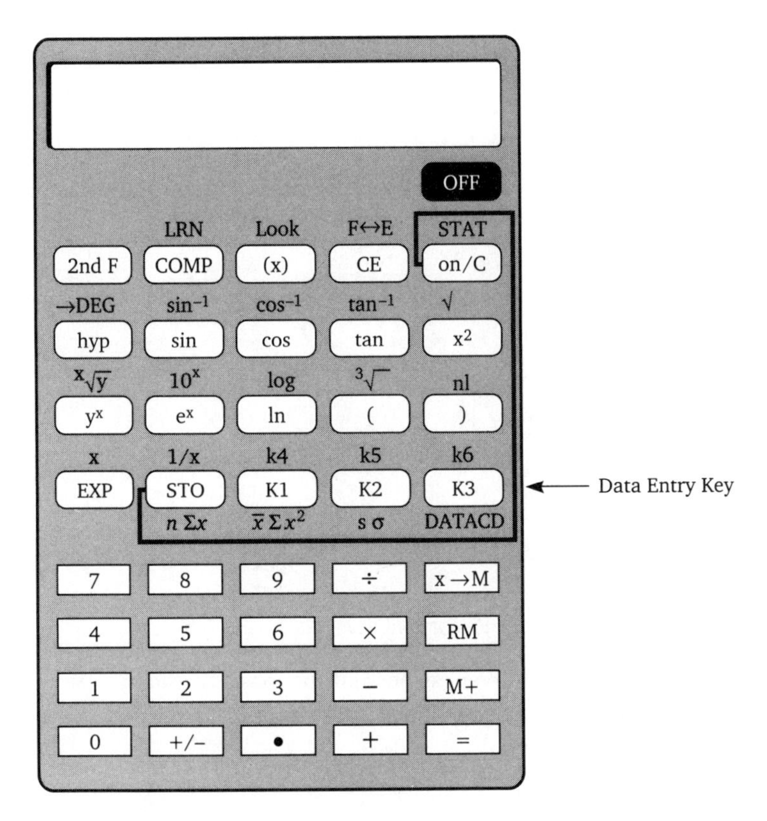

Figure 5.8
A typical calculator with a statistics mode for finding the mean and standard deviation.

Consult your calculator's instruction manual for the specific details on activating the stat mode and entering data. Once the stat mode is activated, data entry can take place. Enter the first number in your data set; then push the data entry key. Enter the second number and again push the data entry key. Keep doing this sequence until all the data points are entered. Finally, to find the mean, push the key marked "mean" or '$\bar{x}$'. To find the standard deviation, push the key marked "s" or "std dev" or "σ" or "σ_{n-1}". The process is represented schematically in Figure 5.9.

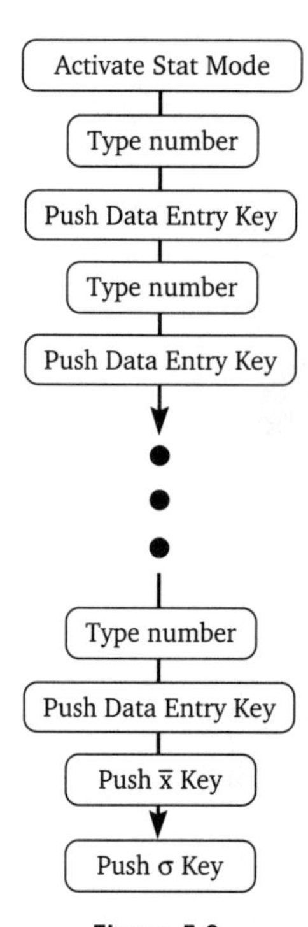

Figure 5.9
Data entry process for finding the mean and standard deviation.

Some calculators give you a choice of using two different standard deviations: σ_n usually means the standard deviation is normalized to N rather than to $N - 1$; i.e., it uses the formula

$$\sigma_n = \left(\frac{\sum_{i=0}^{N} \delta_i^2}{N} \right)^{1/2} . \tag{5-15}$$

If possible, use the correctly normalized standard deviation in Equation (5-14). However, the difference between the results returned by Equations (5-14) and (5-15) becomes negligible when the data sets are large.

EXAMPLE 5-2

Estimate the actual value and the absolute error for the six observed values of a period: 18.1 ms, 18.2 ms, 18.3 ms, 18.5 ms, 18.6 ms, 18.7 ms. Do this two ways: first by using Equation (5-11) to estimate the actual value and then use Equation (5-14) to estimate the error. Next use the statistics package on a calculator to calculate the mean and standard deviation.

Solution:

Using the Equation (5-11) for the mean

$$\overline{T} = \frac{18.1+18.2+18.3+18.5+18.6+18.7}{6} = 18.4 \text{ ms}$$

There are two deviations each of 0.1, 0.2, 0.3 from the mean. So the standard deviation is

$$s = \left(\frac{2(0.1)^2+2(0.2)^2+2(0.3)^2}{6-1} \right)^{½} = 0.24 \text{ ms}$$

We need to round the error to agree with the proper number of significant figures to $0.24 \simeq 0.2$.

$$T = 18.4 \text{ ms} \pm 0.2 \text{ ms}$$

On your calculator you should obtain 18.4 exactly for the mean and 0.2366...for the standard deviation.

5.5 ERROR PROPAGATION AND PREDICTED VALUES

Now that we have the data and the error, what do we do with it? Sometimes you will compare the measured quantity directly. Often, however, it is necessary to combine the measurements in a function. For instance, if you are measuring the length of a table top to obtain its area, and you want to know how the uncertainty in the two length measurements will translate into an uncertainty in the area. In the following discussion, we explore how an error propagates through mathematical calculations.

If we have a simple function of one dimension $f = f(x)$, and we know x and the absolute error—i.e., $x \pm \Delta x$—then the task of constructing the error in f is relatively simple. It is the difference in the maximum value of f and the expected value of f,

$$\Delta f = f(x + \Delta x) - f(x) \quad (5\text{-}16)$$

where

$$f = f(x) \pm \Delta f. \quad (5\text{-}17)$$

Normally $f(x + \Delta x) - f(x)$ and $f(x) - f(x - \Delta x)$ should be about the same. The problem with a technique like the one used in Example (5-2) arises when there are wide disparities between the two. In those cases, you should use the one that yields the larger value of Δf.

EXAMPLE 5-3

Find the absolute error in $f = 3x^2$ for $x = 5.0 \pm 0.1$.

Solution:

Using Equation (5-16):

$$\Delta f = 3(5.1)^2 - 3(5.0)^2 =$$

$$78.03 - 75.00 =$$

$$\Rightarrow \Delta f = 3.0$$

$$\Rightarrow f = 75.0 \pm 3.0$$

For this problem $f(x + \Delta x) - f(x) \simeq f(x) - f(x - \Delta x)$, but try $f = e^{x^2}$ for the same value and error, $x = 5.0 \pm 0.1$. Then:

$$f(x + \Delta x) - f(x) = 1.26 \times 10^{11}$$

and

$$f(x) - f(x - \Delta x) = 5.5 \times 10^{10}$$

The values obtained for the error in the second case is a factor of 3 less than that for the first case. You should be conservative in reporting errors and use the first result, giving a larger error.

Using the simple difference is fine for functions of one dimension, but what if the function depends on two variables, $f = f(x, y)$? Then should we use $\Delta f = f(x + \Delta x, y + \Delta y) - f(x, y)$? Here we must be careful because the error in two variables could easily cancel each other, leading us to underestimate the error. For example, try $f = \frac{x}{y}$ where the effect of increasing x and increasing y will tend to cancel. This is just a simple example, but there could be far more complicated situations where canceling effects would not be so apparent. If you are getting the idea that error is not so simple, you are correct. You must never use statistics blindly, because if abused, even unintentionally, it can lead you to have inordinate confidence in your results.

Suppose we look at the variations in x and y independently, so that we have two pieces of information to use.

$$\Delta f_x = f(x + \Delta x, y) - f(x, y),$$

and

$$\Delta f_y = f(x, y + \Delta y) - f(x, y). \tag{5-18}$$

In Equation (5-18) we have two uncertainties for f. Which do we use? Using just one would underestimate the uncertainty, but simply adding the two errors would overestimate the uncertainty. The solution is to look at how we treat independent coordinates in two dimensions. In a Cartesian coordinate system the x and y coordinates are independent meaning the motion in the x direction can be completely independent of the motion in the y direction. For instance, you can walk due east and then walk due north. The walk east involves no northerly walking and is therefore independent of north walk. The total distance traveled is not just the sum of the Δx and Δy distances because they occurred at right angles to one another. The total distance traveled is given by the Pythagorean theorem.

$$d = \sqrt{\Delta x^2 + \Delta y^2} \tag{5-19}$$

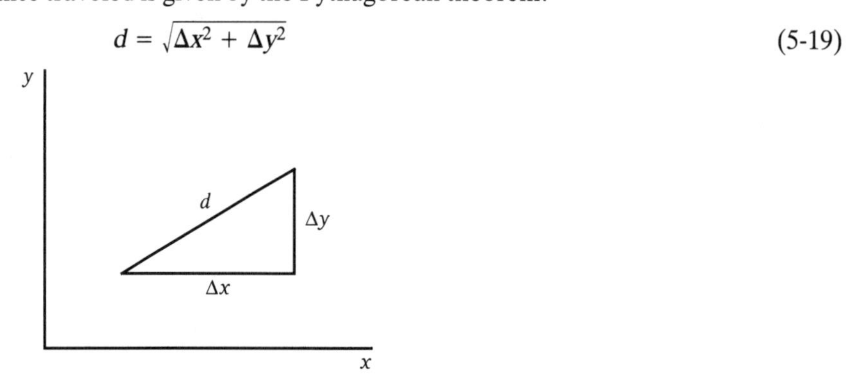

Figure 5.10
The total distance traveled after walking east and then north is given by the Pythagorean theorem.

Of course the x and y displacements are independent of one another because the x and y directions are orthogonal, and that's why we must use the Pythagorean theorem to find the magnitude of the displacement. Since the Δy and Δx in Equation (5-19) come from independent measurements of two distinct quantities, they are orthogonal to one another in the sense that there is no overlap between the two measurements. There is no way the measurement of x can affect the measured value of y and vice versa. When this property exists, we should also use the Pythagorean theorem to combine the two uncertainties

$$\Delta f = \sqrt{(\Delta f_x)^2 + (\Delta f_y)^2} \tag{5-20}$$

Using the Pythagorean theorem for combining the uncertainties is known as quadrature.

EXAMPLE 5-4

Calculate the area and absolute error of a table whose dimensions have been measured to be

$$L = 3.00 \pm 0.04 \text{ m}, \; w = 2.00 \pm 0.02 \text{ m}.$$

Solution:
The area is $A = L \times w = (3 \text{ m})(2 \text{ m}) = 6.00 \text{ m}^2$. Next, calculate the uncertainties in the independent variables. Here L and w are independent. Use Equation (5-20).

$$\Delta A_L = w(L + \Delta L) - wL$$

$$\Delta A_L = w\Delta L = (2.00 \text{ m})(0.04 \text{ m}) = 0.08 \text{ m}^2 \tag{5-21}$$

Similarly for ΔA_w

$$\Delta A_w = L(w + \Delta w) - wL \Rightarrow$$

$$\Delta A_w = (3.00 \text{ m})(0.02 \text{ m}) = .06 \text{ m}^2$$

Now combining the uncertainties in quadrature:

$$\Delta A = \sqrt{(\Delta A_L)^2 + (\Delta A_w)^2}$$

$$\Delta A = \sqrt{(0.08 \text{ m}^2)^2 + (0.06 \text{ m}^2)^2} = 0.1 \text{ m}^2$$

Thus,

$$A = 6.0 \pm 0.1 \text{ m}^2$$

The least significant 0 on the 6.00 was dropped because the error is 0.1 so there is no longer any reason to keep the last zero on the 6.00 m^2.

5.6 USING GRAPHS

Other people will be interested in the results of your experiment, so you must be able to effectively communicate both your experiment and the results to people. In addition, before you end your experiment, you must understand the data and the relationships implied by your data. In both these endeavors, it is good to remember the old Chinese proverb, "*A picture is worth a thousand words.*" When your data is displayed in graphic form, relationships that you might otherwise have overlooked become apparent. Also, pictures can communicate ideas more effectively than words or tables. Graphs are particularly good at conveying relative sizes, and including the errors in a graph can efficiently show how big the error is relative to your data.

A problem can arise during data analysis when you discover *exceptional data points*—points that vary drastically from neighboring data points. You might be tempted to throw them out because you never want to use data that contain illegitimate error. However, when you are analyzing data, you have no idea whether the exceptional point represents good data or a major error. To get around this problem, it is good practice to make a quick graph during data collection because exceptional data points strikingly stand out from the rest, and you can then immediately retake that data, thus checking whether it is an error or a legitimate point. More than a few times, researchers have thrown out data because they assumed that an error occurred, when in fact it was a real effect, and the prize-winning discovery went to the person who took the time to check it out. Making a quick graph as you collect your data can save you from making major experimental errors.

There are several rules for using graphs either in a report or in a presentation.

Rules for Graphing

1. Always use graph paper!
2. Title the graph.
3. Label the axes, including units.
4. Clearly mark the data points.
5. Include error bars.
6. Draw a smooth curve through the data set.

For lab work, regular notebook paper is not acceptable because it is too hard to read values from the graph, and hence is not accurate. Always use ruled graph paper. Always title the graph with a title that conveys the major ideas presented. It should be simple. For example, "Length versus Period for a Simple Pendulum" would convey the idea for a graph to an audience or to yourself 6 months hence, when you will have long forgotten what this graph is. For your own use while taking data, you might use a simple "*L vs T*, run 3" as a title, but always include a title. You will be surprised at how quickly you forget what you were doing. At the time you make the graph, it may seem very obvious, but even in 1 week, it can become a nightmare to figure out what in blazes you were doing. You should always have a title because it can make the graph clear and can save many hours; even more important, however, labeling the axes is something that absolutely must be done, even on a quick-and-dirty graph. A graph with unlabeled axes is useless. Not labeling your axes on an exam or a lab report would be considered a major mistake.

Data points should be clearly marked with a small filled circle bold enough to be clearly seen. If you need to distinguish among sets of data points, then use triangles, squares, or other marker shapes, as simulated in Figure 5.11.

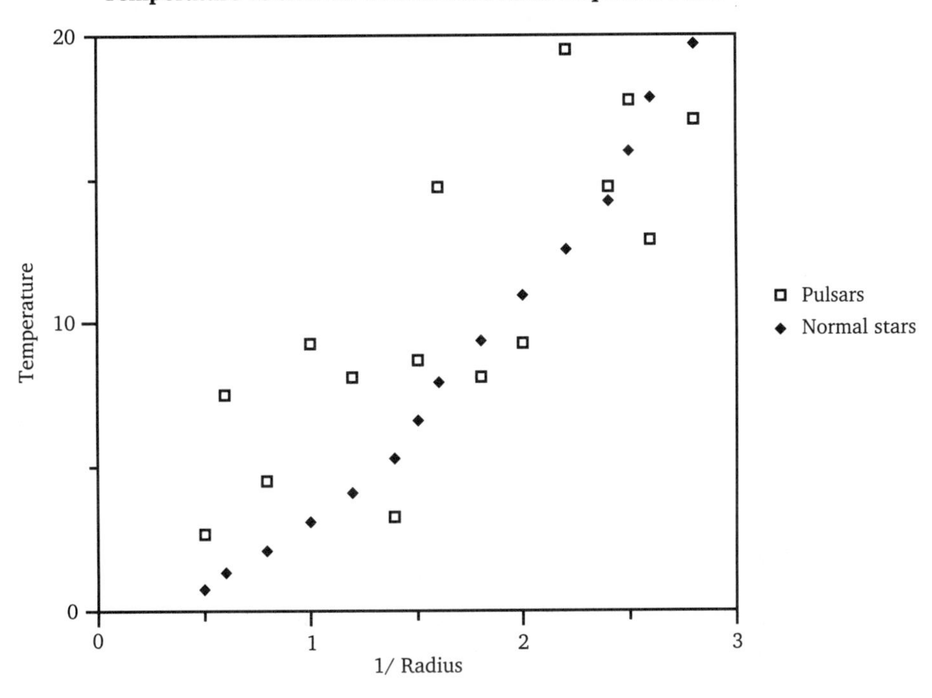

Figure 5.11
A graph comparing two different data sets.

One of the things that should strike you about this graph is the strange units used. What could they possibly be? Because they were left off the graph, it's impossible to determine them. The units should be marked in parenthesis next to the axis label. For such a graph the temperature might be in units of 1000 degrees Kelvin. The inverse radius might be in inverse solar radii where one solar radius equals the radius of our sun. These units are included in the next graph, Figure 5.12.

The graph should convey some sense of how large is the error in the measurements. Error bars are, just as the name implies, bars representing the absolute error $(\pm\Delta x)$ in each data point. If you do not include error bars, people will assume that the error was so small as to be unrepresentable on the graph. If you have not included error bars and you have significant error, your graph will mislead people, which is unethical, and other scientists will judge you harshly by this if you are discovered. So when you have significant error, always include error bars. A graph with error bars is shown in Figure 5.12.

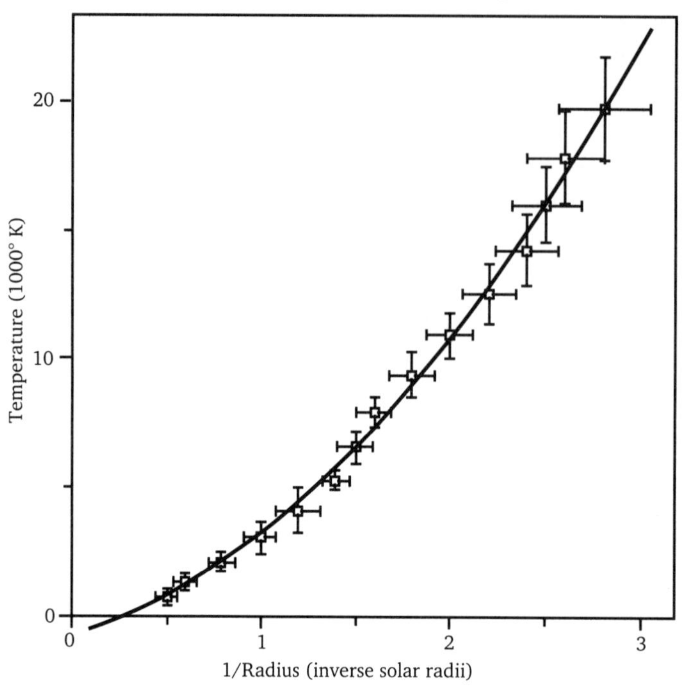

Figure 5.12
A graph of well-behaved data with error bars.

Notice in Figure 5.12 that the smooth line drawn through the data does not go through the center of each data point. The curve is to represent our best guess of the actual behavior, and the data, because it contains error, should fluctuate with roughly the same number of points on either side of the curve. The curve should be smooth and drawn with the aid of a French curve or drawn very carefully freehand. It should not be jagged or sloppy. This is also not a connect the dots game, so do not connect the data point with straight lines. Many computer plotting programs offer several choices for fitting the data with a curve.

The data in Figure 5.12 is well-behaved and represents a classic scientific graph, but your data may not be particularly well-behaved, as in Figure 5.13. Note in Figure 5.13 that there are data points with error bars that are not large enough to contain the curve. This is natural because the error was calculated statistically, representing one standard deviation either way from the mean. Therefore, it's statistically logical that some of the data should have error bars that do not contain the curve. Error bars where only two thirds of the taken data lie within the stated error are called "one-sigma" error bars. Figure 5.13 very efficiently conveys how much confidence should be placed in the curve. There is far too much scatter of the data points to accurately predict a curve; however, when you are at the cutting edge of science, you may find yourself with just such data. At the cutting edge, the equipment will always be pushed to the extremes of what is possible.

Figure 5.14 presents a graph as an example of what *not* to do.

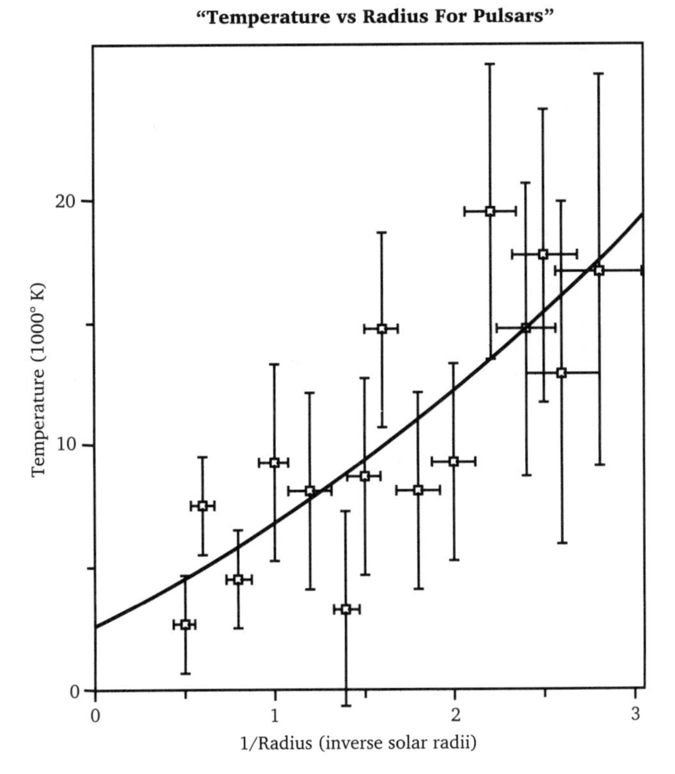

Figure 5.13
A graph of not-so-well-behaved data.

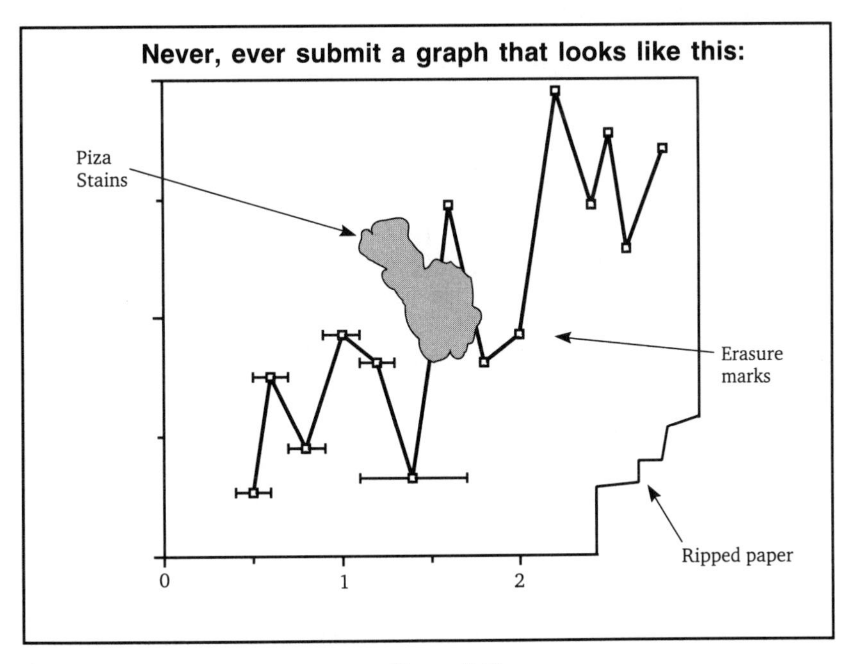

Figure 5.14
The wrong way to present data.

Chapter 5 Problems

The following data applies to Problems 1–4.

Twenty-five engineering students independently measured the engineering float for Picnic Day to the nearest inch with a 12" ruler. Their results are shown in Table 5.1.

Table 5.1

Number of Students	Result
2	32' 10"
4	32' 11"
5	33' 00"
6	33' 01"
4	33' 02"
2	33' 03"
1	35' 03"

[5.1] What type of error is involved here? Note that it may involve more than one type.

[5.2] Express the length of the float as a length and an absolute uncertainty.

[5.3] Express the uncertainty in the length as a relative uncertainty.

[5.4] Express the uncertainty in the length as a percent uncertainty.

* * * *

[5.5] The radius of a sphere is known to be $r = 4.52 \pm 0.03$ cm. The volume of a sphere is given by $(4/3)\pi r^3$. Find the absolute uncertainty and the percent uncertainty in the volume of the sphere.

• **[5.6]** The sides of a slab of concrete are measured to be 10.0 ± 0.1 m, 5.0 ± 0.1m, and 0.50 ± 0.03m. Find the absolute and the percent uncertainty in the volume.

[5.7] List all of the things wrong with the graph shown in Figure 5.14.

[5.8] Give the absolute, relative, and percent error in the following rational approximations for π:

a) 3 **b)** $\frac{22}{7}$ **c)** $\frac{333}{106}$ **d)** $\frac{355}{113}$ **e)** $\frac{103993}{33102}$.

[5.9] Two perfect six-sided dice are thrown on a table (Lost Wages, here we come). **a)** Make a table of all the possible outcomes. **b)** What is the probability of rolling double fours? **c)** What is the probability of any double being rolled? **d)** What is the probability of rolling two unlike numbers? **e)** What is the combined probability of rolling a 7 or an 11?

• **[5.10]** There are 21 particles moving along the x axis with the following distribution of velocities: 3 have −8 m/s, 4 have −7 m/s, 3 have −6 m/s, 1 has 4 m/s, 1 has 5 m/s, 5 have 6 m/s, and 4 have 7 m/s. **a)** Make a histogram of the velocity distribution, find **b)** the average velocity, **c)** the rms velocity, and **d)** the standard deviation in the velocity. Mark all these quantities on your velocity distribution graph.

e) Make a histogram of the speed distribution, find **f)** the average speed, **g)** the most probable speed, and **h)** the rms speed; mark these quantities on your speed distribution graph.

[5.11] Sally makes 10 independent measurements of the time it takes a pendulum to complete a cycle. This time is called "the period of the pendulum." She obtains 1.5, 1.6, 1.5, 0.9, 1.6, 1.7, 1.6, 1.5, 1.4, and 1.5 seconds. She noted in her lab book that her finger slipped off the timer during the fourth reading. **a)** Find the best estimate of the period. **b)** Find the uncertainty in the period. **c)** Assuming that Sally measured each period separately, find a more accurate way to measure the period using the same equipment as Sally.

• [5.12] José makes 10 independent measurements of the length of a pendulum: 1.515 m, 1.513, 1.510, 1.511, 1.512, 1.514, 1.500, 1.512, 1.513, 1.508 meters. **a)** Find the best estimate of the length. **b)** Find the absolute error in the length. **c)** Find the percent error in the length. **d)** If the period is related to the length by the formula $T = 2\pi\sqrt{\frac{L}{g}}$, then find the best estimate of the period where g = 9.8 m/s^2. **e)** Find the best estimate of the absolute error in the period.

CHAPTER 6
Return to LineLand

(Picture this . . .)

Submitted for your approval—one reader, sitting in the same easy chair that once took him to Lineland. A strange, vaguely familiar feeling comes over you and once again you find yourself in Lineland, but this time there is a difference. You are equipped with visualization and graphing skills, and here you learn to use them to your advantage. This gives you great power over other Lineland inhabitants. In this chapter, you will learn to geometrically visualize the equations of kinematics. You will not return from this journey the same person who left.

6.1 AREA UNDER THE CURVE AND SLOPE

Usually, the position of an object is known at some particular moment, and the acceleration can be found by summing the forces acting on the object. We want to find the position at some time in the future, and this is the *kinematic problem.* Finding the position from the acceleration and initial position is the inverse of finding the velocity from the position.

Consider the formal process of solving for the velocity from the acceleration. From Chapter 2, we learned a silly trick

$$\Delta \mathbf{v} = \frac{\Delta \mathbf{v}}{t} t \tag{6-1}$$

where $\Delta \mathbf{v} = \mathbf{v} - \mathbf{v}_o$ so that

$$\mathbf{v} = \mathbf{v}_o + \frac{\Delta \mathbf{v}}{t} t. \tag{6-2}$$

The $\Delta \mathbf{v}/t$ is the acceleration giving

$$\mathbf{v} = \mathbf{v}_o + \mathbf{a}t. \tag{6-3}$$

Because $\Delta \mathbf{v}/t$ is also the slope of the velocity vs time curve, the acceleration is the slope if the velocity vs time curve. However, we are interested in another interpretation; namely, given the acceleration, how do we find the velocity.

An interesting note is that if we plot $\mathbf{a}$ vs t, then $\mathbf{a} \cdot t$ is the area under the $\mathbf{a}$ vs t curve. Figure 6.1 represents a plot of a constant acceleration vs time. It is clear that the area under a segment of the curve is given by

$$\text{Area} = \mathbf{a}\Delta t = (\text{height} \times \text{width}). \tag{6-4}$$

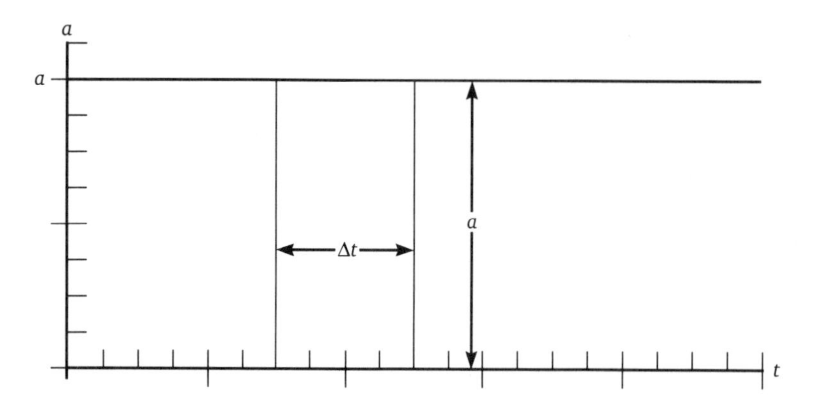

Figure 6.1
The acceleration vs time curve. The "area" under the curve is $a\Delta t$.

To be sure, the area we have defined is generalized from the usual definition. It doesn't even have the usual units of m^2, but rather has units of $(m/s^2)s = m/s$, i.e., units of velocity. Thus, the velocity has a very nice interpretation of being the area under the acceleration vs time curve.

This relationship between slope and the area under the curve is a general one. Suppose that m is the slope of a function $f = f(x)$ where

$$m = \frac{\Delta f}{\Delta x}. \tag{6-5}$$

Then, f is given by the area under the m vs x curve. This point is important enough that we should consider the general case.

If we have a more interesting acceleration, one that changes with time, then choosing a wide Δt will only give us a crude approximation to the area underneath the curve. A better approximation would be to let Δt become small, as shown in Figure 6.2, and the smaller Δt becomes, the better the approximation becomes as shown in Figure 6.2.

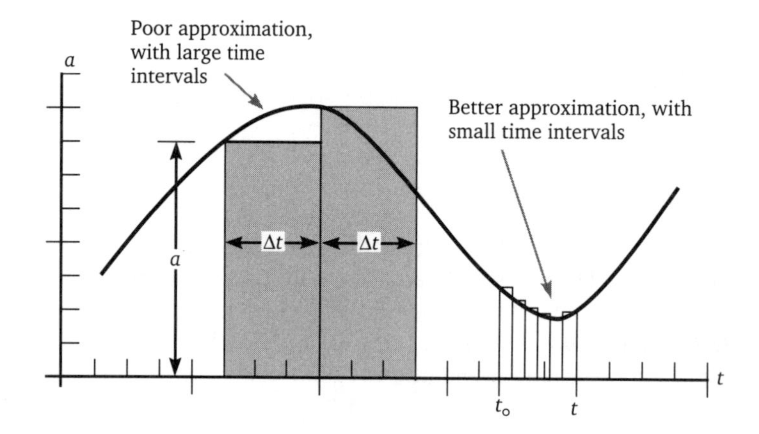

Figure 6.2
Acceleration vs Time curve. The smaller the time interval Δt, the better the approximation to the area under the curve.

Suppose we take the six small time intervals shown in Figure 6.2 between t_o and t, and then, sum the areas associated with each where individually,

$$\mathbf{v}_1 - \mathbf{v}_o = \mathbf{a}_o\Delta t,$$

$$\mathbf{v}_2 - \mathbf{v}_1 = \mathbf{a}_1\Delta t, \ldots$$

$$\mathbf{v} - \mathbf{v}_5 = \mathbf{a}_5\Delta t. \tag{6-6}$$

Summing all the equations,

$$(\mathbf{v} - \mathbf{v}_5) + (\mathbf{v}_5 - \mathbf{v}_4) + (\mathbf{v}_4 - \mathbf{v}_3) + (\mathbf{v}_3 - \mathbf{v}_2) + (\mathbf{v}_2 - \mathbf{v}_1) + (\mathbf{v}_1 - \mathbf{v}_o) = \mathbf{a}_o\Delta t + \mathbf{a}_1\Delta t + \mathbf{a}_2\Delta t + \mathbf{a}_3\Delta t + \mathbf{a}_4\Delta t + \mathbf{a}_5\Delta t. \tag{6-7}$$

Note that all the intermediate velocities on the left hand side of Equation (6-7) cancel leaving

$$\mathbf{v} - \mathbf{v}_o = \mathbf{a}_o\Delta t + \mathbf{a}_1\Delta t + \mathbf{a}_2\Delta t + \mathbf{a}_3\Delta t + \mathbf{a}_4\Delta t + \mathbf{a}_5\Delta t. \tag{6-8}$$

Thus, summing the small areas gives us a geometrical interpretation of velocity as the area underneath the acceleration vs time curve where small intervals of Δt must be used if there is much curvature to the curve.

$$\mathbf{v} - \mathbf{v}_o = \sum_i \mathbf{a}_i\Delta t \tag{6-9}$$

EXAMPLE 6-1

A 5 kg object, initially at rest, is subjected to a constant force of 6 N for 5 s, then the force decreases linearly to zero in 3 s, as shown in Figure 6.3. Find the total change in momentum, Δp, of the mass m if $F = \frac{\Delta p}{\Delta t}$.

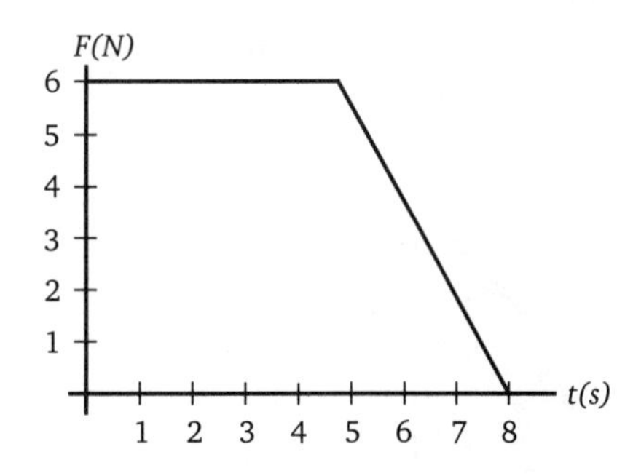

Figure 6.3
The force as a function of time.

Solution:

If F were constant, we could just solve $F = \frac{\Delta p}{\Delta t}$ for Δp; however, F changes, which means for each small Δt, $\Delta p = F\Delta t$ so that for a changing force, the change in momentum is the area under the force vs time curve.

$$\Delta p = \text{area under } F \text{ vs } t \text{ curve.}$$

$$\Delta p = \text{area of box} + \text{area of triangle}$$

$$\Delta p = (6\text{ N}) \times (5\text{ s}) + \tfrac{1}{2}(6\text{ N}) \times (3\text{ s}) = 39\text{ N}\cdot\text{s}.$$

6.2 GRAPHICAL TECHNIQUES IN KINEMATICS

We have now reached the point where we can formally define the rules of motion called *kinematics*. The kinematic problem is to start with a given acceleration and find the position of an object. Consider first a constant acceleration.

Constant Acceleration:

First we want to find the velocity, which we have already seen is given by

$$\boldsymbol{v} = \boldsymbol{v}_o + \boldsymbol{a}t \tag{6-10}$$

where $\boldsymbol{v}$ is the area under the $\boldsymbol{a}$ vs t curve.

Once we have found the velocity, we need to find the position $\boldsymbol{x}$. As we did in solving for the velocity, first use the chain rule for $\Delta\boldsymbol{x}$, and then use the definition of the velocity.

$$\Delta\boldsymbol{x} = \frac{\Delta\boldsymbol{x}}{\Delta t}\Delta t \tag{6-11}$$

so that

$$\Delta\boldsymbol{x} = \boldsymbol{v}\,\Delta t. \tag{6-12}$$

Because $\boldsymbol{v}$ is the slope of $\boldsymbol{x}$ vs t, it must be true that $\boldsymbol{x}$ is the area under the curve of $\boldsymbol{v}$ vs t, just as with the acceleration, $\boldsymbol{a}$, is the slope of the $\boldsymbol{v}$.

$$\boldsymbol{x} - \boldsymbol{x}_o = \sum_i \boldsymbol{v}_i \Delta t \tag{6-13}$$

A plot of $\boldsymbol{v}$ vs t for the $\boldsymbol{v}$ given in Equation (6-10) is made in Figure 6.4. Because the graph is a straight line with slope $\boldsymbol{a}$, the area is easy to determine. It is the area of the light gray box plus that of the darker gray triangle.

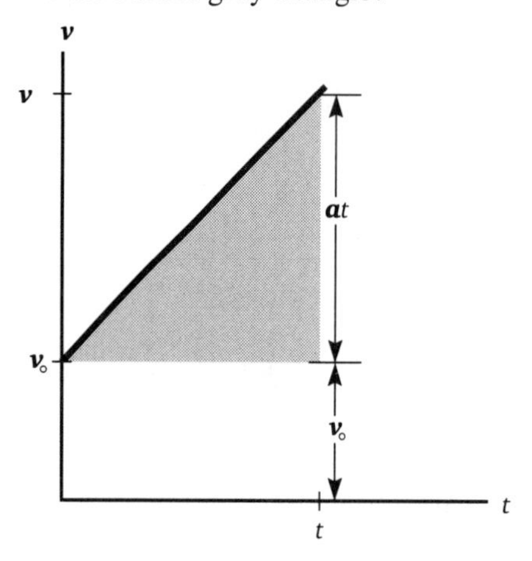

Figure 6.4
A graph of $\boldsymbol{v} = \boldsymbol{v}_o + \boldsymbol{a}t$ vs t for constant $\boldsymbol{a}$.

The area of the light gray box is

$$A_{\text{box}} = \boldsymbol{v}_{\text{o}}t \tag{6-14}$$

and the area of the triangle is

$$A_{\text{triangle}} = \frac{1}{2}(t)(\boldsymbol{a}t) = \frac{1}{2}(\text{base})(\text{height}). \tag{6-15}$$

Hence, the total area under the curve is:

$$\boldsymbol{x} - \boldsymbol{x}_{\text{o}} = \boldsymbol{v}_{\text{o}}t + \frac{1}{2}\boldsymbol{a}t^2 \tag{6-16}$$

$$\boldsymbol{x} = \boldsymbol{x}_{\text{o}} + \boldsymbol{v}_{\text{o}}t + \frac{1}{2}\boldsymbol{a}t^2, \tag{6-17}$$

which is the result found in Section 2.3. $\boldsymbol{x}_{\text{o}}$ is the initial position of the object, $\boldsymbol{v}_{\text{o}}t$ is the distance the object would travel if it were falling with a constant velocity, and $\frac{1}{2}\boldsymbol{a}t^2$ is the additional distance the object travels because its velocity is changing. We have found the kinematic results from Section 2.3 which we derived from geometrical considerations alone. In the next section we give you practice with these techniques by applying them to the motion of a bead on a frictionless wire.

6.3 A BEAD ON A WIRE

To describe a ball thrown into the air, we have only used the equations for motion in one dimension, yet a ball can travel horizontally as well as vertically. Should we also consider horizontal motions: What determines the dimensions we use in a given problem?

A ball thrown straight up moves only vertically. There is no horizontal force on the ball so there is no horizontal acceleration or velocity. Thus the x coordinate for the ball is constant and does not change, so we need not consider it. That is, we need not use two dimensions because only one coordinate is changing. Actually, there are two horizontal coordinates, neither of which change. This approach forms the basis for deciding how many dimensions to use. If you can find some constraint that causes one of the coordinates not to change, then you needn't consider that coordinate. For instance, how many dimensions should we use to describe a pendulum that consists of a small mass m hung on a string of length l, as shown in Figure 6.5?

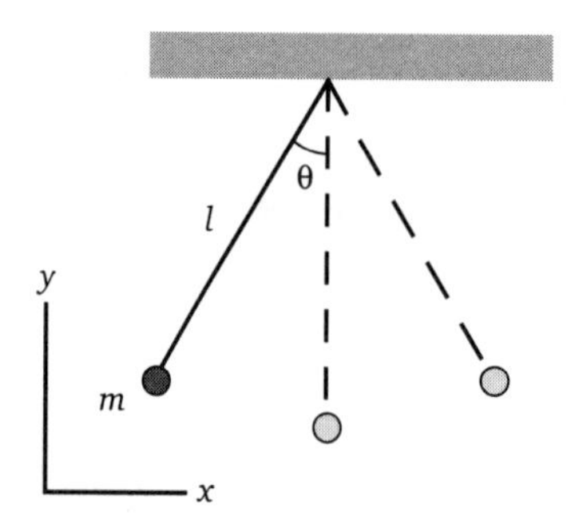

Figure 6.5
Motion of a simple pendulum.

"Ah–h," I hear you say, "here is a case of two dimensions because the mass moves through both x and y!" The string, however, constrains the mass to move along an arc of a circle of constant radius. Thus, x and y are not independent of one another but are related by

$$l = \sqrt{x^2+y^2} \ . \tag{6-18}$$

This relation means that given x, we could find the y, and thus we have a one–dimensional problem for the position. It is easier to see in polar coordinates (r,θ). Only the θ varies: $r = l =$ constant. The key in this example is that the string constrains the particle to follow a given path, which reduces the dimensions needed to describe the magnitudes of the position, velocity, and acceleration to one dimension. Later chapters show that the directions of all these quantities are not the same, and we must ultimately use two dimensions to completely describe this problem. The point, however, is to use the symmetry of the problem to simplify the mathematics.

In other problems, constraints work similarly to reduce the dimension of the problem. For example, a ship exists in three dimensions, but because gravity and the buoyant force of the water constrain a ship to move along the ocean's surface, only two dimensions are needed to uniquely determine any ship's position, the rolling of the ship being negligible. (Anyone who has been seasick on a ship is well aware of the three–dimensional motion of the ship.) Another example is a bead sliding along a wire track. Suppose the track were to bend and twist in three dimensions, as shown in Figure 6.6. How many dimensions are needed to uniquely determine the position of the bead?

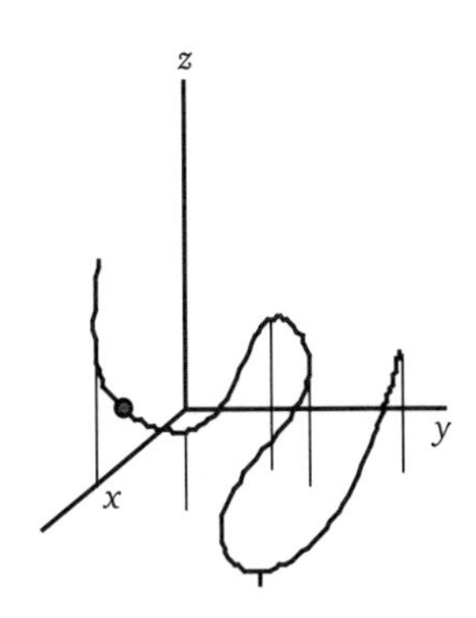

Figure 6.6
A bead sliding along a twisted wire track.

The answer is one. Because the bead is always constrained to be on the wire track, we only need specify how far along the track the bead has moved to find where it is. One end of the track is usually chosen to be the origin, $x = 0$. A positive velocity moves the bead away from the origin, and a negative velocity moves the bead toward the origin. The direction of the acceleration is more difficult because for curved sections of the track, it needn't be along the track at all. Because of this complication, we consider straight sections of track in this chapter and discuss the acceleration for curved motion later.

If we allow a bead to slide down a straight frictionless wire track, the acceleration will be less than that for a bead in free-fall. i.e., a steeper inclination of the track will yield a greater value of acceleration than a small inclination, as shown in Figure 6.7. In fact, $\boldsymbol{a} = g \sin \theta$ which will be demonstrated in Chapter 9.

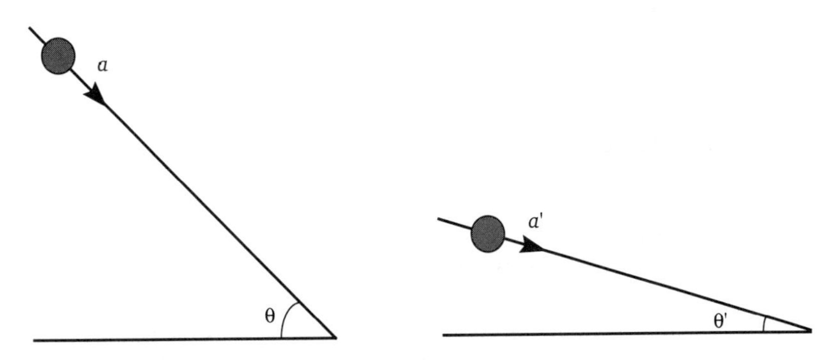

Figure 6.7
The greater the inclination θ, the greater the acceleration $\boldsymbol{a}$.

Note that we can represent a vector such as the acceleration graphically with an arrow that has a length that is proportional to the magnitude.

Because the acceleration is constant for any straight segment of track, the velocity increases linearly. Inclined tracks provide excellent examples on which to exercise your geometry–handling muscles. For example, suppose we are given the velocity–vs–time curve shown in Figure 6.8. Could we obtain the acceleration curve and the position curve?

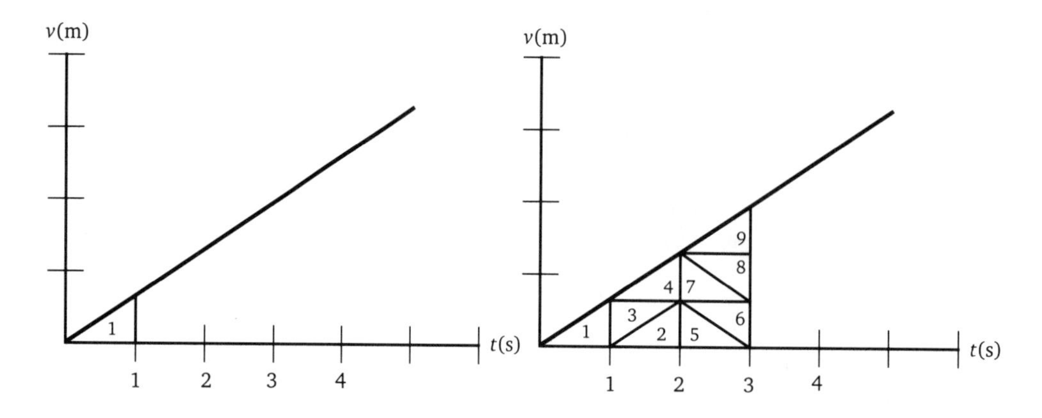

Figure 6.8
Velocity as a function of time for an inclined track. The area under the velocity curve is the displacement which is obtained by counting the triangles. There is 1 triangle after 1*s*, 4 triangles after 2*s*, 9 triangles after 3*s*.

To draw the acceleration vs time curve is trivial because the acceleration is the slope of the velocity curve $\boldsymbol{a} = dv/dt$, which is constant, as shown in Fig. 6.9.

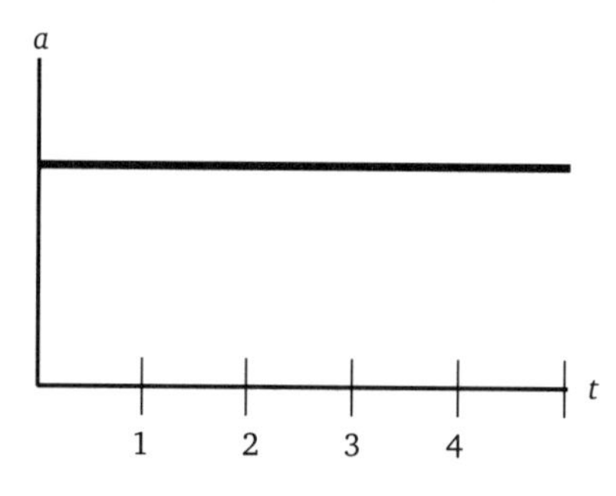

Figure 6.9
A constant acceleration versus time.

To find the position curve we must integrate the velocity curve $\boldsymbol{x} = \int \boldsymbol{v}\, dt$, which means finding the area underneath the curve. Let's allow the first triangle in Figure 6.8 to define one unit of area. Then after 2 time units, we have a total of 4 triangles under the curve so we have 4 units of area. After 3 time units, there are 9 triangles under the curve. How many will there be after 4 time units? This obviously gives us a curve for a parabola, shown in Figure 6.10 where we assume $\boldsymbol{x}_o = 0$.

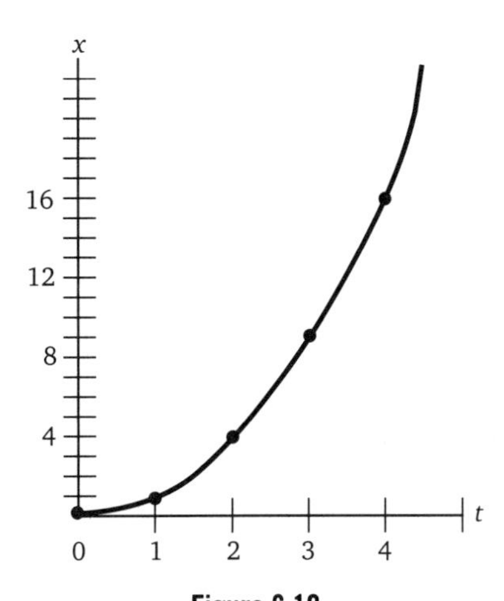

Figure 6.10
Position versus time for a constant acceleration is a parabola.

By using bent wire tracks we can produce quite varied results. Consider the track shown in Figure 6.11. Assume that all the segments [1,2], [2,3], [3,4] have the same length, and the bead slides freely without friction along the wire and past any kinks in the wire. Let s be the position along the track, and let gravity act in the negative y direction.

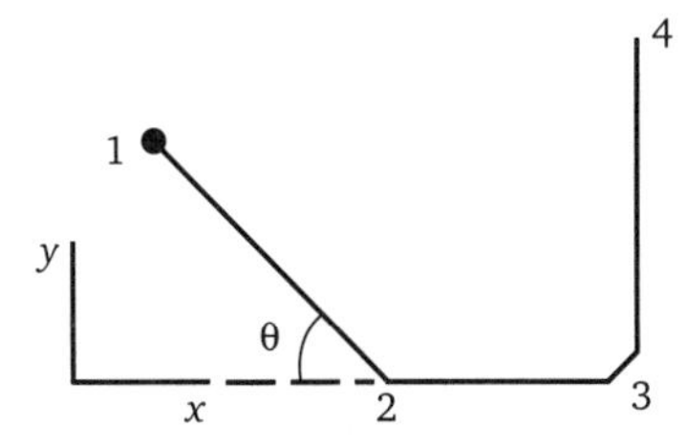

Figure 6.11
A complicated track.

As the bead slides along the track, the behavior will be more complex than in the previous case. We first need to set up a time axis to graph the velocity (v), acceleration (***a***), and position (s). Because each segment is straight and the acceleration along a given segment will be constant, we can best analyze the ball's behavior by looking at each segment individually. Because the bead will be moving at varying speeds along the track, the time taken to traverse each segment will be different— i.e., *not uniform.*

To find how the time intervals will compare to one another, we need to investigate how the speeds compare. Generally, the lower the bead gets, the faster it will be moving, due to gravity. The first segment from 1 to 2 is the same as the preceding problem, where the speed will increase linearly under a constant acceleration. Because the slope of this segment is less than from 3→4, the bead will have less acceleration along 1→2, $\boldsymbol{a}_{12} = g \sin \theta$. When the bead reaches the flat section from 2 to 3, it will not accelerate, so its velocity will remain constant. This segment, 2→3, is lowest, so the ball will travel the fastest here, which means that the ball will take the least amount of time to cover this segment. Over the final segment from 3 to 4, the bead will slow down with a constant acceleration g equal to that in free-fall until its velocity decays to its zero. This segment has the greatest slope, and thus the acceleration is largest. Because the acceleration is back toward the origin, it is negative.

Note that point 4 is slightly above point 1. If the bead initially starts from rest, it will not have enough energy to reach the pinnacle at 4. You can think of the bead as being like a pendulum that does not swing higher than the initial height, and that repeats its motion over and over again. The bead will thus start to move back toward point 1. Since it is moving back towards the origin, the velocity will now be negative. Based on the foregoing arguments, the longest time interval will be from 1 to 2, the shortest from 2 to 3. The bead will not reach 4, so let the time the bead reaches a height equal to point 1 be t_o'. The bead will then turn around and begin to slide backward. The remaining intervals on the swing back correspond to the time intervals on the initial run. The axes for one complete oscillation where the bead returns to point 1 at time t_1' are shown in Figure 6.12. The velocity axis is labeled with the highest speed, v_{23}, attained on segment 2→3.

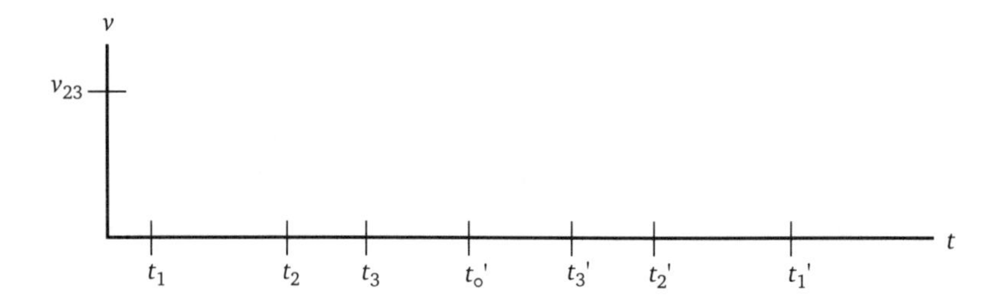

Figure 6.12
The axes for a bead on a track. Note that the time intervals are not equal. The lengths are covered more quickly where the bead is moving fastest. The greatest speed is attained at the lowest point on the track.

Now we can plot the velocity vs time. During times when the bead is moving to the right away from the initial point, the velocity will be positive; when it moves to the left, back toward the origin, the velocity is negative. Based on the preceding discussion, the graph of the velocity is shown in Figure 6.13 where from t_1 to t_2 the slope is g sin θ and from t_3 to t_3' the slope is $-g$.

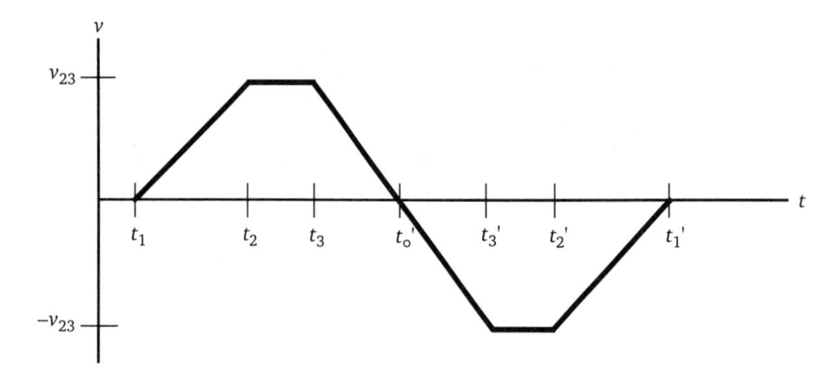

Figure 6.13
Velocity vs time. The bead slides from point 1, almost up to 4, then back to point 1. This completes one oscillation.

Note that the velocity vs time differs significantly from the shape of the track. There are two common mistakes students make here. One is to just repeat the shape of the track for the graph, as shown in Figure 6.14. This signals to your instructor a total lack of understanding and analysis on your part.

Don't make this mistake!

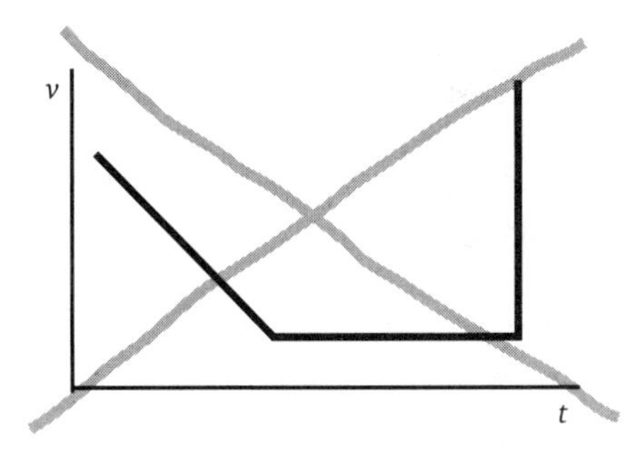

Figure 6.14
Don't repeat the shape of the track in your time graphs. It shows a total lack of understanding!

The other most common mistake is to make all the time intervals equal, ignoring the fact that the velocity is not constant.

Next, we can find the acceleration from the velocity–vs–time curve. Because the acceleration is the slope of the velocity, it is constant during the first segment, and positive. The curve has zero slope over the second segment, so the acceleration is zero. Then over the third segment, the slope is again constant but negative with a greater magnitude than the first segment because the acceleration is the full acceleration due to gravity $(-g)$. Gravity acts as the sole source of acceleration the entire time the bead is on the vertical segment of the track. The total time on the vertical segment is from t_3 to t_3'. The acceleration vs time is plotted in Figure 6.15. Note that the acceleration from t_2' to t_l' is positive even though the bead is slowing down as it returns to Point 1. The velocity is negative, so a positive acceleration opposes the velocity, thus slowing the bead.

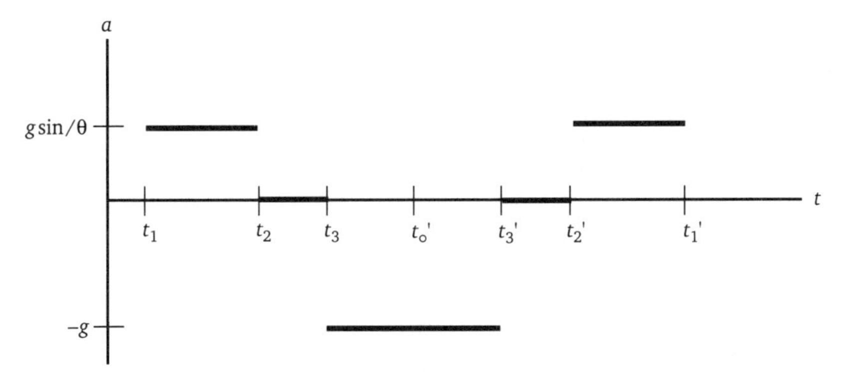

Figure 6.15
Acceleration vs time for a bead on a double dipper track.

Compare Figure 6.15 with Figure 6.13 for the velocity. Note that at the moment t_o' when the bead reaches its highest point, the velocity is zero but the acceleration is not! Remember that acceleration is the slope of the velocity. Just because the value of v is zero at some moment in time does not mean that the slope is zero! There can be a nonzero acceleration at a moment when the velocity is zero.

To find the position as a function of time we need to sum the area under the velocity vs time, curve. We've already seen that the area under a straight line is parabolic, so the first segment in Figure 6.17 is parabola. Be careful though when the velocity flattens out because the position does not remain constant. The position increases linearly when the velocity is constant. This can be seen from Figure 6.16. As we take equal chunks in time, the area increases at a constant rate. Thus, during the second segment in Figure 6.17 the position continues to increase linearly.

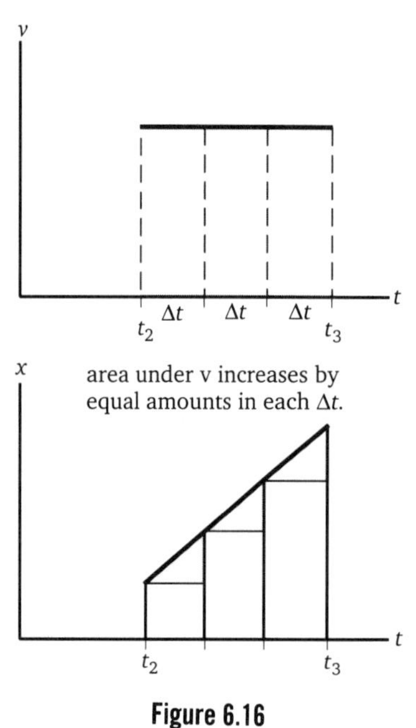

Figure 6.16
Area increases linearly under a curve with zero slope.

Therefore, in the flats between t_2 and t_3 the position must increase linearly. In the third segment in Figure 6.17, between t_3 and t_4, the position does not decrease, but rather continues to increase as more area continues to be added, but at a slower and slower rate. Even though the acceleration is negative, the velocity is still positive. A negative acceleration means that the bead is slowing down as it moves up the vertical hill i.e., the position graph will have negative curvature between t_3 and t_o'. Then as the bead slides back down the vertical wire, the bead speeds up as it falls, but because it is moving back toward the origin, the position graph curves down. Note that in the entire region from time t_3 to t_3', the positive graph has negative curvature because the acceleration is negative in this region. For t_3' to t_2' the graph is a straight line as the bead coasts back to the origin. Then on the final segment in Figure 6.17, the bead slows to zero velocity as it approaches the original position.

To review, consider all three graphs for acceleration, velocity, and position, as shown in Figure 6.18.

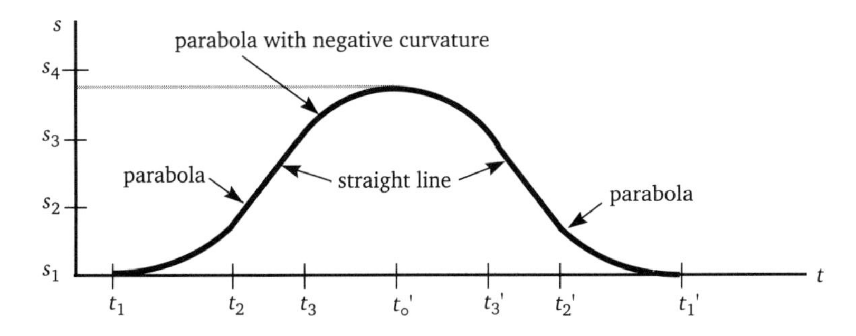

Figure 6.17
Position vs time for a bead on a track.

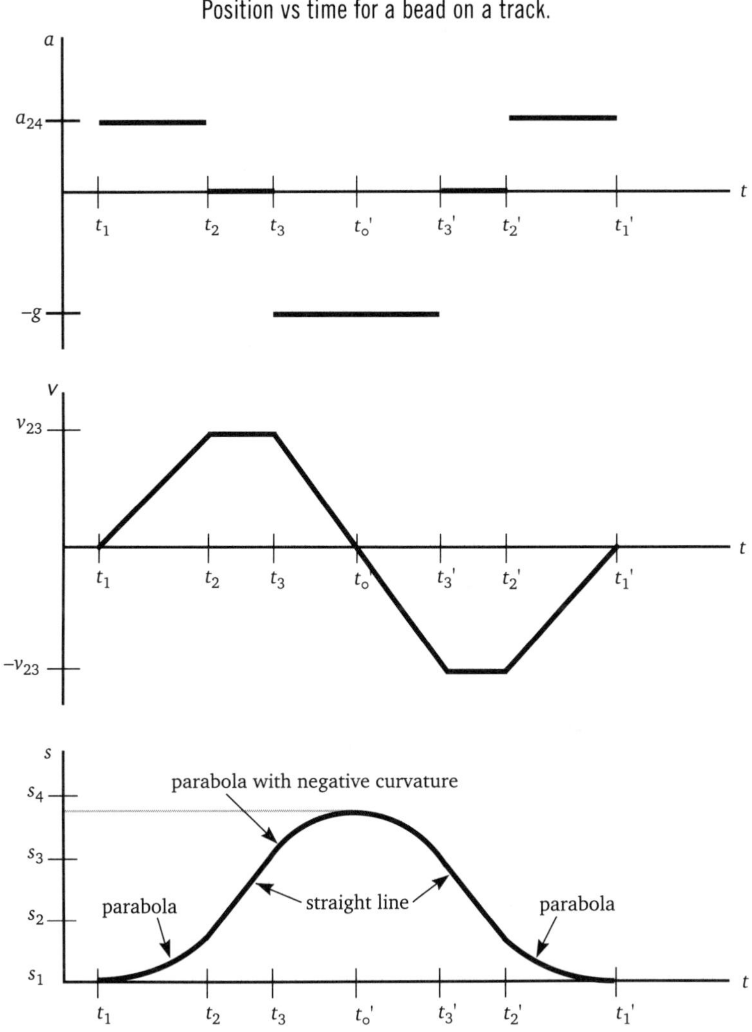

Figure 6.18
Acceleration, velocity, and position vs time for a bead on a double dipper track.

Before going on, you should compare the velocity, position, and acceleration during each time interval. Be sure you understand how they relate to one another.

• 6.4 DENSITY

Another example where slope and area under a curve become very useful is with a substance distributed throughout some region of space. In this section, we will consider only mass distributed along a line, but the analysis will apply to any "substance," be it mass, electric charge, or energy. The average linear mass density λ is defined by

$$\overline{\lambda} = \frac{M}{L} = \frac{\text{Total Mass}}{\text{Total Length}} \qquad \text{linear mass density} \qquad (6\text{-}19)$$

The density relates the mass to the geometry, i.e., it tells us how the mass is distributed in space. If we had mass distributed over a surface, then it would make sense to use the mass per unit area σ, and for a volume, the mass per unit volume ρ.

$$\overline{\sigma} = \frac{M}{A} \qquad \text{surface mass density} \qquad (6\text{-}20)$$

and

$$\overline{\rho} = \frac{M}{V} \qquad \text{volume mass density} \qquad (6\text{-}21)$$

All are average densities.

An important point about density is that even though it is formed from two additive quantities, the mass and volume—the density itself is not additive for the same reason $1/2 + 1/3 \neq 1/5$. To find the average density of a composite object, we must be careful to use the definition of average density, $\dfrac{\text{total mass}}{\text{total volume}}$. The following one dimensional example illustrates the principle.

EXAMPLE 6-2

A 3 cm rod with density 6 g/cm is glued to a 2 cm rod with density 8 g/cm. Find the average density of the composite rod.

Solution:

The definition of the average density is the total mass divided by the total length. Densities do not add!

$$\lambda_{tot} \neq \lambda_1 + \lambda_2$$

but mass does add so

$$M = M_1 + M_2$$

$$M = \lambda_1 L_1 + \lambda_2 L_2$$

$$\Rightarrow M = (6\text{ g/cm})(3\text{ cm}) + (8\text{ g/cm})(2\text{ cm}) = 34\text{ g}.$$

The total length is $L = L_1 + L_2 = 5$ cm. So the average density is $\overline{\lambda} = M/L$

$$\bar{\lambda} = \frac{34\text{ g}}{5\text{ cm}} = 6.8\text{ g/cm}.$$

Note that this is far different than the addition of the densities, which would give us 14 g/cm, very much in error.

If the mass of an object is distributed uniformly along a line, then we can use

$$M = \bar{\lambda} L \tag{6-22}$$

to find the mass. It becomes interesting when the mass is not distributed uniformly, but varies with position. Then each little piece of the object will have a different density, such that, λ becomes a function of position, $\lambda = \lambda(x)$. In general, the only way to handle the situation is to divide the object into small pieces ΔL, and to define the density for each piece. Note that if the length of the piece is ΔL, then only a small amount of mass Δm can exist in that small piece of space, ΔL, so that

$$\Delta m = \frac{\Delta m}{\Delta L}\Delta L.$$

where

$$\lambda = \frac{\Delta m}{\Delta L} \tag{6-23}$$

is the density at a point on the object. Each Δm is related to its length ΔL by its density. Note that the density relates Δm to its geometry—i.e., how Δm is distributed in space. Here Δm is distributed over a line ΔL, as shown in Figure 6.19. When using density, always draw a picture of what the small piece of mass looks like so you will know what geometry to use.

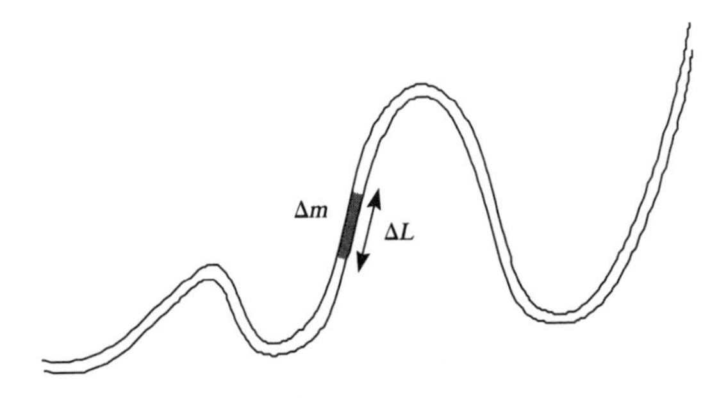

Figure 6.19
An infinitesimal piece of mass Δm has a length ΔL.

The mass of any macroscopic piece L can be found by using the chain rule and summing the small pieces Δm. This is of course nothing more than we have just discussed in the last three sections, i.e., adding up the area under the $\lambda(x)$ vs x curve.

$$M = \sum_{i=0}^{n} \frac{\Delta m_i}{\Delta L_i}\Delta L_i \tag{6-24}$$

so that

$$M = \sum_{i=0}^{n} \lambda_i \Delta L_i. \tag{6-25}$$

This technique can be used to find the mass when the density varies. In general, Equation (6-25) is tedious to use; however, we can treat linear variations in density quite easily.

EXAMPLE 6-3

A thin wire has a length of 12 m. The mass density varies linearly from 14 g/m at one end to 20 g/m at the other. Find the mass and average density of this length of wire.

Solution:
The easiest method is to graph the density as shown in Figure 6.20.

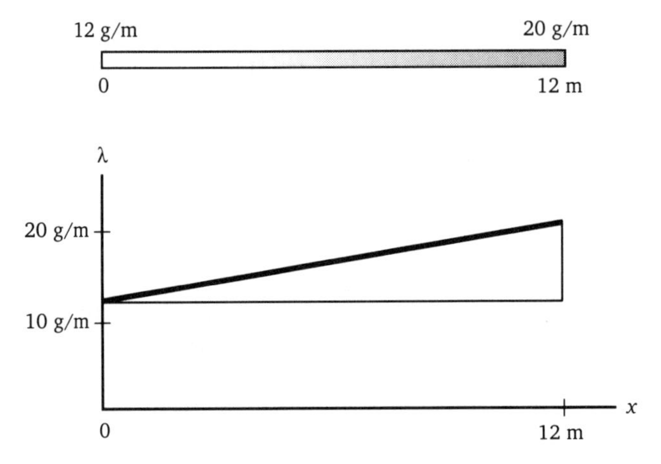

Figure 6.20
The mass of a wire whose density increases linearly with position can be found from the area under the density vs × curve which is the area of the shaded triangle + the (12 g/m × 12 m) box below the triangle.

The area under the λ vs x curve is the area of the triangle + (12 × 12) box. In

$$M = \frac{1}{2}(12\text{ m})(8\text{ g/m}) + (12\text{ m})(12\text{ g/m})$$

$$M = 48\text{ g} + 144\text{ g} = 192\text{ g}.$$

The average density is the total mass divided by the total length.

$$\overline{\lambda} = \frac{M}{L} = \frac{192\text{ g}}{12\text{ m}} = 16\text{ g/m}.$$

Can this be right? The average density, 16 g/m, is halfway between 12 g/m at one end, and 20 g/m at the other end! The average density is the value of the density of the bar at its middle. Under what conditions is the average density equal to the density at the midpoint? The proof is left as an exercise, Problem 6.19.

We could also handle special shapes. For instance, consider a wire bent along the arc of a circle. In this case, the length of the wire L becomes the arc length

$$L = r\,\Delta\theta \tag{6-26}$$

where $\Delta\theta$ is the angle subtended by the length L as shown in Figure 6.21.

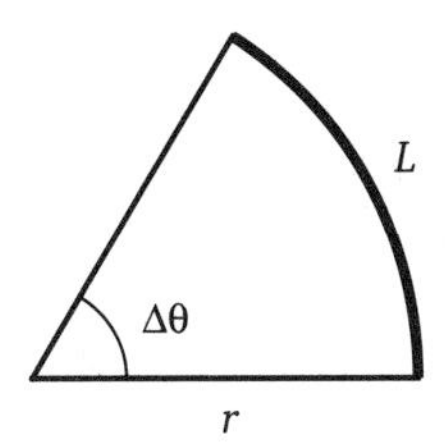

Figure 6.21
A wire of length L bent into the arc of a circle.

In this case, the total mass is given by

$$M = \lambda L = \lambda r \Delta\theta. \tag{6-27}$$

Always be aware of the geometry of the shape of the object you are trying to describe.

EXAMPLE 6-4

A wire, bent into a circular arc of radius 45 cm, has a uniform linear mass density of 6 g/cm. Find the mass of the wire for the angular interval $\left[0, \frac{\pi}{6}\right]$.

Solution:
Since the wire has uniform density, we can use $M = \lambda L$. But in this case L is the arc of a circle, so $L = r\theta$.

$$M = \lambda(r\theta)$$

where $\theta = \pi/6$. Substituting in the values:

$$M = (6\text{ g/cm})(45\text{ cm})\left(\frac{\pi}{6}\right) = 141\text{ g}.$$

Chapter 6 Problems

[6.1] The position of a particle is described by the function

$$x = 5\text{ m} + (8\text{ m·s})\,t - (4\text{ m·s}^2)t^2$$

for all positive t. **a)** Draw a graph of x vs t labeling all intercepts and extrema **b)** Find the position, velocity, and acceleration each at $t = 3$ s. **c)** During what time interval is the velocity positive? **d)** Is the initial acceleration positive or negative? **e)** At the moment the velocity is zero, is the acceleration positive, negative, or zero?

• **[6.2]** The velocity of a rocket is given by

$$v = (25\text{ m/s}) + (50\text{ m/s}^3)t^2.$$

a) Draw a graph of v vs t. **b)** Find the acceleration at $t = 4$ s, and **c)** find the position at $t = 4$ s. Assume the rocket starts from $x = 0$.

[6.3] For the track shown in Figure 6.23, a bead starts from rest at point A and slides to point E where it momentarily comes to rest and then slides back toward point A. Drawing arrows at the points A through E, show the direction of the velocity and acceleration for both the journey to point E and the return trip back to point A.

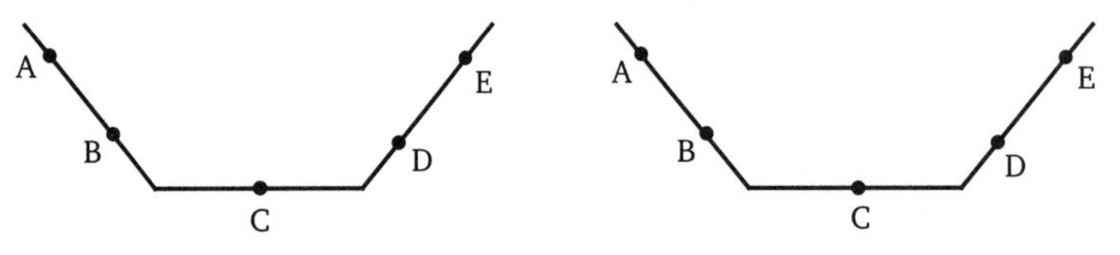

(a) bead moves from A to E (b) bead moves from E back to A

Figure 6.22

For Problems 4–7, given the graph for the velocity versus time, use graphical methods to find the position and acceleration versus time.

[6.4]

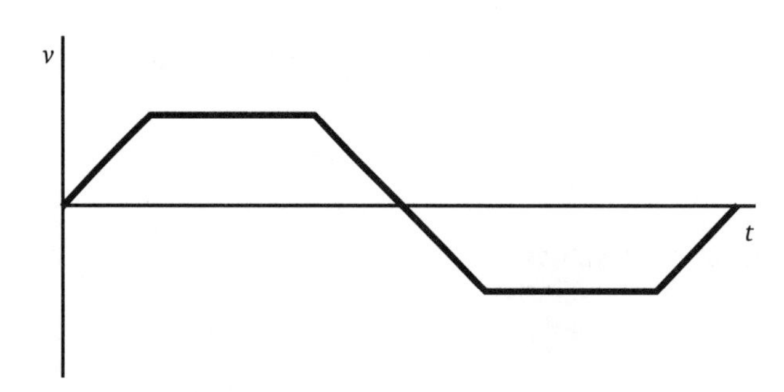

[6.5]

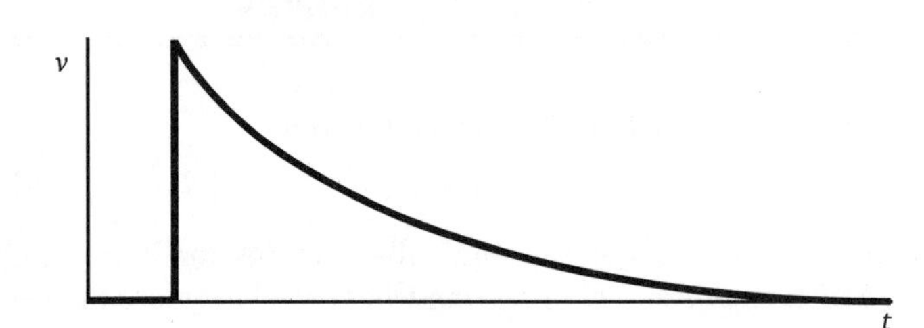

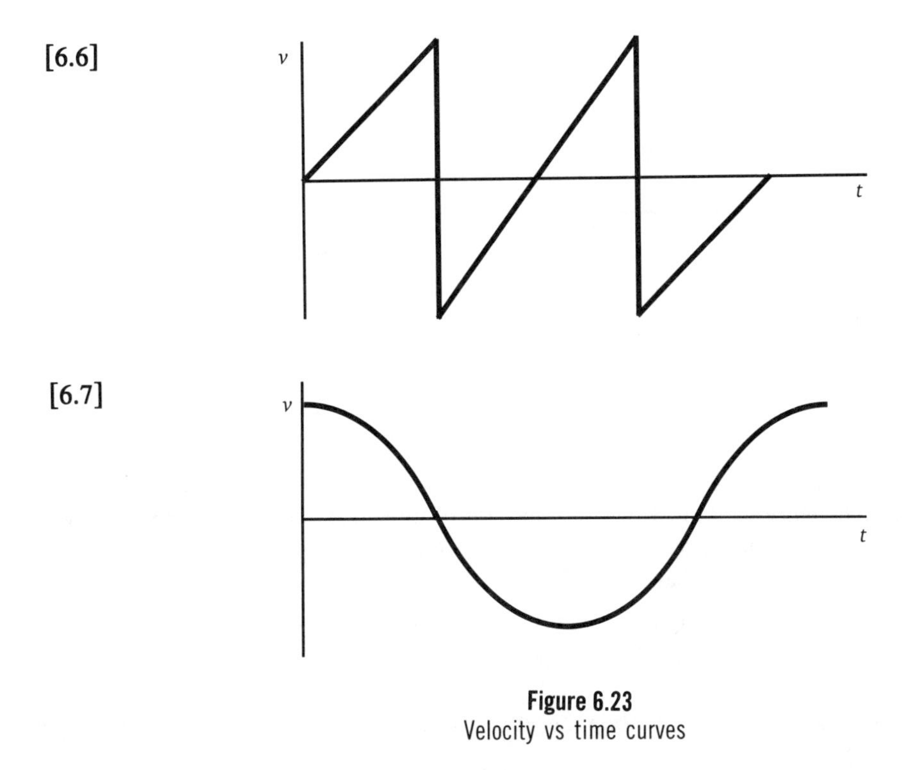

Figure 6.23
Velocity vs time curves

Given the wire tracks with beads shown in Problems 8–11, make a graph of the position s, velocity v, and acceleration a in time. Be sure to label the points t_1 through t_6 on the abscissa and the corresponding points on the ordinate. Assume s is the distance along the track, and all the track segments are of the same length. The tracks exist in a vertical plane with x being horizontal and y being vertical. Gravity acts in the negative y direction.

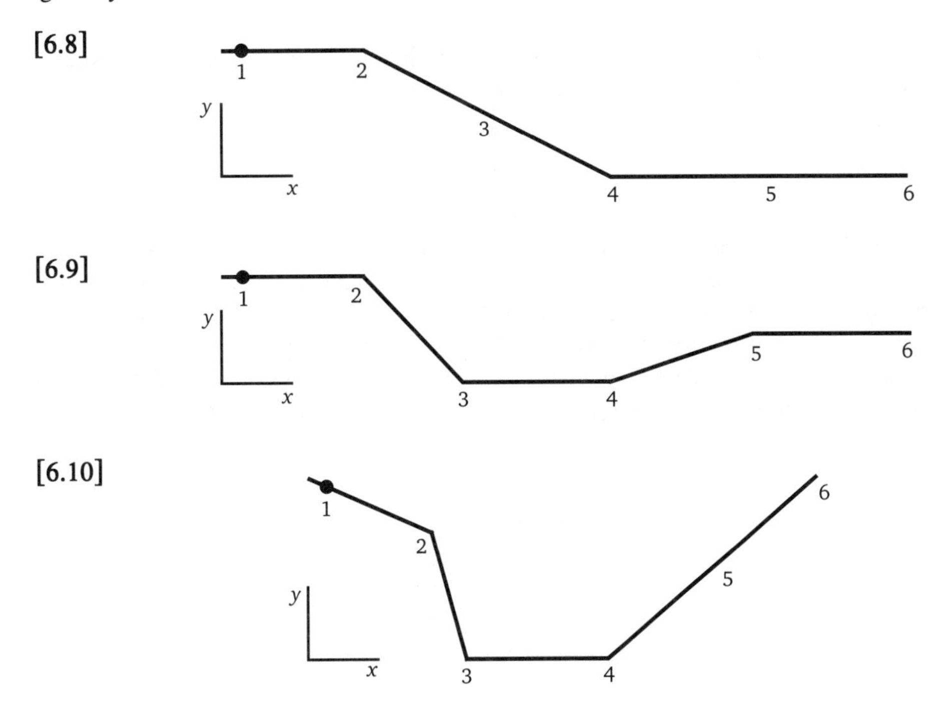

• **[6.11]**

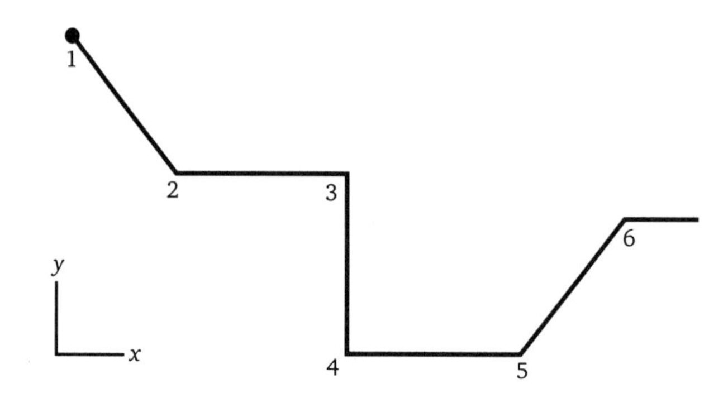

[6.12] A block of length 12.0 cm and cross-sectional area 7.0 cm2 consists of three pieces of different wood that have been glued together end to end. The densities of the three individual pieces are 0.60 g/cm3, 0.73 g/cm3, and 0.80 g/cm3, and their respective lengths are 4.7 cm, 3.5 cm, and 3.8 cm (all of cross sectional area 7.0 cm2). Find the density of the whole block.

[6.13] A thin wire of length L, has a density of 43 g/m at one end and decreases linearly to 73 g/m at the other end. **a)** Find the average density. **b)** Find the mass of the wire.

• **[6.14]** A thin wire of length L, stretched along the x axis from $x = -\frac{L}{2}$ to $x = +\frac{L}{2}$, has a density that increases from λ_o at $x = -\frac{L}{2}$ to $2\lambda_o$ at $x = +\frac{L}{2}$. **a)** Draw a graph of the density **b)** Find the average density. **c)** Find the mass of the wire.

• **[6.15]** A thin wire of length L, stretched along the z axis beginning at $z = +a$, has a density of λ_o and increases linearly to $5\lambda_o$ at $z = a + L$. **a)** Find the average density **b)** Write an equation for the density as a function of x.

• **[6.16]** A thin uniform wire is bent into a circular arc subtending one third of a circle of radius 6 cm. The density of the wire is $\lambda = 15$ g/cm. Find the mass of the wire.

• **[6.17]** A section of a circular wire of radius R has a density which increases linearly from λ_o at $\theta = 0$ to $3\lambda_o$ at $\theta = 60°$. **a)** Find the average density. **b)** Write a formula (in terms of λ_o, π, θ, and appropriate numbers) for the density of this segment of wire as a function of θ.

• **[6.18]** A bead slides down a frictionless wire that is made of equal lengths of straight wire that has bends at the junctions of the lengths. Assume s is the distance along the track, and all the track segments are of the same length. The tracks exist in a vertical plane with x being horizontal and y being vertical. Gravity acts in the negative y direction. The labeled points on the time axis of the graph of the velocity vs time shown in Figure 6.24 correspond to times when the bead reaches the end of each segment. Given this graph, sketch: **a)** a graph of the acceleration vs time, **b)** a graph of the positive vs time, **c)** the approximate shape of the track.

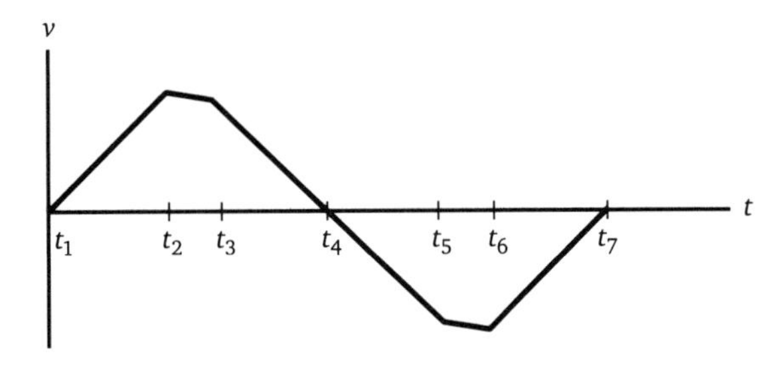

Figure 6.24

• **[6.19]** Prove that for a long thin straight object whose density increases linearly along its length that the average density of the object equals the value of the density at the object's midpoint.

[6.20] Ride a bike down a fairly steep hill in the following manner. Choose a hill with a smooth, paved road straight down the hill. Start from rest near the top and allow yourself to coast down the hill. Try to stay bent over like a bike racer to minimize the friction effects. Draw a picture of the hill and mark on the picture: **a)** where your speed is least, **b)** where your speed is greatest, **c)** where your speed is changing at the greatest rate.

[6.21] In the graph shown in Figure 6.25, the force, F, in newtons acting on a mass m is plotted along the vertical axis, and the time in seconds is along the horizontal axis. Find the total change in momentum, Δp, of the mass m if $F = \dfrac{\Delta p}{\Delta t}$.

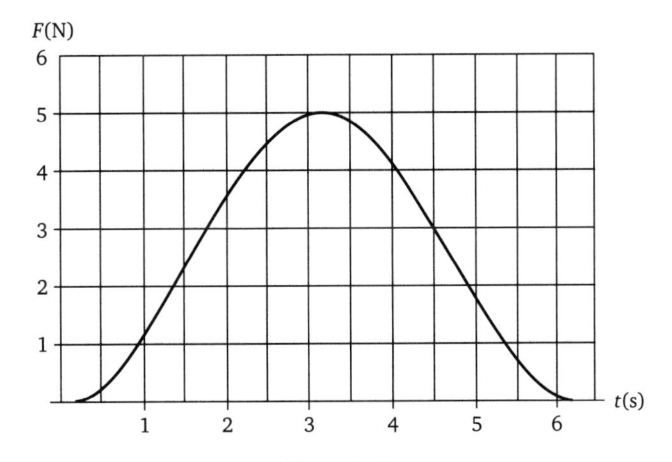

Figure 6.25
Graph of F vs t.

CHAPTER 7
Vectors, Displacement, and Velocity

(Has your life been without direction?)

Have you ever watched a swallow fly? It darts this way and that, changing its direction in the wink of an eye. Unencumbered by a large mass, it can travel up, down, left,or right and can quickly turn around and travel back from where it came. How many pieces of information do we need to describe the motion of the swallow? Would we need the same number to describe the motion of a garbage truck? In this chapter, we tackle the problem of motion in more than one dimension. At first, it might seem hopelessly complicated, but some powerful techniques make it straightforward. First, you must memorize those techniques and practice them until they become natural.

7.1 DIMENSION

Not all problems require the same number of dimensions to solve the problem. For instance, the aforementioned swallow can move through three dimensions, but the garbage truck is constrained to the surface of the earth, so we only need two dimensions to uniquely describe its position. Likewise, the bead sliding along a wire is constrained by the wire to a linear path, so it moves in only one dimension. It may be that the wire has loops and twists and turns, but if we know the path, then the position of the bead is uniquely given by the distance along the wire. We saw this when we treated circular motion. A wire bent into a circle exists in two dimensions, but for a bead sliding along the wire, we only need one dimension to uniquely locate its position, and that is the angular coordinate θ. The radius is constant and therefore doesn't change. If the wire the bead slides along is not bent into a circle but is some crazy shape such as in the cases considered in Chapter 6, we still only need to specify the distance along the track to locate the bead's position. Even though the track may bend around and travel up and down, all the other coordinates are dependent on how far along the track we have traveled.

Constraints determine how many dimensions are required to describe the motion because constraints limit the motion, thus reducing the number of dimensions required to describe a system. Constraints do this by providing equations that link two or more of the coordinates. Whenever two coordinates are linked by an equation, we lose a dimension because one of the coordinates is dependent on the other. For example, the position of an object in uniform circular motion is determined by one coordinate because the radius is a constant, and this provides an equation that links x and y:

$$r = \sqrt{x^2 + y^2} \tag{7-1}$$

If we know the x position, then the y position is determined by the preceding equation. Therefore, y is not independent of x. Generally, a system constrained to move along a path needs only one dimension to describe its position, while a system constrained to move along a surface needs two dimensions to describe its position. Remember, it is the number of *independent* coordinates that determines the number of dimensions needed to solve any problem.

Although the position of a bead constrained to a track is uniquely given by one coordinate, Section 8.1 shows that the calculation of the acceleration is more difficult. For circular motion, the acceleration points toward the center of the circle, perpendicular to the velocity, so describing the acceleration of a particle moving in uniform circular motion requires two dimensions after all.

For motion in a plane, two coordinates are necessary to describe even the position of an object. There are a variety of ways to choose the particular coordinate system, but the two most common are the Cartesian and the polar coordinate systems.

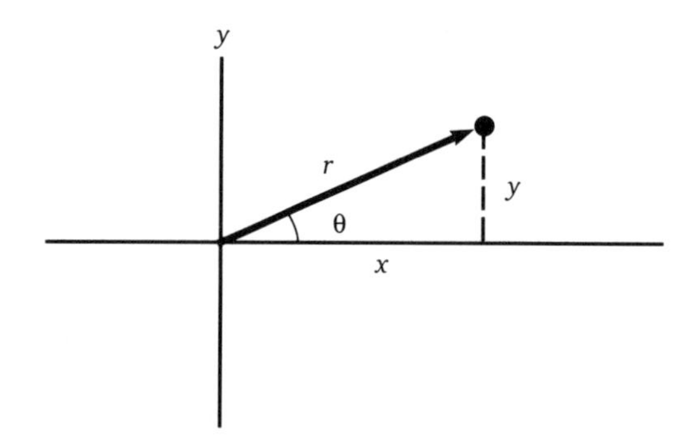

Figure 7.1
Coordinates of an object constrained to move in the x–y plane.

In the Cartesian coordinate system, points in the plane are assigned to a rectangular grid (see Figure 7.1). Any point is then labeled by its distance both along the x axis of the grid and along the y axis of the grid. These two pieces of information are stored in an ordered pair (x,y). The x coordinate is always given first. If the plane is flat, and the motion of the system is along straight lines, then there is an inherent advantage to using a rectangular grid to find the position of an object. However, if the path is circular, it is usually easier to use the polar coordinate system. The position is again represented by two pieces of information (r,θ), which are again given as an ordered pair. That is, r is the distance from the origin to the object, and θ is the angle with the x axis made by a line drawn from the origin to the point. Positive θ is measured in the counterclockwise sense. Thus, θ gives the direction you must walk to reach the point (r,θ) from the origin.

Note that there is a certain economy in the polar system: (r,θ) represent a set of instructions for finding the object. They say start at the origin, and turn to an angle θ, with respect to the x axis. Now walk a distance r and there you find the object. If you followed that set of instructions, you would take the shortest path to the object. Instead, if you followed the Cartesian instructions you would walk along the x axis a distance x. and then walk along the y axis a distance y. The polar instruction set is giving information in "as the crow flies" form, and the Cartesian instruction set is more like being constrained to follow city streets.

In two dimensions for either the Cartesian or the polar coordinate system, two pieces of information are required to uniquely label any point in the system. Much of what follows is an advanced bookkeeping system for keeping track of two or three pieces of information.

7.2 VECTORS: POSITION AND DISPLACEMENT

A *vector* is a quantity that has both *magnitude* and *direction* and is denoted by writing an arrow above the letter. For instance, if we wished to write (A) as a vector, it would appear $\vec{A}$. In most other books, a vector is written in bold-faced type, **A**. You will probably see a variety of techniques for denoting vectors. Sometimes, the writing arm gets tired, and the arrow sinks to the bottom or a tilde $\tilde{A}$ is used instead of an arrow. The $\tilde{A}$ sometimes finds its way into classroom use.

The magnitude is a measure of the size of the vector. When the vector measures position or a change in position, the magnitude is the length, but in general, magnitude need not be a length. For example, force is a vector, and the magnitude of the force is a measure of how strong the force is, but the magnitude of the force is not a length, so magnitude is a broader concept, having much greater applicability than length. The magnitude is a *scalar*—i.e., is always an unsigned non-negative number, and it is denoted by

$$\text{magnitude of } \vec{A} \equiv |\vec{A}|.$$

The direction is much harder to construct mathematically. A very useful way is through unit vectors. All *unit vectors,* by definition, have a magnitude of 1, and hence their name *unit,* which follows from *unity.* Their sole purpose is to mathematically denote the direction of a vector. Unit vectors are written by placing a caret over a letter (e.g. â). The direction of the vector A can be denoted by $\hat{a}$. We can always find the unit vector for a given vector by dividing the vector by its magnitude.

$$\hat{a} \equiv \frac{\vec{A}}{|\vec{A}|} \qquad \text{unit vector} \qquad (7\text{-}2)$$

For example in one dimension, the direction of the position vector $\mathbf{x} = -5$ is $(-5)/|-5| = -1$, means that the vector points in the negative-x direction. The positive-x direction is given by unit vector $\hat{\imath}$ and is the direction of increasing x. Similarly, the other Cartesian unit vectors are denoted by $\hat{\jmath}$ for the y direction and $\hat{k}$ for the z direction. Some will also use $\hat{x}$, $\hat{y}$, $\hat{z}$ for the Cartesian unit vectors.

Using unit vectors, a vector can be written just as its definition is written: a magnitude times a direction.

$$\vec{A} = |\vec{A}|\,\hat{a} \qquad \text{vector} \qquad (7\text{-}3)$$

We can represent vectors graphically as in Figure 7.2 by drawing an arrow, the length of which is proportional to the magnitude, and the tip of which points in the direction $\hat{a}$.

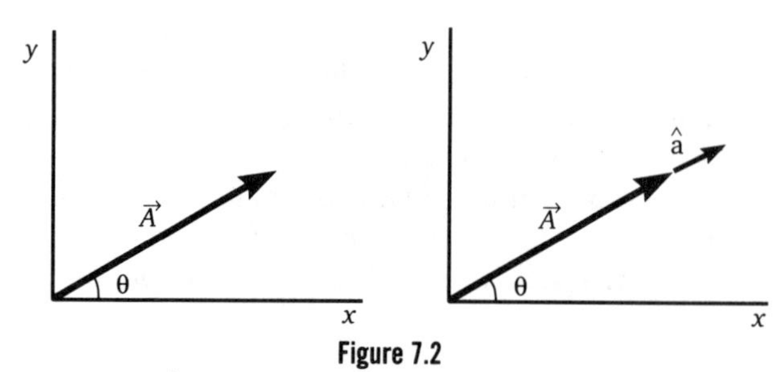

Figure 7.2
A vector $\vec{A}$, and a unit vector $\hat{a}$ that points in the direction of $\vec{A}$.

In general, a vector need not have its tail attached to the origin but can be moved as needed, so long as the length and direction are preserved. The one exception to this rule is the position vector $\vec{r}$, which always has the tail at the origin and is discussed next. All the vectors shown in Figure 7.3 are equivalent.

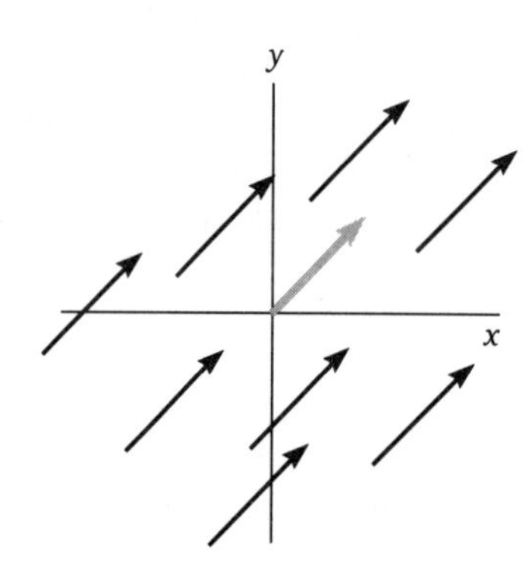

Figure 7.3
All the vectors shown are equivalent because they have the same length and direction, but only the gray vector, with its tail at the origin, can be a position vector.

The simplest vector we can create is the position vector, which points from the origin to any point in our coordinate system. The position vector is denoted as r and can be written as

$$\vec{r} = r\hat{r}, \qquad \text{position vector} \qquad (7\text{-}4)$$

where

$$r \equiv |\vec{r}|. \qquad \text{magnitude} \qquad (7\text{-}5)$$

Because the direction from the origin to any point in the coordinate system is radially out from the origin, $\hat{r}$ always points in the radial direction. Position vectors for various points in the x–y plane are shown in Figure 7.4. Note that they all point radially away from the origin.

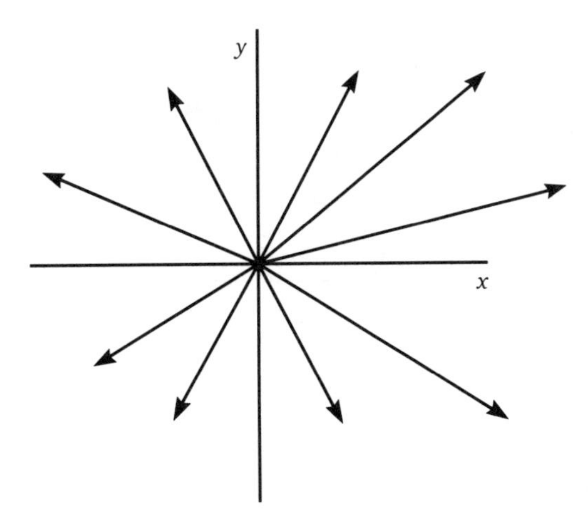

Figure 7.4
All position vectors point radially away from the origin.

Usually, we will need to change the position of some object—i.e., displace the object from an initial position $\vec{r}_0$ to a new position $\vec{r}$. Label the *displacement vector* $\Delta\vec{r}$. Then, we will get to the new position by starting at the tip of $\vec{r}_0$ and walking in the direction of $\Delta\vec{r}$ for a distance $|\Delta\vec{r}|$. Graphically, we can reconstruct this set of instructions by placing the tail of the displacement vector at the tip of $\vec{r}_0$, as in Figure 7.5.

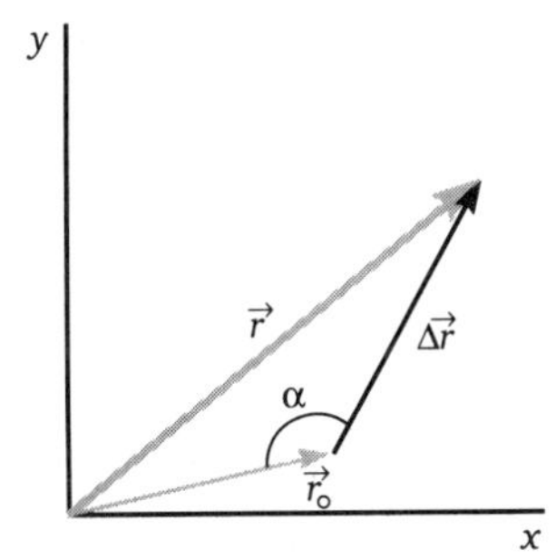

Figure 7.5
A displacement from r_0 to r by moving through Δr.

We can find the resultant vector by using some basic trig. We can find the length of r by using the law of cosines. To do this, we need the angle subtended* by r, which is α in Figure 7.5. Assume we know both the angle θ that $\vec{r}_0$ makes with the x direction, and the angle β that $\Delta\vec{r}$ makes with the x direction. Then the angle α is $\pi - (\beta - \theta)$, as shown in Figure 7.6.

* A line segment or an arc of a circle *subtends* a given angle when the end points of the segment lie on the sides of the angle. The angle α in Figure 7.5 has sides Δr and r_0, and these sides intercept the end points of r, so r is subtended by the angle α.

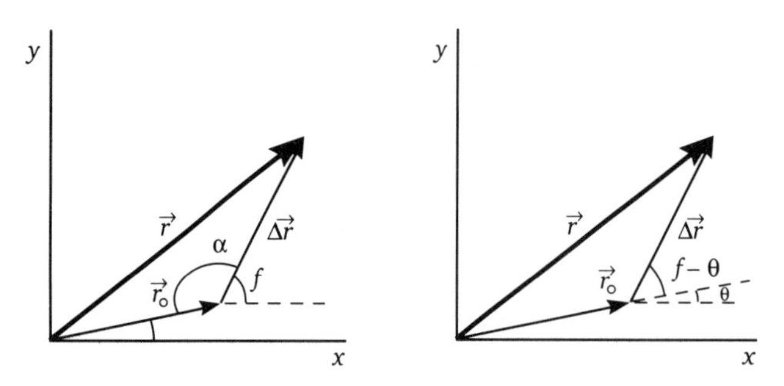

Figure 7.6
The angle subtended by r is $\alpha = \pi - (\beta - \theta)$.

The law of cosines then gives

$$r = r_0 + \Delta r - 2r_0 \Delta r \cos \alpha, \qquad \text{law of cosines} \qquad (7\text{-}6)$$

the length of the new position vector r. Note that the law of cosines is not some strange trig identity to be memorized only for tests; it is the generalization of the Pythagorean theorem to non-right triangles, and this makes it a very useful law, indeed. If $\alpha = 90°$, then the triangle shown in Figure 7.6 will be a right triangle with a hypotenuse of r and legs r_0 and Δr. Because *cos* $90° = 0$, the law of cosines in Equation (7-6) is then $r^2 = r_0^2 + \Delta r^2 - 0$ which is the Pythagorean theorem. Thus, the law of cosines contains the the Pythagorean theorem as the special case for $\alpha = 90°$.

We also need the direction of the new position vector. If we can find the angle γ in Figure 7.7, we'll be able to find the angle $\vec{r}$ makes with the x axis by adding it to θ

$$\phi = \gamma + \theta. \qquad (7\text{-}7)$$

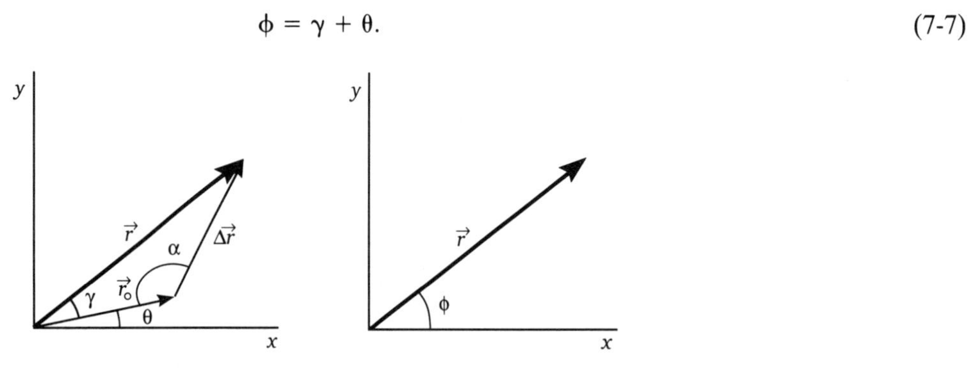

Figure 7.7
The direction of r is given by $\phi = \gamma + \theta$.

We only know one interior angle of the triangle formed by $\Delta\vec{r}$, $\vec{r}$, and $\vec{r}_0$, so we can't simply sum the interior angles to find γ, but we can find γ by using the law of sines. Use α and the side opposite α, r, and γ and its side opposite, Δr:

$$\frac{\sin \alpha}{r} = \frac{\sin \gamma}{\Delta r} \qquad \text{law of sines} \qquad (7\text{-}8)$$

$$\gamma = \sin^{-1}\left(\frac{\Delta r}{r}\sin\alpha\right) . \tag{7-9}$$

As a general procedure, use the law of cosines to find the magnitude of the new position, and the law of sines to find the direction.

EXAMPLE 7-1

The initial position vector is 5 m at 30° with the x axis. A displacement of 4 m is made at 63° with the x direction. Find the new position.

Solution:
$\vec{r}$ is subtended by an angle

$$\alpha = 180° - (63° - 30°) = 147°,$$

so the law of cosines gives

$$r^2 = 25 + 16 - 2(5)(4)\cos(147°) \Rightarrow$$
$$r = 8.6 \text{ m}$$

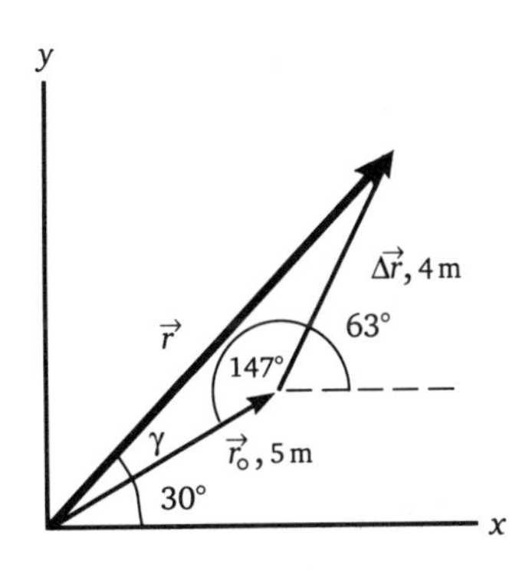

Figure 7.8
Finding the final position vector.

Using the law of sines to get γ,

$$\frac{\sin 147°}{8.6\text{m}} = \frac{\sin\gamma}{4\text{m}} ; \quad \gamma = \sin^{-1}\left(\frac{4}{8.6}\sin 147°\right) = 14.6°.$$

This gives the direction of $\vec{r}$:

$$\phi = 14.6° + 30° = 44.6°.$$

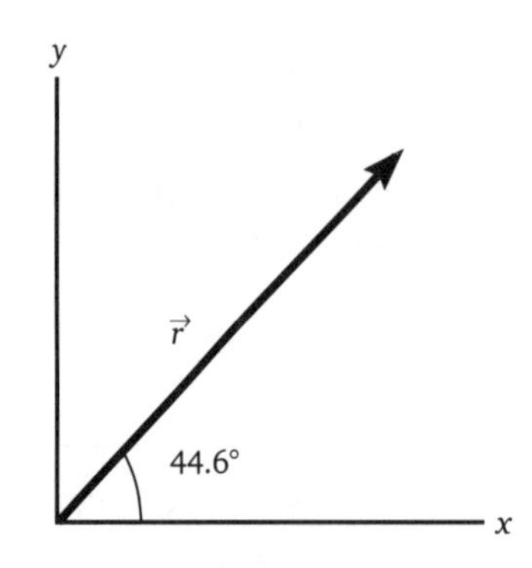

Figure 7.9
The direction of *r* is the angle made with the *x* axis.

This method of adding vectors is *coordinate free*, in the sense that we have not made use of any particular type of coordinate system. Often, particularly when more than three vectors are involved, it is easier to resolve the vectors into Cartesian components when adding than to use trigonometry. We tackle this method next.

7.3 VECTOR ALGEBRA

Although we can manipulate vectors using the techniques in section 7.2, we really need to use vectors in formulas. For example, we will want to derive a formula for the position vector when the position changes in time as in Chapter 2. Therefore, we need to develop a vector algebra—i.e., we must be able to add vectors, subtract vectors, and multiply vectors. Furthermore, the algebra we construct should be as much like regular algebra as possible. As many of the rules of scalar algebra as possible should carry over to vector algebra. There are some differences, particularly for vector multiplication, but we strive to make vector algebra as much like regular algebra as possible. To aid us in this endeavor, we introduce the idea of *resolving a vector into Cartesian components.* In many ways, components form the heart of our vector algebra.

Resolving a Vector into Cartesian Components

A *component* is the projection of a vector along a particular coordinate direction. Cartesian components are generally the easiest to find. We start by dropping a perpendicular line from the tip of the vector onto one of the coordinate axes. In Figure 7.10, we've done this for the *x* axis.

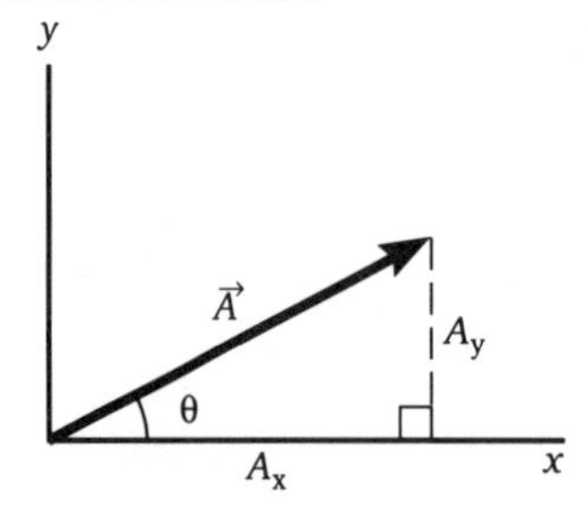

Figure 7.10
Resolving a vector into its Cartesian components.

We can use trig on the triangle formed by A_x, A_y, and $\vec{A}$. The length of the hypotenuse is

$$|\vec{A}| = A.$$

Therefore,

Cartesian Components of the Position Vector

$$A_x = A\cos\theta$$
$$A_y = A\sin\theta. \tag{7-10}$$

θ = angle $\vec{A}$ makes with the x axis

Given the components, the original vector can be retrieved by polar components:

Polar Components

$$|\vec{A}| = \sqrt{A_x^2 + A_y^2}, \text{ and}$$
$$\theta = \tan^{-1}\left(\frac{A_y}{A_x}\right) \tag{7-11}$$

We wish to put the vector in a form that may be handled algebraically. We can do this by using the components and the unit vectors that define the x and y directions. Literally, the x component is the amount of the vector that lies in the x direction, so using the definition of a vector,

$$\vec{A} = A\,\hat{a}, \tag{7-12}$$

but applied to the components gives

$$\vec{A} = A_x\hat{\imath} + A_y\hat{\jmath}. \tag{7-13}$$

That is, A_x of $\vec{A}$ lies in the x direction and A_y lies in the y direction. This notation forms the heart of the algebra. Note that there is one important difference between the definition of the vector $\vec{A} = A\,\hat{a}$, and its component representation. A, being the magnitude of $\vec{A}$, is an unsigned number, but A_x is a signed number, and therefore, the components (A_x, A_y) carry an implicit $(+/-)$ direction. This is an important point.

Although some textbooks refer to the components as scalars, strictly speaking, they are not. A scalar, such as a mass of 5 kg, is totally independent of the orientation of the coordinate system. The components, on the other hand, are not. For example, suppose we rotate the x and y axes through an angle such that the x axis now lies along the vector $\vec{A}$ (or see Problem 7.8). Now A_x, $= A$ and $A_y = 0$. Thus, the components are not independent of the orientation of the coordinate system. This sensitivity to direction that a component displays is manifest in the sign of the component. An x-component of $(+5)$ means that part of the vector points in the negative x-direction as shown in Figure 7.11, in contrast to an x component of (-5). It is important to correctly find the sign of the component as well as its magnitude.

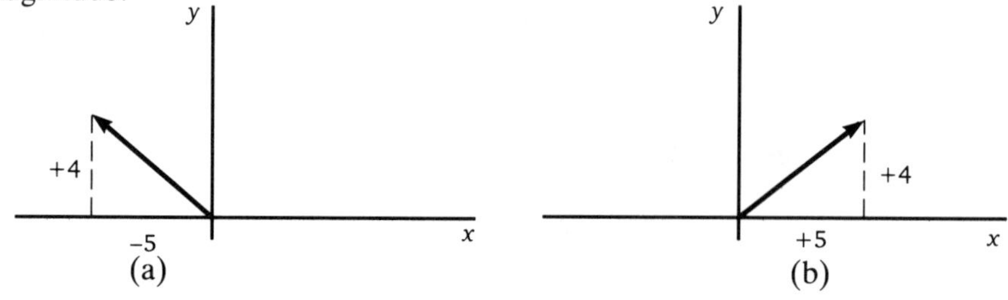

Figure 7.11
(a) A vector with a negative-x component. (b) A vector with a positive-x component.

One final point is made by comparing the coordinate-free representation with the Cartesian representation

$$\vec{A} = A\,\hat{a},$$
$$\vec{A} = (A\cos\theta)\hat{\imath} + (A\sin\theta)\hat{\jmath}.$$

On pulling out the A, it is clear that

$$\hat{a} = \cos\theta\,\hat{\imath} + \sin\theta\,\hat{\jmath} \tag{7-14}$$

and thus, specifying θ does indeed give the direction of the vector.

EXAMPLE 7-2

Resolve into Cartesian components the vector for which the magnitude is 5, and the angle with the negative x axis is 30°.

Solution:
From a graph of the vector, Figure 7.12, it is clear that the x component is negative.

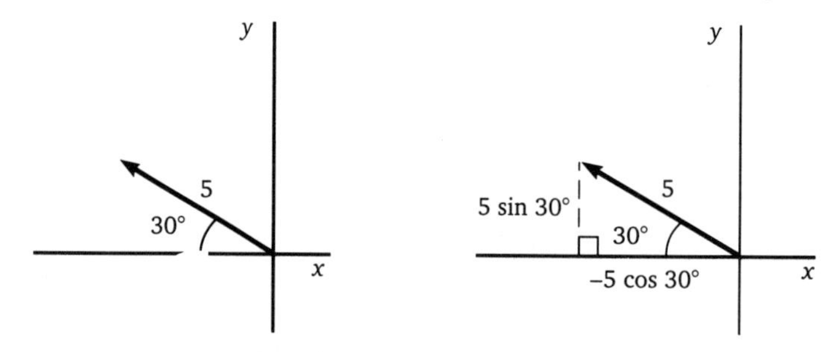

Figure 7.12
The x component is in the negative-x direction.

The very first thing to do is to drop a perpendicular from the tip of the vector onto one of the axes so that the components can be clearly seen. You should always draw a graph of the vector, so that you can clearly see the angles involved and can see which components are positive, which are negative. The angle the vector makes with the positive x axis is 150°, so the x component is

$$x = 5\cos 150^\circ = -5\cos 30^\circ, \text{ and}$$
$$y = 5\sin 150^\circ = 5\sin 30^\circ.$$

In their book *Don't Be a Dodo, Improve Your Physics Grade,* Ronald and Robin Aaron make the point, "if your mother were to awaken you at 2:00 AM, flash Figure 7.13 in your face, and shout,

"What is the *x*-component of *A*?"

you must immediately respond,

"10 cos 50°, Mom!"

If you cannot, you will probably earn a C or even a lower grade in your Physics course."

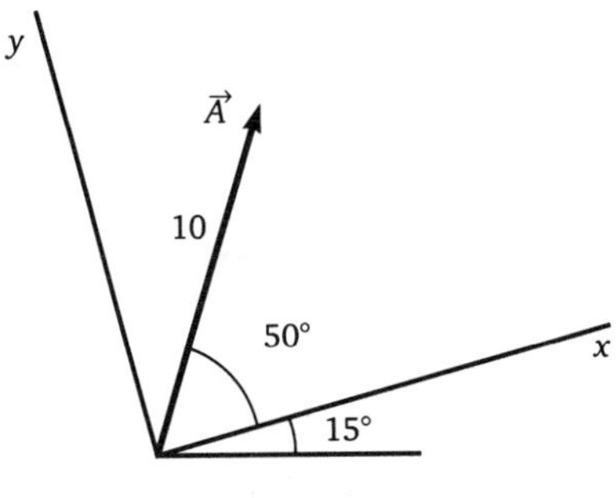

Figure 7.13
Quick, its 2:00 A.M.!

Believe it, they are not exaggerating. If you must spend time on an exam fumbling around because you can't immediately break a vector up into its components, there will be little chance of your reaching the real physics, where most of the partial credit resides, and thus, little chance of passing the exam.

Now that we have cast the vector into a form that definitely looks algebraic, we must see whether it indeed behaves algebraically.

Vector Addition

If we were to follow the rules of algebra and add two functions,

$$\begin{array}{r} y_1 = x^2 + 3x + 4 \\ +\quad y_2 = 2x + 3 \\ \hline y_1 + y_2 = x^2 + 5x + 7, \end{array}$$

we'd note that the two coefficients of the first power of x can be added, and the two constant terms can be added. Thus, we come to a rule: Only the coefficients of like powers may be added. Similarly, if we add two vectors, we'd expect to be able to add only those things that lie in the same direction. Thus,

$$\vec{r}_1 = x_1\hat{\imath} + y_1\hat{\jmath}$$
$$+\ (\vec{r}_2 = x_2\hat{\imath} + y_2\hat{\jmath})$$

gives

$$\vec{r}_1 + \vec{r}_2 = (x_1 + x_2)\hat{\imath} + (y_1 + y_2)\hat{\jmath}. \tag{7-15}$$

The rule of algebraic addition generalized to vectors follows:

Vector Addition

Only those components that lie along the same axis may be added or subtracted.

Note that Equation (7-15) is the distributive property. What other properties have we incorporated into the algebra? Although we've endeavored to follow the rules of algebra, we must check to see whether this makes sense geometrically. First drop a perpendicular from each vector onto the x axis, resolving the vectors into x and y components, as in Figure 7.14.

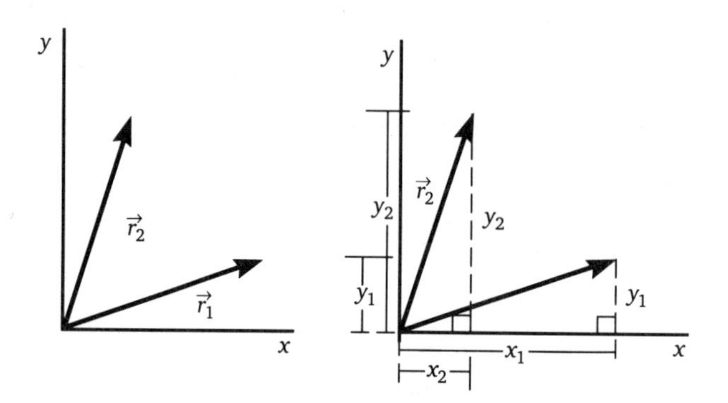

Figure 7.14
Resolving two vectors into their Cartesian components.

Following our rule for vector addition, we add $x_1 + x_2$, and $y_1 + y_2$.

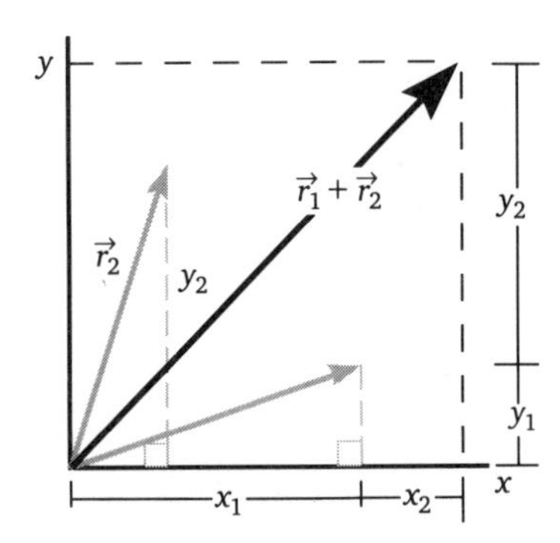

Figure 7.15
Add the like components to get the components of the resultant vector.

The resultant must have components $(x_1 + x_2)$ and $(y_1 + y_2)$. The components to the upper right in Figure 7.15 are just the components for r_2, so it looks as though $\vec{r}_2$ has been moved to the head of $\vec{r}_1$. This procedure forms our first rule for adding vectors graphically:

> **Head-to-Tail Vector Addition**
>
> To add the two vectors $\vec{A}$ and $\vec{B}$ graphically:
>
> - Move the tail of $\vec{B}$ to the tip of $\vec{A}$
> - Draw the resultant vector $\vec{A} + \vec{B}$ from the tail of $\vec{A}$ to the tip of $\vec{B}$.

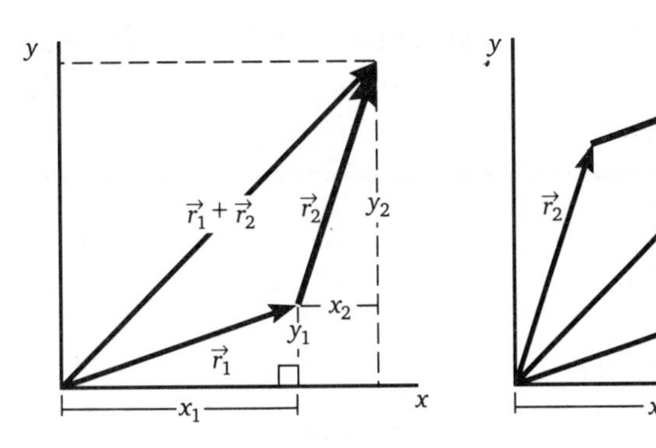

Figure 7.16
By looking at the components we can move $\vec{r}_2$ to the head of $\vec{r}_1$ and vice versa.

Also, the top left can be completed by moving $\vec{r}_1$ to the head of $\vec{r}_2$ so that the two vectors to be added form a parallelogram, as shown in Figure 7.16. We have just graphically proved the commutation property of addition—i.e.,

$$\vec{r}_1 + \vec{r}_2 = \vec{r}_2 + \vec{r}_1.$$

In Figure 7.17, the parallelogram is shown without all the constructions which become distracting.

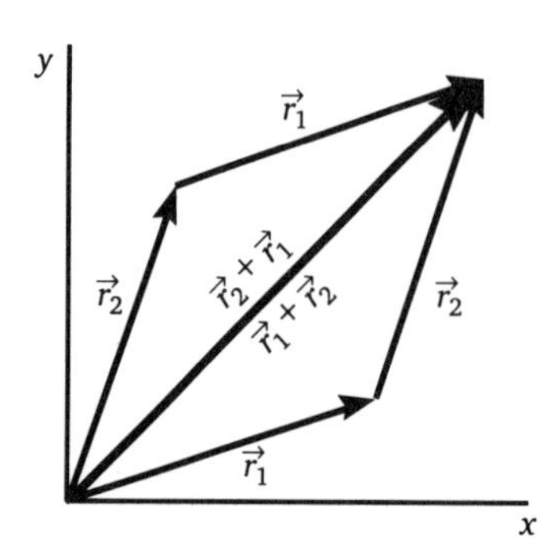

Figure 7.17
The parallelogram method for adding two vectors.

The simple elegance of Figure 7.17 is striking and powerful. To graphically construct the addition of two vectors, we don't need to go back to our analytical method for adding two vectors; rather, we can lay the vectors head-to-tail to form a parallelogram. The resultant vector is the diagonal of the parallelogram.

Note that the parallelogram method is a geometrical proof that vector addition is commutative—that is, it doesn't matter in which order we add the vectors. The sides below the diagonal of the parallelogram in Figure 7.17 form $\vec{r}_1 + \vec{r}_2$, and above the diagonal form $\vec{r}_2 + \vec{r}_1$, both giving the same resultant vector, proving commutation.

Parallelogram Method of Vector Addition

To graphically add two vectors, move the tail of one vector to the head of the other vector and vice versa to form a parallelogram. The resultant vector is the diagonal of the parallelogram.

Analytically, subtraction is an easy extension of addition, following the same form we have already developed.

$$\vec{r}_1 - \vec{r}_2 = (x_1 - x_2)\hat{\imath} + (y_1 - y_2)\hat{\jmath}. \qquad \text{subtraction} \qquad (7\text{-}16)$$

Note that a common mistake is to try to put a minus sign between the components of the resultant vector:

$$\vec{r}_1 - \vec{r}_2 \neq (x_1 - x_2)\hat{\imath} - (y_1 - y_2)\hat{\jmath}. \qquad \text{wrong}$$

Like addition, we must only subtract like components. Do not subtract components that lie along different axes!

As with regular algebra, subtraction can be considered to be the addition of a negative vector. Geometrically, we can add a negative of vector $\vec{r}_2$ to $\vec{r}_1$ as in Figure 7.18, where $\vec{r}_1$ has been moved to the head of $(-\vec{r}_2)$, so the addition is $(-\vec{r}_2) + \vec{r}_1$.

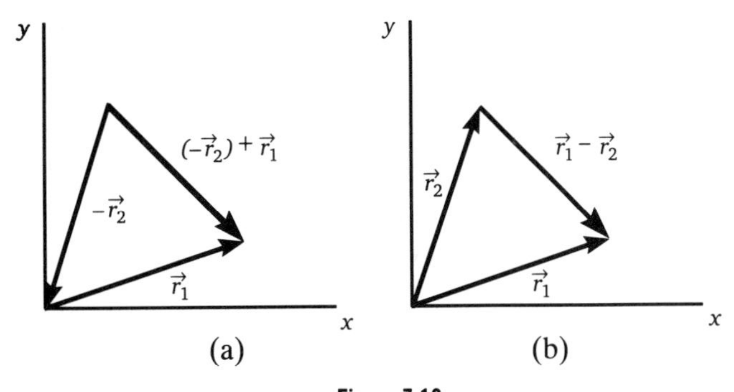

Figure 7.18
The result of subtracting $\vec{r}_1 - \vec{r}_2$ goes from the head of $\vec{r}_2$ to the head of $\vec{r}_1$.

Note that the final resultant vector $\vec{r}_1 - \vec{r}_2$ in Figure 7.18(a) is the same as a vector going from the head of $\vec{r}_2$ to the head of $\vec{r}_1$ as shown in Figure 7.18(b). For subtracting two vectors graphically, we have found the following rule:

Subtraction of Vectors

Place the tails of the two vectors together. For $\vec{A} - \vec{B}$, draw the resultant from the head of $\vec{B}$ to the head of $\vec{A}$—remember: from the head of $\vec{B}$ to the head of $\vec{A}$!

As an example, let's return to the displacement problem found in Section 7.2.

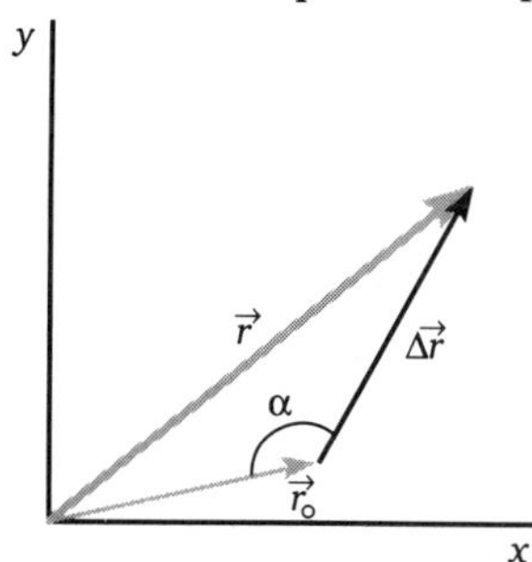

Figure 7.5
A displacement from $\vec{r}_0$ to $\vec{r}$ by moving through $\Delta\vec{r}$.

With our newfound algebra, it should be easy to tell which vector is the addition of two vectors and which is the subtraction of two vectors. Such skills are vital to your success in physics. Of course, $\vec{r}$ is the addition of $\vec{r}_0 + \Delta\vec{r}$ because of the head to tail arrangement of $\vec{r}_{01}$ and $\Delta\vec{r}$. Similarly, $\Delta\vec{r}$ is the subtraction of $\vec{r} - \vec{r}_0$ because $\Delta\vec{r}$ starts at the head of $\vec{r}_0$ and goes to the head of r. Thus,

$$\vec{r} = \vec{r}_0 + \Delta\vec{r}, \tag{7-17}$$

and

$$\Delta\vec{r} = \vec{r} - \vec{r}_0. \qquad \text{displacement} \qquad (7\text{-}18)$$

Note that the equation we obtained from Figure 7.5 for $\Delta\vec{r}$ could have been obtained just as easily by algebra from the equation just above it!

$$\Delta\vec{r} = \text{final position vector} - \text{initial position vector}$$

We have an algebra that works for addition and subtraction. We'll take up multiplication at a later time. For the moment, we need to polish our skills in vector addition and subtraction.

EXAMPLE 7-3

Find the magnitude and direction both analytically and graphically of **a)** $\vec{A} + \vec{B}$ and **b)** $\vec{A} - \vec{B}$ when $\vec{A}$ is a vector of length 4, and a direction of 30° with the positive x axis, and $\vec{B}$ is of length 3 and a direction 60° with the positive x axis.

Solution:
a) What should the answer be? (i) First, draw a picture of the vectors, and graphically resolve them into components, as in Figure 7.19 (a) and (b). Next, graphically add the two vectors so that you know what the resultant looks like. You should always do the graphical addition before you start the analytic solution, even if a graphical addition is not required. It will save you from making mistakes and will give you a good idea what the answer is. From Figure 7.19(c), you can see that the resultant is less than 7 but is probably larger than 6, and it makes an angle slightly closer to vector $\vec{A}$ than to vector $\vec{B}$, so the direction will be between 45° and 40°.

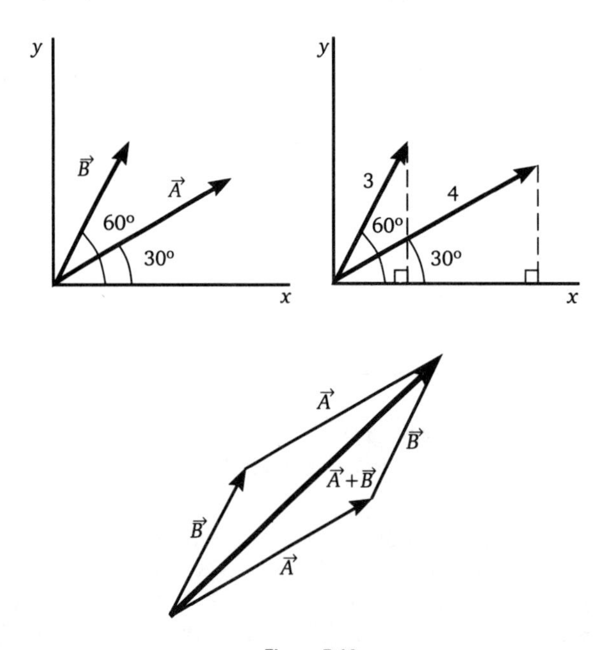

Figure 7.19
First draw the vectors in graph (a). Next resolve the vectors into components (b), then add (c).

(ii) To find the analytical solution, first resolve the vectors into Cartesian components,

$$\begin{array}{rl} \vec{A} &= 4\cos 30°\,\hat{\imath} + 4\sin 30°\,\hat{\jmath} = 3.5\,\hat{\imath} + 2\,\hat{\jmath} \\ + \ \vec{B} &= 3\cos 60°\hat{\imath} + 3\sin 60°\,\hat{\jmath} = 1.5\,\hat{\imath} + 2.6\,\hat{\jmath} \\ \hline \vec{A} + \vec{B} &= \qquad\qquad\qquad\qquad\qquad 5\,\hat{\imath} + 4.6\,\hat{\jmath} \end{array}$$

(iii) Next, add the like components, and finally (iv) find the magnitude and the direction.

$$|\vec{A} + \vec{B}| = \sqrt{x^2 + y^2} = \sqrt{5^2 + 4.6^2} \Rightarrow |\vec{A} + \vec{B}| = 6.8$$

$$\theta = \tan^{-1}\frac{y}{x} = \tan^{-1}\frac{4.6}{5} \Rightarrow \theta = 42.6°$$

Before we finish, it is important to resolve one final point. (v) What angle is θ? It's the angle the resultant vector $\vec{A} + \vec{B}$ makes with the x axis. Mark it on your diagram, as shown in Figure 7.20. It is important that you understand what the angle θ is. Readers may deduct points from exams if θ is not clearly marked or described!

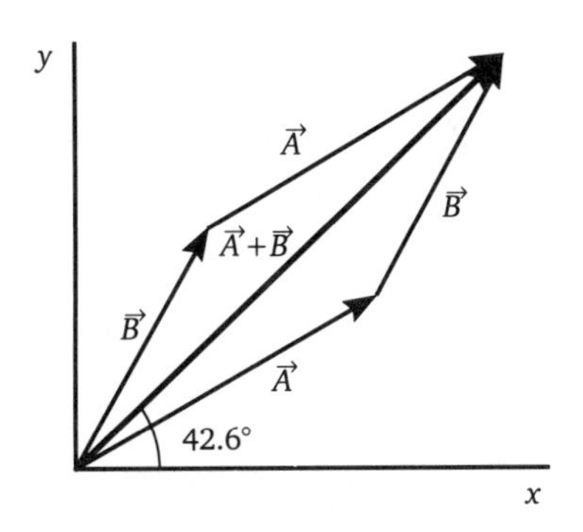

Figure 7.20
$\theta = 42.6°$ is the direction of the resultant vector.

b) To find $\vec{A} - \vec{B}$, graphically subtract the vectors first. As shown in Figure 7.26, follow the rule for graphically subtracting vectors.

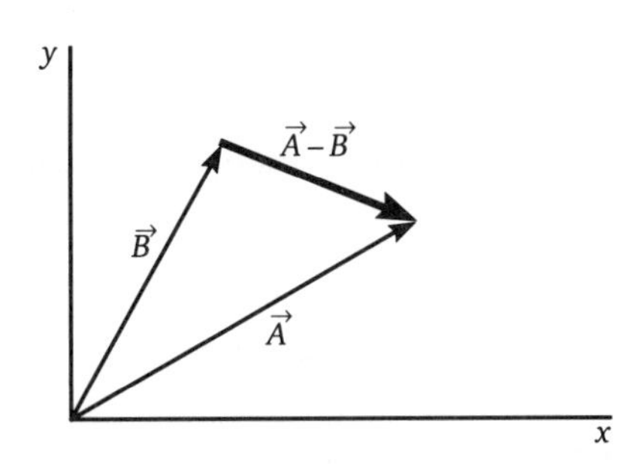

Figure 7.21
The vector $\vec{A} - \vec{B}$ goes from the head of $\vec{B}$ to the head of $\vec{A}$.

Analytically,

$$\begin{array}{l} \vec{A} = 4\cos 30°\,\hat{\imath} + 4\sin 30°\,\hat{\jmath} = 3.5\,\hat{\imath} + 2\,\hat{\jmath} \\ -\ (\vec{B} = 3\cos 60°\,\hat{\imath} + 3\sin 60°\,\hat{\jmath} = 1.5\,\hat{\imath} + 2.6\,\hat{\jmath}) \\ \hline \vec{A} - \vec{B} = \qquad\qquad\qquad\qquad 2\,\hat{\imath} - 0.6\,\hat{\jmath}, \end{array}$$

so the magnitude and direction are

$$|\vec{A} - \vec{B}| = \sqrt{2^2 + 0.6^2} = 2.1$$

$$\theta = \tan^{-1}\frac{-0.6}{2} = -16.7°.$$

The negative angle means that θ is measured in the clockwise direction from the x direction. We can show the angle by drawing in a dotted line at the tail of the vector to represent the x axis as shown in Figure 7.22.

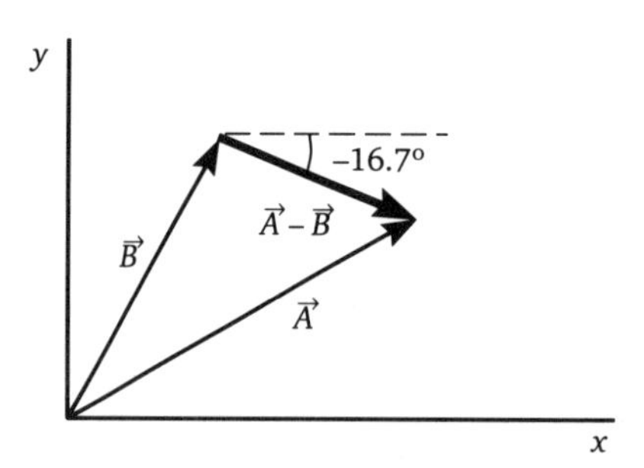

Figure 7.22
The direction of $\vec{A} - \vec{B}$ is with respect to the x direction.

7.4 DISPLACEMENT, VELOCITY, AND ACCELERATION

Consider again the displacement vector Equation (7-18), which represents a change in position and is formed by subtracting the initial-position vector from the final-position vector.

$$\Delta\vec{r} = \vec{r} - \vec{r}_0 \qquad (7\text{-}18)$$

To calculate the displacement, we subtract like Cartesian components so that the displacement vector has the form

$$\Delta\vec{r} = \Delta x\,\hat{\imath} + \Delta y\,\hat{\jmath}. \qquad (7\text{-}19)$$

Velocity must be a vector. We can form the average velocity by dividing the displacement vector by the time that it takes to make the displacement:

$$\vec{v} = \frac{\Delta x}{\Delta t}\hat{\imath} + \frac{\Delta y}{\Delta t}\hat{\jmath}. \qquad \text{average velocity} \qquad (7\text{-}20)$$

The $\Delta x/\Delta t$ is the x component, and similarly, $\Delta y/t$ is the y component.

$$v_x = \frac{\Delta x}{\Delta t}$$

$$v_y = \frac{\Delta y}{\Delta t} \qquad \text{velocity components} \qquad (7\text{-}21)$$

Note that neither Δt not dt have any direction associated with them, so the average velocity vector must point in the same direction as the total-displacement vector.

EXAMPLE 7-4

A ship starts 5 nautical miles, 60° northeast of a lighthouse. Fifteen minutes later ,the ship is 9 miles, 30° northeast of the lighthouse. Find the average speed of the ship, and find its direction of travel.

Solution:
We work this example two ways. First, we use the law of cosines to find the net displacement and the law of sines to find the direction of the displacement. Second, we resolve the vector into Cartesian components and find the net displacement and direction.

I. Using the law of cosines applied to the acute triangle in Figure 7.23(a),

$$\Delta r^2 = 5^2 + 9^2 - 2(5)(9)\cos(60° - 30°).$$

$$\Delta r = 5.3 \text{ nautical miles}$$

Also using the law of sines to find ϕ

$$\frac{\sin(60° - 30°)}{\Delta r} = \frac{\sin\phi}{5}$$

$$\Rightarrow \phi = \sin^{-1}\left(\frac{5}{5.3}\sin 30°\right) = 27.8°$$

Using the two horizontal lines and the 9 mi bisector in Figure 7.23(b)

$$\theta + \phi = 30° \Rightarrow \theta = 2.2°$$

II. Start by resolving the position vectors into components. Remember to graphically solve the problem before starting on the analytical part.

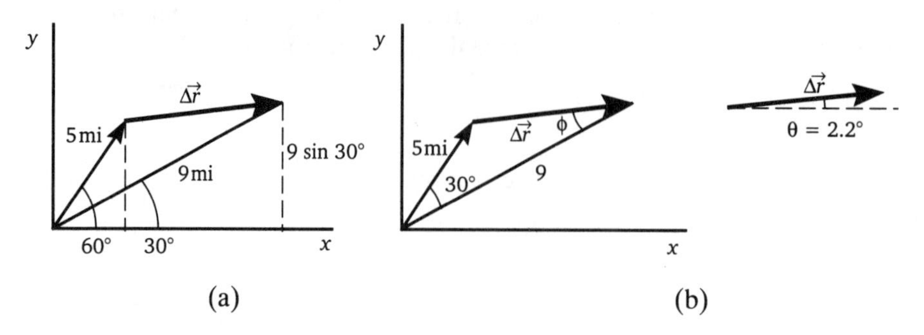

Figure 7.23
The displacement of the ship.

The components for the position vectors are

$$\vec{r} = 9\cos 30°\,\hat{\imath} + 9\sin 30°\,\hat{\jmath} = 7.8\,\hat{\imath} + 4.5\,\hat{\jmath}$$
$$\vec{r} = 5\cos 60°\,\hat{\imath} + 5\sin 60°\,\hat{\jmath} = 2.5\,\hat{\imath} + 4.3\,\hat{\jmath}$$
$$\Delta\vec{r} = \vec{r} - \vec{r} \qquad 5.3\,\hat{\imath} + 0.2\,\hat{\jmath}$$

The average speed is

$$v = \frac{|\Delta\vec{r}|}{\Delta t} = \frac{\sqrt{(5.3)^2 + (0.2)^2}}{0.25\text{ hour}} = 21.2\text{ nautical miles per hour}$$

which is $v = 21.2$ knots. Again because 0.2 is so much less than the 5.3, we could have ignored it for this part. However, we cannot ignore the 0.2 miles for the direction which is

$$\theta = \tan^{-1}\frac{0.2}{5.3} = 2.2°,$$

and from Figure 7.23 is 2.2° north of east.

The acceleration also becomes a vector and is defined by taking the change of the velocity in time.

$$\vec{a} = \frac{\Delta\vec{v}}{\Delta t} = \frac{\Delta v_x}{\Delta t}\,\hat{\imath} + \frac{\Delta v_y}{\Delta t^2}\,\hat{\jmath} \qquad \text{acceleration} \qquad (7\text{-}22)$$

The second derivative of x with respect to time is the x component of the acceleration, and the second derivative of y with respect to time is the y component. Note that the direction of the acceleration is the same as the direction of the change in the velocity, and so it is not necessarily in the same direction as the velocity.

7.5 CARTESIAN DISPLACEMENTS IN THREE DIMENSIONS

The first obstacle to working in three dimensions is one of drawing. Paper is a two-dimensional medium, so it makes the representation of three dimensions difficult. Typically, the y and z axes are drawn perpendicular to one another in the plane of the paper. The line representing the x axis is then drawn at a 45° angle, with the y axis as in Figure 7.24, and the figure formed is supposed to represent a perspective drawing of three coordinate axes. Remember that all three axes are mutually perpendicular, so the x axes is actually coming straight out of the page.

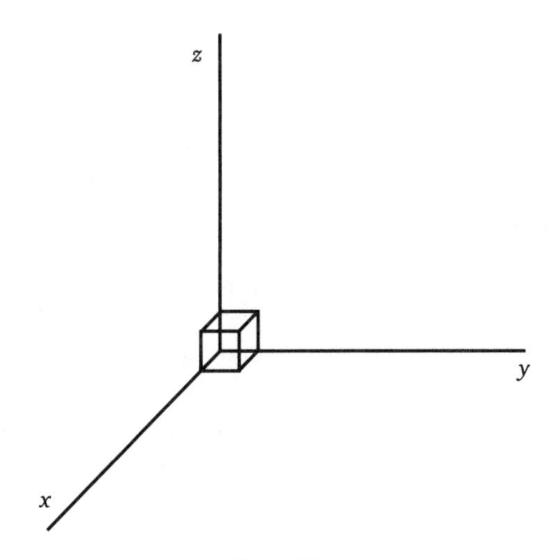

Figure 7.24
The three axes are mutually perpendicular. The x axis actually is perpendicular to the page.

The Cartesian coordinate system is also defined to be a right-handed system. By this, we mean that once the x and y axes are set, the z axis must be determined by a right-hand rule:

Right-Hand Rule

Point the fingers of your right hand along the x axis (perpendicular to the page) with your palm facing the y axis, so that your fingers curl toward the y axis. (Your fingers can cross the x axis into the y axis.) Your thumb will point in the z direction. (See Figure 7.25.)

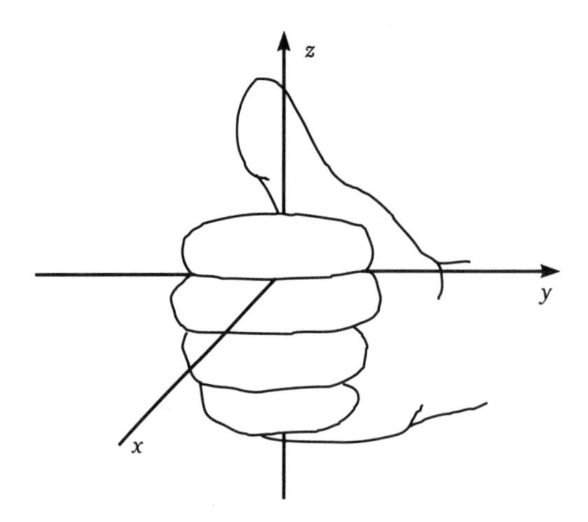

Figure 7.25
The right-hand rule. When the fingers of the right hand curl from *x* to *y*, your thumb points along the *z* axis.

One of the problems with representing three-dimensional systems on a two-dimensional page is that it is difficult to tell in which plane a point lies. One way to uniquely show the plane is to drop a perpendicular to the *x–y* plane, and show the *x* and *y* components as in Figure 7.26. The perspective is enhanced by drawing a line from the point of intersection with the *x–y* plane to the origin. Note that this line is a shadow for the line drawn from the origin to the point *P*. You must be able to accurately make such drawings, and you should practice drawing them for points in various octants. You must also be able to tell which of the angles in Figure 7.26 are right angles. In a perspective drawing, it is often hard to pick out the right angles, but it is crucial to the rest of the discussion that you can do this quickly. On Figure 7.26, the right angles are marked.

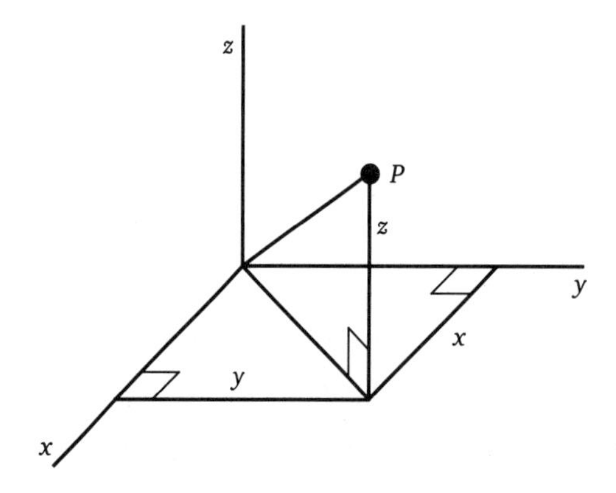

Figure 7.26
A representation of a point on a three-dimensional drawing.

The position vector of a point in a three-dimensional coordinate system is a simple extension from two dimensions:

$$\vec{r} = x\hat{\imath} + y\hat{\jmath} + z\hat{k}. \tag{7-23}$$

Addition and subtraction of vectors are handled using the same algebraic system we used in the preceding chapter, where only the components along like axes may be added or subtracted. The displacement vector from point $\vec{r}_0$ to $\vec{r}$ is shown in Figure 7.27.

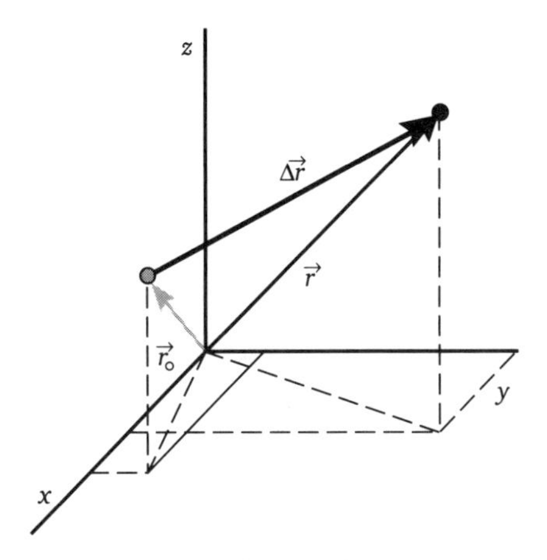

Figure 7.27
A displacement in three dimensions from $\vec{r}_0$ to $\vec{r}$.

In terms of Cartesian components the displacement is

$$\Delta\vec{r} = (x - x_0)\hat{\imath} + (y - y_0)\hat{\jmath} + (z - z_0)\hat{k}. \tag{7-24}$$

As in two dimensions, the three-dimensional position vectors all point in the positive radial direction, but the displacement vectors do not necessarily do so.

7.6 VELOCITY ADDITION AND REFERENCE FRAMES

Undoubtedly, the most famous theory in physics is Einstein's theory of relativity. Relativity is one of those great ideas, such as, Darwin's theory of evolution that finds application to areas outside the original field. Physicists and astronomers borrowed Darwin's idea and applied it to the interiors of stars which culminated in the theory of the evolution of stars. Stars start with hydrogen-rich cores and then progress through various fusion cycles until the core is iron–rich. Iron nuclei, being the most stable, cannot be combined to form more complex nuclei, and thus, the star has run out of fuel to burn and either collapses under its own weight to form a white dwarf, or if it is big enough, becomes unstable and explodes in a nova leaving a very dense neutron star. This is the end of the evolutionary cycle for the star; however, the material that was carried off the star in the explosion may become fresh material for the formation of a new star. A great idea always has applicability beyond the specific field in which it was developed. With the possible exception of the entropy form of the second law of thermodynamics, there is probably no theory that is misrepresented more than relativity. Very often the application of the concept of relativity outside the realm of physics usually begins, "Einstein proved that *everything* is relative to the observer's point of view. Therefore. . . . ." The critical error is that Einstein did not prove that everything is relative. In fact, one of the fundamental assumptions of relativity is that any two observers of an event must agree on the event took place, that is, they must see the same event. Two observers may not agree on the specific details of the event, such as the velocity of a thrown ball. But if one observer sees the ball land in a baseball glove, any other observer must also see the ball land in the glove, i.e., they will agree that the ball was caught. A classical form of relativity worked out by Galileo was known long before Einstein and to this day is called Galilean relativity. Before you will be able to understand Einstein's relativity, you must understand the simpler Galilean relativity. The best way is to start with some examples of Galilean relativity. Here, as in Einstein's relativity, a basic tenet will be that all observers must agree on the event that took place and the forces involved, but they can disagree on the specific details, such as the velocity of a given object.

EXAMPLE 7-5

Two people are playing catch while standing on a flatcar of a train moving at 30 m/s. Observer O is standing on the ground, and observer O′ is sitting on the train. a) If observer O′ sees player A throw the ball toward player B with a horizontal component of velocity 20 m/s, as shown in Figure 7.28, find the velocity of the ball with respect to observer O. b) If observer O′ sees player B throw the ball toward player A with a horizontal velocity of − 20 m/s, find the velocity of the ball with respect to observer O.

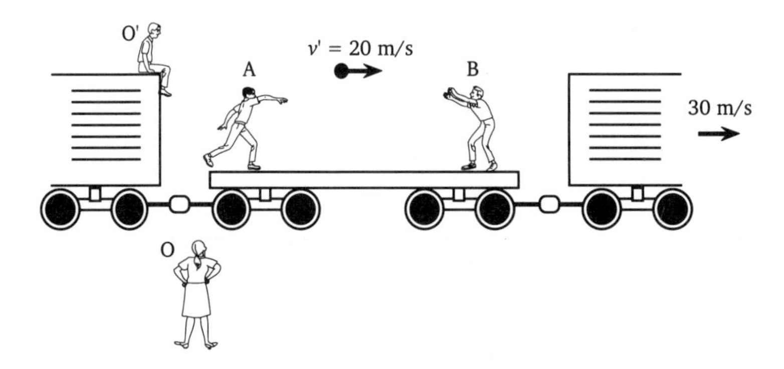

Figure 7.28
Playing catch on a train.

Solution:

a) Note that both observer O and observer O′ agree on the event. They both see player A throw the ball and observer B catch the ball; however, they do not agree on the details. Observer O′ sees both players at rest and sees the ball move with a uniform horizontal component of the velocity.

$$v_{\text{ball}\to\text{O}'} = 20 \text{ m/s}$$

Note that subscript notation on v reads, "the velocity of the ball relative to O′." Observer O, however, sees the velocity of the ball added to the velocity of the train. So O sees the ball travel at a speed of

$$v_{\text{ball}\to\text{O}} = 20 \text{ m/s} + 30 \text{ m/s} = 50 \text{ m/s}.$$

Does this mean O sees player B get hit really hard with the ball? How hard player B is hit depends on the change in velocity of the ball, i.e., on the acceleration he must supply to the ball. O′ sees the velocity of the ball change by 20 m/s.

$$\Delta v_{\text{ball}\to\text{O}'} = \text{final } v - \text{initial } v = 0 - 20 \text{ m/s} = -20 \text{ m/s}$$

On the other hand, O sees player B moving away from himself at 30 m/s so he concludes that when B catches the ball, the ball's velocity will only change by

$$\Delta v_{\text{ball}\to\text{O}} = 30 \text{ m/s} - 50 \text{ m/s} = -20 \text{ m/s}$$

Thus, O sees the same change in velocity, − 20 m/s, that O′ sees when B catches the ball. We can conclude that both observers see the same accelerations of the ball, even though they don't agree on the individual velocities.

b) Because the ball is thrown against the motion of the train, we need to subtract the velocity of the ball from the velocity of the train to find the velocity relative to O. So

$$v_{\text{ball}\to\text{O}} = 30 \text{ m/s} - 20 \text{ m/s} = 10 \text{ m/s}$$

Note that in this case observer O and O′ do not even agree on the direction in which the ball is traveling. O′ sees the ball traveling to the left with velocity

$$v_{\text{ball}\to\text{O}'} = -20 \text{ m/s } \hat{\imath},$$

(remember velocity is a vector), but O sees the ball travlling to the right with velocity

$$v_{\text{ball}\to\text{O}} = +10 \text{ m/s } \hat{\imath},$$

as shown in Figure 7.29(b). How can this be? How will O ever see A catch the ball?

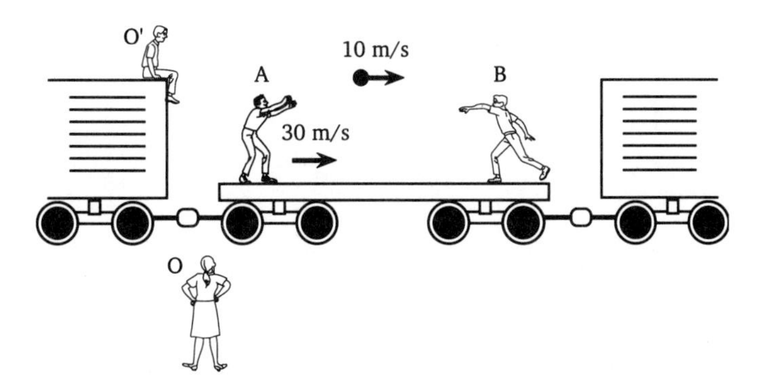

Figure 7.29
When B throws the ball back to A, O sees the ball move to the right!

Of course O sees A moving at $v_{\text{A}\to\text{O}} = 30$ m/s, so that, although O sees the ball always moving to the right, he also sees A moving to the right at a much higher rate of speed. Thus, O sees A overtake the ball. As in part (a), the two observers O and O′ do not agree on the details of what happened. O′ sees the ball race towards A, and O sees A race towards the ball. But they both agree on what happened. A catches the ball changing the ball's velocity by 20 m/s in the positive x-direction.

In Example 7-5, it might at first seem like we have used two different rules to find the velocity in the unprimed frame of reference; however, if we use vector notation, we find the same rule being applied in each case.

$$\vec{v}_{\text{ball}\to\text{O}} = \vec{v}_{\text{ball}\to\text{O}'} + \vec{V}_{\text{O}'\to\text{O}} \tag{7-25}$$

This is the rule to transform the velocity from frame O′ to frame O. In both parts (a) and (b), $\vec{V}_{\text{O}'\to\text{O}} = 30\text{ m/s }\hat{\imath}$, but $\vec{v}_{\text{ball}\to\text{O}'}$ changes sign. For part (a) $\vec{v}_{\text{ball}\to\text{O}'} = 20\text{ m/s }\hat{\imath}$,

$$\vec{v} = 20\text{ m/s }\hat{\imath} + 30\text{ m/s }\hat{\imath}$$

and for part (b) $\vec{v}_{\text{ball}\to\text{O}'} = -20\text{ m/s }\hat{\imath}$,

$$\vec{v} = -20\text{ m/s }\hat{\imath} + 30\text{ m/s }\hat{\imath}$$

i.e., when B throws the ball back to A only the direction changes, and it is opposite to the direction in part (a); hence, in part (b) we subtract the speeds by adding the velocities. Note that $\vec{V}_{\text{O}'\to\text{O}}$ is the velocity of frame O′ relative to frame O. Let's apply the Galilean velocity transformation equation to a problem in two dimensions to see if it works.

EXAMPLE 7-6

A boat is headed at top speed straight across a river with a current of 3 mph. If the top speed of the boat in still water is 4 mph, find the direction and speed of the boat with respect to the ground.

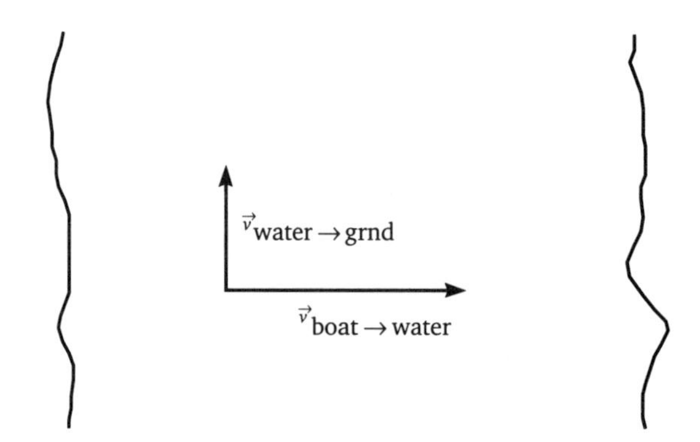

Figure 7.30
A boat travels across a river, but the river carries the boat downstream.

Solution:
Although the boat is headed across the river, it will not travel straight across the river because the river will carry the boat downstream. In Figure 7.30, downstream is pointed toward the top of the page. *Downstream* refers to the direction in which the current will carry a drifting object.

During the same time that the boat travels Δx, the stream will carry the boat a distance Δy downstream, as shown in Figure 7.31. The net displacement vector $\Delta \vec{r}$ is the vector sum of the individual displacements

$$\Delta \vec{r} = \Delta x \hat{\imath} + \Delta y \hat{\jmath}. \tag{7-26}$$

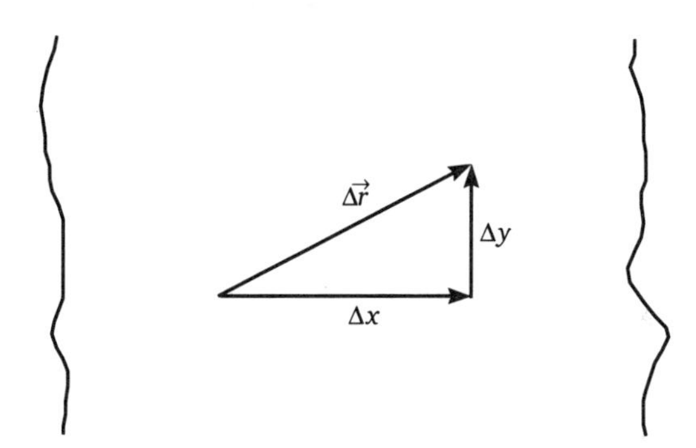

Figure 7.31
During the same time that the boat travels Δx, the stream will carry the boat a distance Δy.

Because both displacements occur in the same time Δt, the resultant velocity of the boat with respect to the ground will be the vector addition of the velocity of the boat in still water and the velocity of the boat with respect to the water as shown in Figure 7.32.

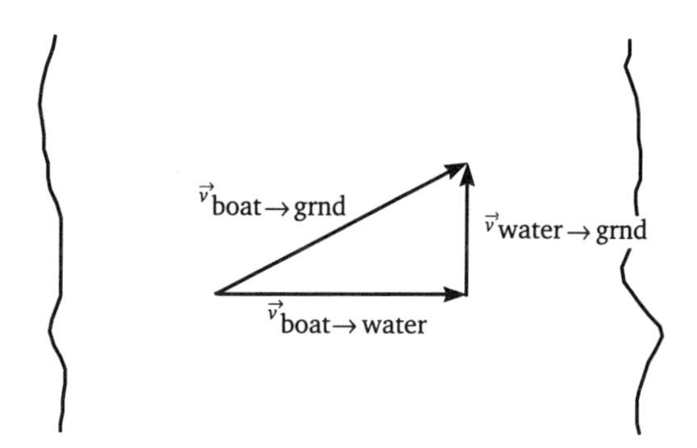

Figure 7.32
The resultant velocity of the boat is given by a vector addition.

Vectorially, we have

$$\vec{v}_{\text{boat}\rightarrow\text{gmd}} = \vec{v}_{\text{boat}\rightarrow\text{water}} + \vec{v}_{\text{water}\rightarrow\text{gmd}}.$$

Note that the ordering on the subscripts is the same as we found for the Galilean transformation:

$$\vec{v}_{\text{ball}\rightarrow\text{O}} = \vec{v}_{\text{ball}\rightarrow\text{O}'} + \vec{v}_{\text{O}'\rightarrow\text{O}}$$

where $\vec{v}'$ is the velocity of an object in the primed frame of reference, and v is the velocity of the object in the unprimed frame of reference. So adding $\vec{V}_{\text{O}'\rightarrow\text{O}}$ to the velocity takes the velocity of an object in the primed frame of reference and transforms it to the unprimed frame of reference.

Returning to our example, we find that the speed of the boat is

$$|\vec{v}| = \sqrt{(4\text{ m/s})^2 + (3\text{ m/s})^2} = 5\text{ m/s},$$

and although the heading of the boat is straight across the river, it is actually traveling at

$$\theta = \tan^{-1}\frac{3}{4} = 36.9°$$

downstream from a line drawn straight across. (See Figure 7.33.) Note that we can treat the displacements in the x and y directions as independent because we have been careful to make the x and y components linearly independent. Thus, motion in the x direction is independent of motion in the y direction unless there is an equation to constrain the two to be coupled, e.g., when a bead slides down a wire.

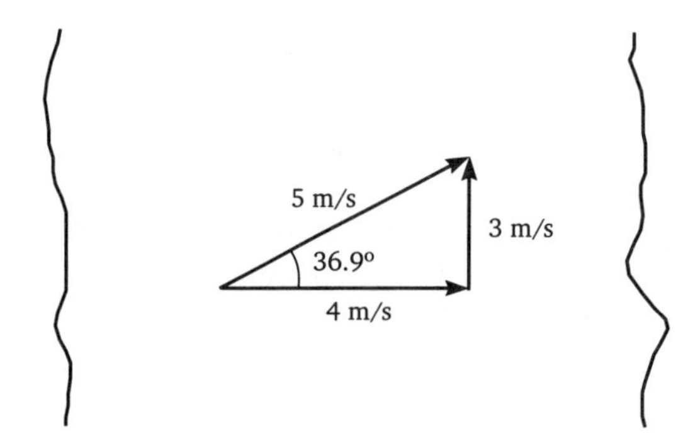

Figure 7.33
The heading is 36.9° with respect to straight across.

Setting up the vector relation between the velocities can be very confusing, but the ordering of the subscripts can help you out of the confusion. Note that the ordering is always of the form:

$$\vec{v}_{\text{object}\to\text{O}} = \vec{v}_{\text{object}\to\text{O}'} + \vec{v}_{\text{O}'\to\text{O}}.$$

To illustrate this, consider the following more difficult example.

EXAMPLE 7-7

The wind is blowing at 50 mph from the north. A pilot wishes to fly a plane at 120 mph 60° south of east with respect to the ground. Find the airspeed and heading of the plane.

Solution:
The heading of the plane is the direction in which the pilot points the plane. This will not be the same direction as the plane is traveling with respect to the ground because the pilot must compensate for the wind. First let's see how the vectors fit together. There are three velocities in this problem: the velocity of the plane with respect to the ground $v_{p\to \vec{g}}$ (ground speed), the velocity of the plane with respect to the air $\vec{v}_{p\to a}$ (air speed), and the velocity of the air with respect to the ground (wind speed)

$\vec{v}_{a\to g}$. Since the wind velocity will transform the air velocity of the plane to the ground velocity, we have

$$\vec{v}_{p\to g} = \vec{v}_{p\to a} + \vec{v}_{a\to g}.$$

Note that this fits the form used in Equation (7-28). Constructing a diagram, as in Figure 7.34, we must add $\vec{v}_{p\to a}$ to $\vec{v}_{a\to g}$.

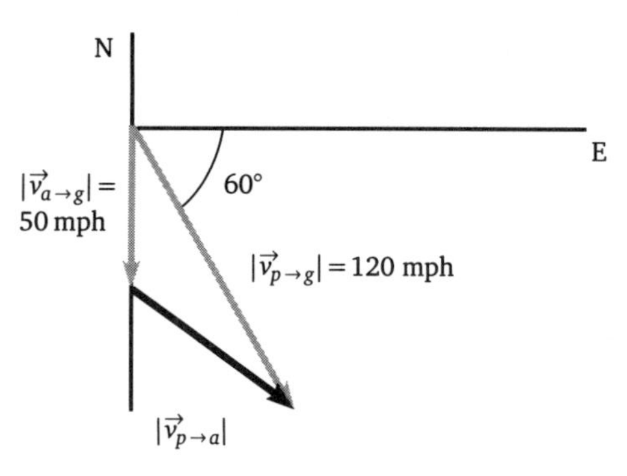

Figure 7.34
$\vec{v}_{p\to g} = \vec{v}_{p\to a} + \vec{v}_{a\to g}$

Solving for $\vec{v}_{p\to a} \Rightarrow$

$$\vec{v}_{p\to a} = \vec{v}_{p\to g} - \vec{v}_{a\to g}.$$

Resolving the vectors into components

$$\begin{array}{rll} & \vec{v}_{p\to g} = 120 \cos 60°\, \hat{\imath} - 120 \sin 60°\, \hat{\jmath} & = 60 \text{ mi/hr}\, \hat{\imath} - 104 \text{ mi/hr}\, \hat{\jmath} \\ - & (v_{a\to g} = 0\, \hat{\imath} - 50 \qquad \hat{\jmath} & = 0\, \hat{\imath} \qquad - 50 \text{ mi/hr}\, \hat{\jmath}) \\ \hline & v_{p\to a} = v_{p\to g} - v_{a\to g}. & = 60 \text{ mi/hr}\, \hat{\imath} - 54 \text{ mi/hr}\, \hat{\jmath} \end{array}$$

Then find the magnitude of the air speed

$$|\vec{v}_{p\to a}| = \sqrt{(60)^2 + (54)^2} = 81 \text{ mi/hr},$$

and the heading

$$\theta = \tan^{-1}\frac{-54}{60} = -42° \text{ south of east.}$$

Note that the plane needs to expend fuel only at the airspeed rate. In this case, since the wind blows partially in the direction the plane wishes to travel, the plane only need expend fuel at the rate required to travel 81 mph through still air, even though the plane is traveling 120 mph. Thus, it is quite an advantage to have the wind at your back. The above discussion, however, neglects to consider whether the plane can actually generate enough lift at 81 mph relative to the air to stay aloft. For a pilot, lift is a very real consideration.

Chapter 7 Problems

[7.1] Resolve the vectors in Figure 7.35 into their Cartesian components.

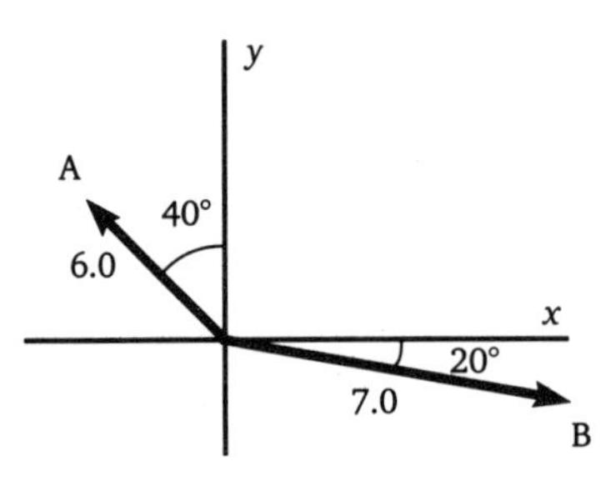

Figure 7.35

[7.2] For the two vectors in Problem [7.1], **a)** graphically find $\vec{A} + \vec{B}$. **b)** Analytically find the magnitude and direction of $\vec{A} + \vec{B}$.

[7.3] For the two vectors in Problem [7.1], **a)** graphically find $\vec{A} - \vec{B}$. **b)** Analytically find the magnitude and direction of $\vec{A} - \vec{B}$.

[7.4] **a)** A boat moves upstream 24 nautical miles (nm) at 10 knots (as measured by a speedometer on the boat) against a 2 knot current. The boat then travels 24 nm downstream at 10 knots with the current. How does the elapsed time for the entire trip compare to the time for a 48 nm trip at 10 knots across a still lake?

[7.5] For three vectors, find the magnitude and direction of the vector $\vec{D} = \vec{A} - \vec{B} + 3\vec{C}$. $\vec{A}$ of length 6.0 and pointing 50° up from the negative-x axis, $\vec{B}$ of length 7.0 and pointing 30° up from the positive-x axis, and $\vec{C}$ of length 3.0 and pointing 60° down from the positive-x axis.

[7.6] A truck starts out at position $x = 3$ mi and $y = 6$ mi. and winds up at position $x = 5$ mi, $y = 2$ mi. Find the magnitude and direction of the truck's displacement.

[7.7] If the truck in Problem [7.6] made the displacement in 10 minutes, **a)** find the average velocity of the truck in miles per hour; b) find the x and y components of the average velocity.

[7.8] A vector A has a magnitude of 5 and makes an angle of 36.9° with the y axis as shown for the rotated coordinate system in Figure 7.36. Find the x and y components of $\vec{A}$.

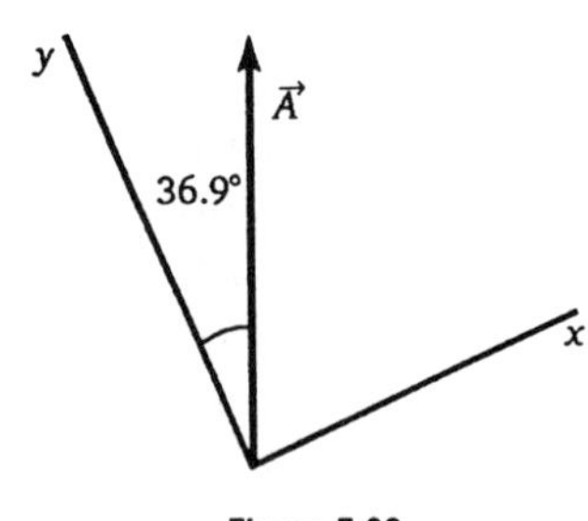

Figure 7.36

[7.9] For each of the arrangements of vectors shown in Figure 7.37 write a vector equation for $\vec{A}$. For example, if adding $\vec{B}$ and $\vec{C}$ gives $\vec{A}$, then write $\vec{A} = \vec{B} + \vec{C}$.

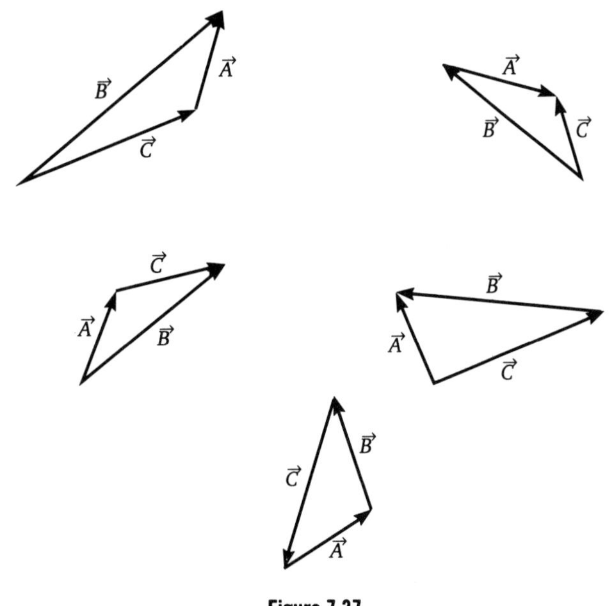

Figure 7.37

[7.10] Which of the coordinate systems shown in Figure 7.38 are right-handed, and which are left-handed?

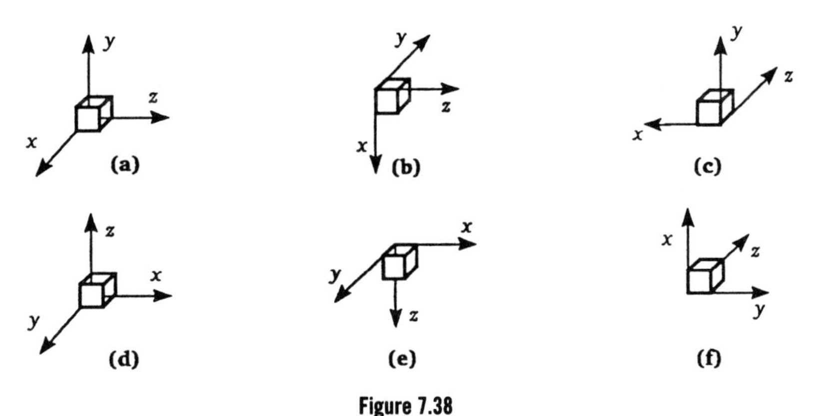

Figure 7.38
Which coordinate systems are right-handed?

[7.11] A bomber flying at 850 km/hr fires a shell with a muzzle speed of 2200 km/hr. What is the speed of the shell with respect to the ground if it is fired **a)** forward; **b)** backward by the tail gunner; or **c)** sideways?

[7.12] A steam engine heads due east at 16 m/s. Smoke particles coming out the smokestack very quickly attain the velocity of the wind. The engineer notes that the trail of smoke makes an angle of 30° in the horizontal plane with respect to the length of the train. If the wind is known to be blowing 60° south of east, find the speed of the wind.

[7.13] An airplane with an air speed of 200 km/hr is pointed due east; however, there is a wind blowing at 70 km/hr headed 45° north of west. Find the velocity of the plane with respect to the ground. Please give the magnitude and direction.

[7.14] An airplane with an air speed of 200 km/hr is pointed due east; however, the pilot notes that her ground speed is 300 km/hr, and she is traveling 30° north of east. Find the velocity of the wind with respect to the ground. Please give the magnitude and direction.

•**[7.15]** A sailboat is moving 40° north of east at a speed of 8.5 km/hr relative to the ground. The wind, measured by an instrument aboard the sailboat, has a speed of 10.0 km/hr and is blowing 30° south of east. Find the speed and direction of the wind relative to the ground.

[7.16] A train is moving east on a straight track with a speed of 12 m/s. Raindrops falling straight down with respect to the ground make tracks on the train windows at an angle of 27° to the vertical. What is the speed of the raindrops with respect to the ground?

•**[7.17]** A ship that has a water speed of 7 km/hr is supposed to sail a straight path 36.9° north of east. A current of 4 km/hr is flowing from the south. What course should be set? (In what direction should the ship head?)

[7.18] Write an equation like Equation (7-28), but instead of finding $v_{\text{ball}\to\text{o}}$, write an equation to find $v_{\text{ball}\to\text{o}'}$ using $V_{\text{o}\to\text{o}'}$. Hint: How is $V_{\text{o}\to\text{o}'}$ related to $V_{\text{o}'\to\text{o}}$?

[7.19] Two vectors with magnitudes a and b have an angle θ between them when their tails are together. **a)** Show that the magnitude of their difference $|\vec{r}| = |\vec{a} - \vec{b}|$ can be written

$$r^2 = a^2 + b^2 - 2ab\cos\theta$$

b) Show that the magnitude of their sum $|\vec{r}| = |\vec{a} + \vec{b}|$ can be written

$$r^2 = a^2 + b^2 + 2ab\cos\theta$$

c) Show graphically that if $\vec{a}$ and $\vec{b}$ are perpendicular, then $|\vec{a} - \vec{b}| = |\vec{a} + \vec{b}|$.

[7.20] **a)** Starting with Equation (7-1) prove that $|\hat{a}| = 1$. **b)** Prove that the vector $\hat{r} = \cos\theta\hat{\imath} + \sin\theta\hat{\jmath}$ is a unit vector. **c)** For the position vector $\vec{r} = 3\hat{\imath} + 4\hat{\jmath}$ find a unit vector that points in the same direction, i.e. find the unit vector's x and y components.

CHAPTER 8
Life on a Sphere

(Life on a sphere—it all goes round and round)

We live constrained to move for the most part on the surface of a huge sphere. However, because the radius is so large, we ignore from day to day the curvature of our world. To us it appears locally flat—so much so that it wasn't until the third century B.C. that the radius of the earth was measured by Eratosthenes from Alexandria, who read that at noon on the twenty-first of June in the outpost city of Syrene, the shadow disappears on a vertical stick and then sent his assistant 500 miles or 800 kilometers to the south to pace the distance from Alexandria to Syrene. Because the shadow from a vertical stick in Alexandria does not disappear, Eratosthenes was able to use simple trigonometry to estimate the radius of the earth (see Problem 8.19). His result, accurate to within a few percent, was eventually lost for many centuries, and the spherical properties of the earth remained a topic of debate until Magellan's proof by brute force—sailing around the world. (Colombus's estimate of the radius of the earth was off by a factor of 2, which fortunately led him to discover North America.)

Those of you who might one day be in a position to decide which research projects get funded, take note. Which project was more cost effective, Magellan's or Eratosthenes? In this chapter, we explore some of the properties of curved paths and curved surfaces.

8.1 TRAVELING IN CIRCLES

In this section we cover the kinematics of circular motion in more detail than in Chapter 3 by using the polar unit vectors and showing how direction is important. You are probably quite comfortable using Cartesian coordinates and wonder why in the world would anyone want to use polar coordinates? The answer lies in simplicity. Consider an object moving in a circle. In Cartesian coordinates there are two pieces of information that are important (x, y). The equation for a circle of radius r is

$$\sqrt{x^2 + y^2} = r. \tag{8-1}$$

If the x coordinate of the object changes, then the y coordinate changes by a corresponding amount that you can calculate from Equation (8-1). Note that both x and y change as the object moves about the circle.

In polar coordinates, we again have two pieces of information (r, θ). But when an object moves in a circle, only θ changes; r stays constant. Having only one coordinate change makes the mathematics much simpler. For instance, finding the displacement Δs from Equation (8-1) in terms of Δx and Δy is a difficult problem requiring the use of calculus. However, in polar coordinates, we can easily find Δs in terms of $\Delta\theta$ as we did in Chapter 3.

$$\Delta s = r\Delta\theta \tag{8-2}$$

The speed is determined from the rate of change in the arc length

$$v = \frac{\Delta s}{\Delta t} = r\frac{\Delta\theta}{\Delta t}\,. \tag{8-3}$$

where $\Delta\theta/\Delta t$ is the angular velocity ω, so that

$$v = r\omega. \tag{8-4}$$

The velocity presents a harder problem. What do we do about the direction? Here, we can define unit vectors that fit naturally into the polar coordinate system. What are the natural displacements in a polar system? As we have seen already, we could change the angle and move along the arc of a circle as in Figure 8.1(a), or we could move from one radius to a new radius by changing r as shown in Figure 8.1(b). The natural displacements are then tangent to the circle, in the $\hat{\theta}$ direction, and radial in the $\hat{r}$ direction. $\hat{\theta}$ is always tangent to the circle, and $\hat{r}$ always lies along the radius to the circle.

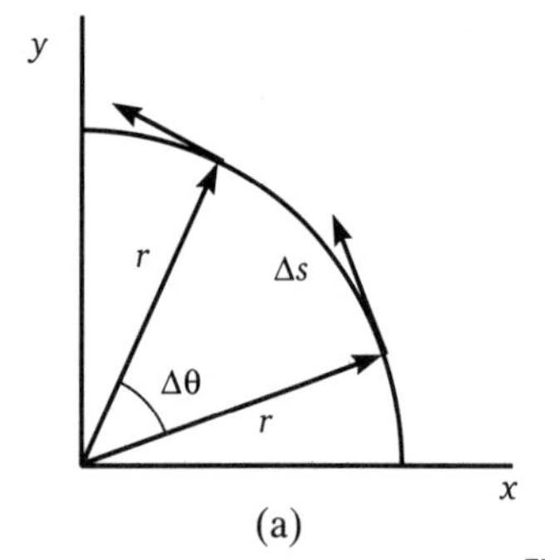

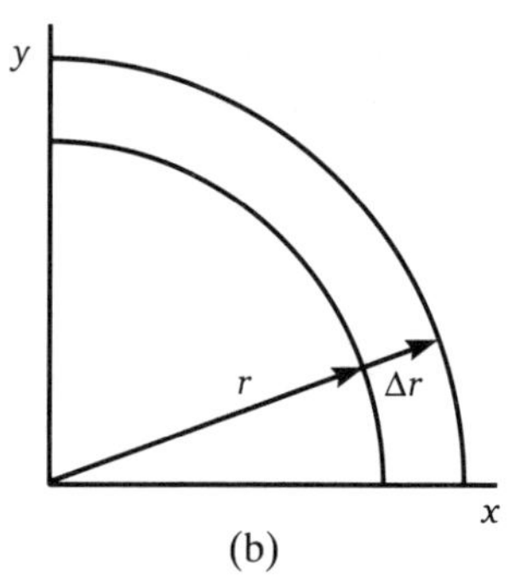

Figure 8.1
In a polar system, you can change your angle, θ, which means you move tangent to a circle, or you can change the radius, *r*, which means you move radially outwards towards a larger circle.

As a first example, can we resolve the position vector into polar coordinates just as we did for Cartesian coordinates? How about

$$\vec{r} = r\hat{r} + \theta\,\hat{\theta}\cdot \qquad \text{Incorrect!!} \tag{8-5}$$

There are several problems with this. First the dimensions are wrong. r has units of length and θ has units of radians, and both terms must have units of length. Second, in what direction does the position vector point? It always has its tail at the origin, and so must always point radially away from the origin. How much of the vector $\vec{r}$ points in the θ-direction? Of course from Figure 8.2, none of it does; all of $\vec{r}$ points in the radial direction. The correct polar form for a position vector is

$$\vec{r} = r\hat{r}. \qquad \text{Correct!!} \tag{8-6}$$

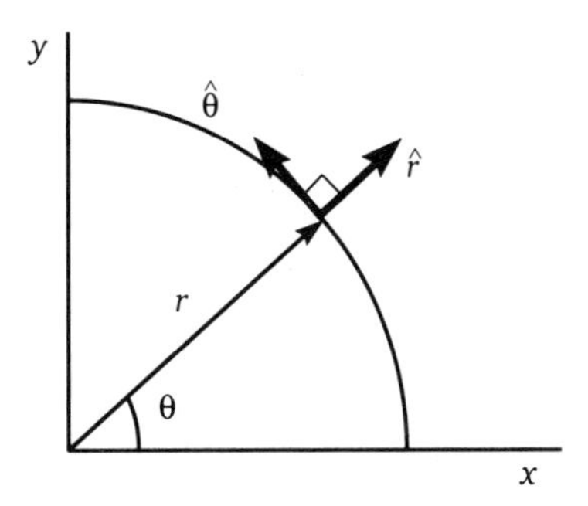

Figure 8.2
Unit vectors $\hat{r}$ and $\hat{\theta}$.

Position vectors cannot have θ components. Can you think of anywhere the θ component might be useful? The Figure 8.2 gives a hint.

Displacement vectors do not have to have their tails at the origin, and therefore, will not necessarily be radial. In fact for circular motion, $\hat{\theta}$ is the direction used for the displacement vector, since it is always tangent to a circle. The velocity is always in the same direction as the displacement and it will also be tangent to the circle. Thus,

$$\vec{v} = r\omega\hat{\theta} \tag{8-7}$$

One word of caution is that unlike the Cartesian unit vectors, that are constant in direction, the polar unit vectors change direction as the position changes as shown in Figure 8.3.

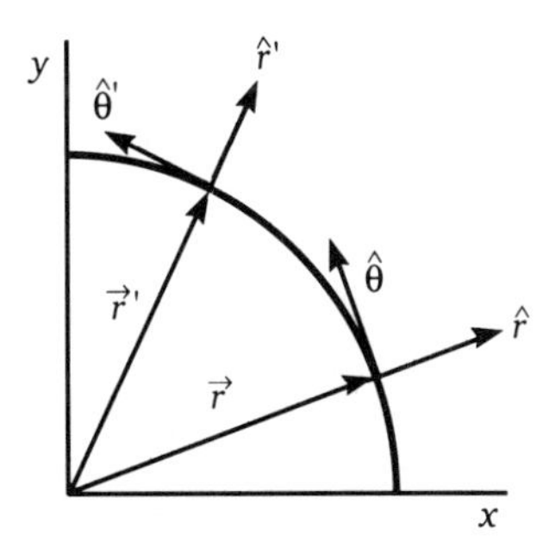

Figure 8.3
$\hat{r}$ and $\hat{\theta}$ change as the position changes.

Before adding polar components always check to be sure that the two r components point in the same direction, and similarly for the θ components. If they do not, then the vectors must be resolved into Cartesian components before adding.

Consider resolving $\hat{r}$ into Cartesian components. At any point (x, y) the unit vector $\hat{r}$ points radially away from the origin as shown in Figure 8.4.

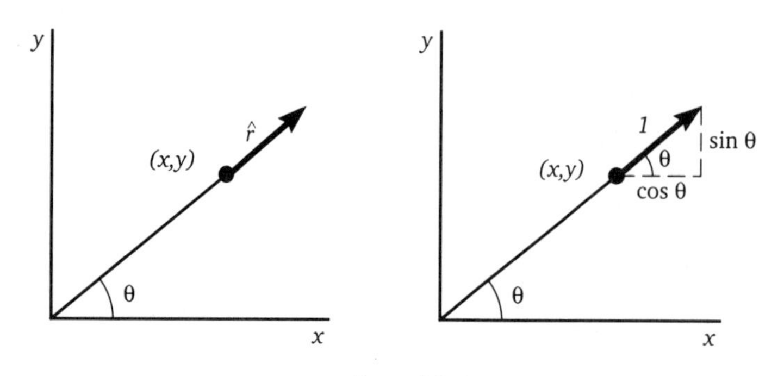

Figure 8.4
The direction of $\hat{r}$ is determined by the angle θ.

From Figure 8.4 we see that the direction of $\hat{r}$ depends on the polar angle θ. Since the magnitude of $\hat{r}$ is unity, it can be easily resolved into its Cartesian components

$$\hat{r} = \cos\theta\hat{i} + \sin\theta\hat{j} \tag{8-8}$$

which fits in nicely with the polar form of the position vector $\vec{r}$

$$\vec{r} = r\hat{r} = r\cos\theta\hat{i} + r\sin\theta\hat{j}. \tag{8-9}$$

So the $\hat{r}$ unit vector is indeed determined by the angle θ, which is the same for any vector, $\vec{A}$.

$$\vec{A} = A\cos\theta\hat{i} + A\sin\theta\hat{j} \tag{8-10}$$

where the direction of the vector is

$$\hat{a} = \cos\theta\hat{i} + \sin\theta\hat{j}. \tag{8-11}$$

Consider an object moving in uniform circular motion where ω is a constant and r is a constant so the speed, $v = r\omega$, must be a constant. If the direction changes but not the speed, then in what direction is the acceleration? It can only be perpendicular to the velocity, in order not to change the speed, because if the acceleration has any component along the direction of the velocity, then the speed will change. For example, the speed will increase if $\vec{a}$ has a component parallel to $\vec{v}$, as shown in Figure 8.5(a). When a ball is thrown down off a bridge, the acceleration due to gravity acts to increase the speed of the ball. What if the acceleration is anti-parallel to the velocity, as in Figure 8.5(b)? The ball should be thrown up into the air, opposite to $\vec{g}$, and the speed will decrease. As long as the acceleration has a component along the velocity, then the speed will change. If, however, the acceleration is perpendicular to the velocity, then the speed will not change, but the direction of the velocity will. We will next show that the acceleration is, indeed, perpendicular to the velocity for circular motion.

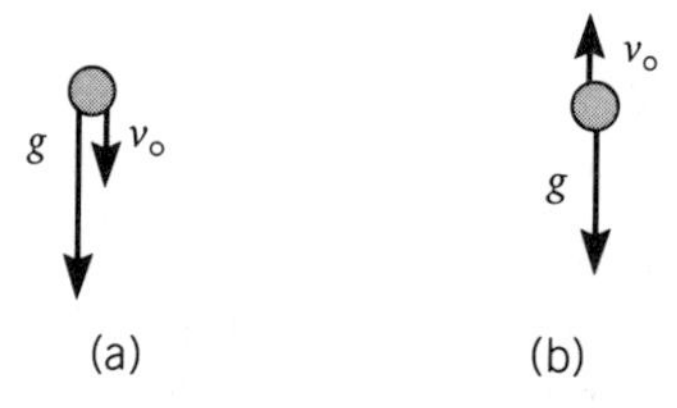

Figure 8.5
In (a) for a ball thrown down, the acceleration is parallel to the velocity, so the speed will increase. In (b) for a ball thrown up, the acceleration is anti-parallel to the velocity, so the speed will decrease.

In Figure 8.6 the velocity is shown at two moments separated in time by Δt.

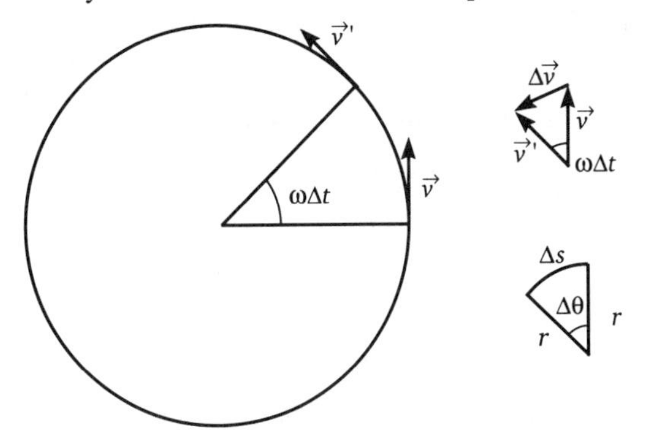

Figure 8.6
The change in the velocity is toward the center of the circle.

The velocity must be pushed towards the center of the circle so that the object will travel in an arc. We can approximate the change in the velocity as part of the arc on a circle. Since

$$\Delta s = r\Delta\theta, \qquad (8\text{-}12)$$

so, too,

$$\Delta v \simeq v\Delta\theta, \qquad (8\text{-}13)$$

where $\Delta\theta = \omega\Delta t$, and $v = r\omega$, giving:

$$\Delta v \simeq r\omega^2\Delta t.$$

The acceleration is thus,

$$a = \frac{\Delta v}{\Delta t} = r\omega^2. \quad \text{centripetal acceleration} \qquad (8\text{-}14)$$

The direction of the acceleration is in the same direction as $\Delta\vec{v}$ in Figure 8.6, toward the center of the circle in the $-\hat{r}$ direction. This acceleration is the *centripetal acceleration*. Thus, in order to produce circular motion, the object must accelerate toward the center of the circle. This acceleration is caused by a force directed in toward the center of the circle. This centripetal force might be gravity in the case of a satellite in a circular orbit about a planet, or the tension in a string for a ball whirled in a circle, or a car door to keep you in when your car rounds a sharp curve.

Using the relation $v = r\omega$, the centripetal acceleration can be written in a very useful form:

$$\vec{a} = -r\left(\frac{v^2}{r^2}\right)\hat{r} = -\frac{v^2}{r}\hat{r}. \quad \text{centripetal acceleration} \qquad (8\text{-}15)$$

EXAMPLE 8-1

Find the linear velocity and centripetal acceleration of a point on the edge of a record album playing at 33⅓ rpm (revolutions per minute).

Solution:
The linear speed is $v = r\omega$ where $\omega = 2\pi f$. The 33⅓ rpm is the frequency, which must be converted to rev/sec.

$$f = 33\tfrac{1}{3}\ \text{rpm}/(60\ \text{sec/min}) = 0.56\ \text{rev/sec}$$
$$\omega = 2\pi f = 3.5/\text{s}$$

Also, the diameter d of the long play album is 12 inches, which must be converted to meters.

$$v = r\omega = (\tfrac{1}{2}\,\text{d})\,\omega$$
$$v = (\tfrac{1}{2}\,12\ \text{in.})(0.0254\ \text{m/in.})(3.5/\text{s}) = 0.53\ \text{m/s}$$

Velocity is a vector, so we must also find its direction. Looking down on a turntable with a tone arm, as shown in Figure 8.7, note the turntable rotates clockwise. Thus, the velocity is tangent to the turntable platter in the $(-\hat{\theta})$ direction.

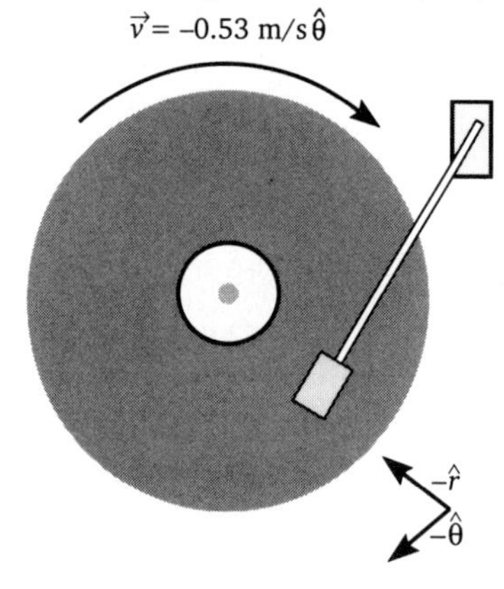

Figure 8.7
Turntable playing a record. The turntable must rotate clockwise so that it doesn't ruin the needle.

The centripetal acceleration is $a = r\omega^2$

$$\Rightarrow a = (\tfrac{1}{2}\,12\ \text{in.})(0.0254\ \text{m/in.})(3.5/\text{s})^2 = 1.86\ \text{m/s}^2$$

The centripetal acceleration is always perpendicular to the velocity and points toward the center of curvature—in the $-\hat{r}$ direction:

$$\vec{a} = -1.86\ \text{m/s}^2\,\hat{r}$$

8.2 CYLINDRICAL POLAR COORDINATES

In three dimensions, we can again define a polar representation consisting of the distance from the origin and two angles (r, θ, ϕ), spherical polar coordinates, or a distance from the z-axis and the angle measured in the plane perpendicular to the z axis, cylindrical polar coordinates. Spherical polar coordinates are the subject of the next section. Note that whatever system we use, we need to specify three independent pieces of information. By convention in a cylindrical polar system, the z–axis is typically taken to be along the central axis of a cylinder. θ is the angle measured in the $x-y$ plane with the x axis. r is the distance from the z axis. The cylindrical polar coordinates are shown in Figure 8.8.

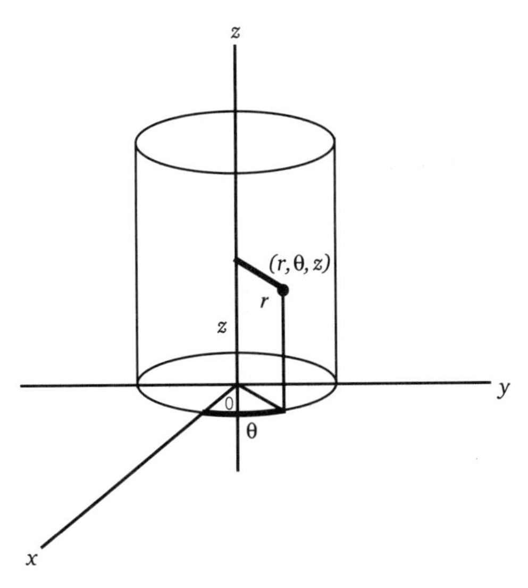

Figure 8.8
Cylindrical polar coordinates.

To find the Cartesian components of the position vector in the cylindrical polar system is straight forward. The z-component is the easiest, since it does not change.

$$z = z \tag{8-16}$$

The x and y components involve the radius r, so that

$$x = r\cos\theta \tag{8-17}$$

$$y = r\sin\theta \tag{8-18}$$

Given the Cartesian components, we can also find the polar components of the position vector. Using the Pythagorean theorem on the $x-$y plane triangle,

$$r^2 = x^2 + y^2. \tag{8-19}$$

The angle is easily found by using the tangent

$$\theta = \tan^{-1}\frac{y}{x}. \tag{8-20}$$

Cylindrical polar coordinates are useful in problems with cylindrical symmetry. Pipes and long wires have cylindrical symmetry and a cylindrical coordinate system is often used to describe physical processes such as waves and electrical currents in these objects.

EXAMPLE 8-2

A right circular cylinder of radius R and length L is spinning with constant angular velocity ω about its central axis. Describe the velocity of points on the surface of the cylinder.

Solution:

Velocity is a vector, so we first need to define a coordinate system. Because we have a cylinder, we should use a cylindrical coordinate system whose z axis lies along the central axis of the cylinder as shown in Figure 8.9. All the points on the curved outer surface, such as point A, have the same radius R, and thus, move with the same speed $v = R\omega$. These points have a tangential velocity and are moving in the $\hat{\theta}$ direction.

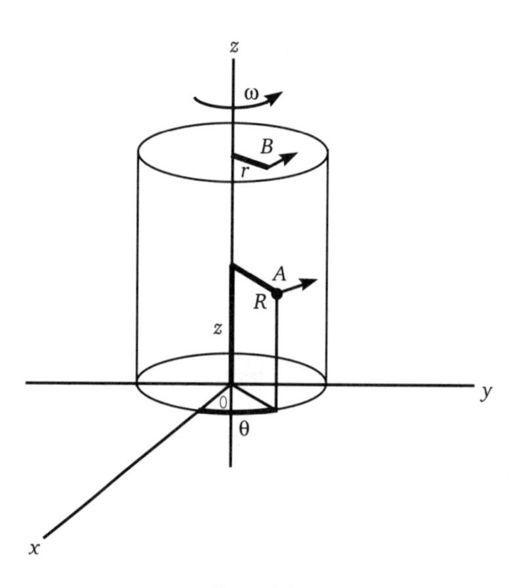

Figure 8.9
The velocity of points on the surface of a spinning cylinder.

$$\vec{v}\Big|_{\text{curved S}} = R\omega\hat{\theta}.$$

On the end caps of the cylinder the points, such as point B, have various radii, and so, they will be moving with various speeds. We can label the radius of one such point with a variable, r, where $0 < r < R$. Then,

$$\vec{v}\Big|_{\text{end cap}} = r\omega\hat{\theta}.$$

8.3 SPHERICAL POLAR COORDINATES

As stated before, we can define a spherical polar representation consisting of the distance from the origin and two angles (r, θ, ϕ). By convention, the z-axis is typically taken to be the north polar axis; i.e., the z-axis runs through the north pole on a sphere whose center is at the center of the coordinate system. θ is the angle the position vector makes with the z-axis, and ϕ is the angle the projection of the vector into the $x-y$ plane makes with the x-axis. The polar coordinates are shown in Figure 8.10

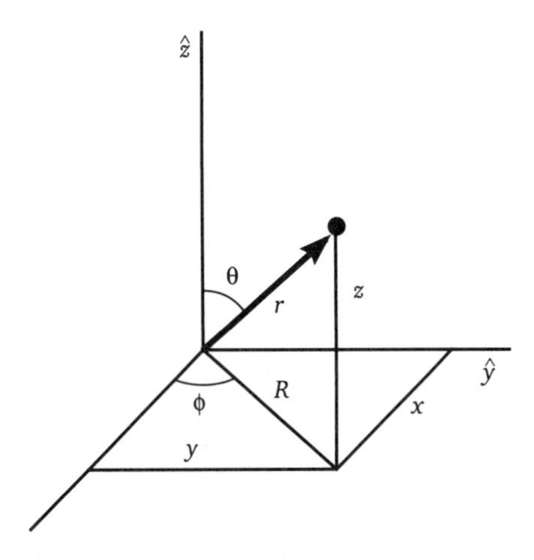

Figure 8.10
Spherical polar coordinates.

To find the Cartesian components of the position vector in three dimensions is somewhat more involved. The z component is the easiest, because it involves only the angle θ.

$$z = r\cos\theta \tag{8-21}$$

The x and y components involve the diagonal R , which is given by

$$\boxed{R = r\sin\theta.} \tag{8-22}$$

Then,

$$x = R\cos\phi$$
$$y = R\sin\phi$$

so that substituting in for R

$$x = r\cos\phi\sin\theta$$
$$y = r\sin\phi\sin\theta\cdot$$
$$z = r\cos\theta\cdot \qquad (8\text{-}23)$$

Given the Cartesian components, we can also find the polar components of the position vector. Using the Pythagorean theorem on the x–y plane triangle,

$$R^2 = x^2 + y^2.$$

The vertical triangle has

$$r^2 = z^2 + R^2,$$

so that upon combining the two

$$r = \sqrt{x^2 + y^2 + z^2}. \qquad (8\text{-}24)$$

The two angles are easily found by

$$\phi = \tan^{-1}\frac{y}{x} \quad \text{and} \quad \theta = \cos^{-1}\frac{z}{r}. \qquad (8\text{-}25)$$

Spherical polar coordinates are useful in problems with spherical symmetry. Because we live on a sphere, polar coordinates can be quite useful. We don't often travel large enough distances to notice or use the spherical polar system; however, pilots, ship navigators, and surveyors are quite familiar with polar systems. Displacements on a spherical surface is the next topic.

8.4 LIFE ON A SPHERE, LONGITUDE, AND LATITUDE

For displacements in three dimensions, we have three pieces of information to keep track of: Δx, Δy, and Δz. When we are constrained to the surface of a sphere, the radial coordinate does not change; therefore, we only need track changes in the two polar angles $\Delta\theta$, and $\Delta\phi$. What strikes you about the circular arc labeled S_θ shown in Figure 8.11? First, you might notice that the radius of the arc is the same as that of the sphere, and that the radius is a constant. Next, the angular coordinate ϕ does not change, but θ does. Finally, note that if we were to slice the sphere with a plane along S_θ, the plane would cleave the center of the sphere. Trajectories for which the plane cleaves the center of the sphere are very special: These *great arcs* have the property that they are the shortest surface arcs between two points on the surface of the sphere.

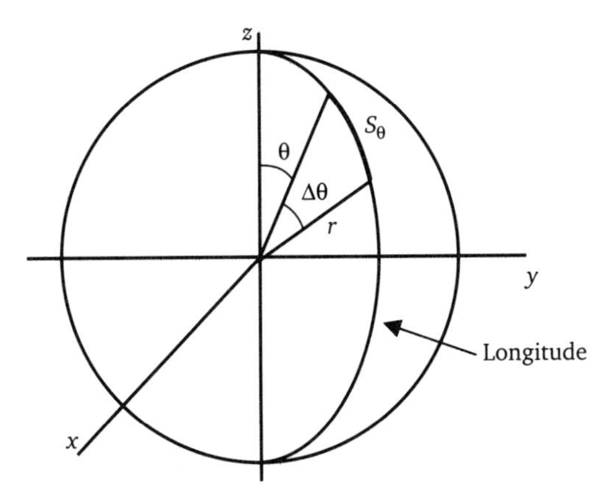

Figure 8.11
A longitude is a great arc through the poles, φ is constant along a longitude.

A displacement through an angle Δθ is along a longitude, and the distance traveled is just the arc length

$$S_\theta = r\,\Delta\theta, \qquad \text{longitudinal arc length} \tag{8-26}$$

where, of course, θ must be in radians.

Not all trajectories are great arcs. The trajectory shown in Figure 8.12 holds θ constant and varies φ, producing a very different result. Such an arc is a *latitude*. The latitudes form horizontal circles around the sphere, and although the line around the equator of the sphere is a great arc, none of the other latitudes are because their centers are not coincident with that of the sphere. A displacement through an angle Δφ along a latitude is shown in Figure 8.12.

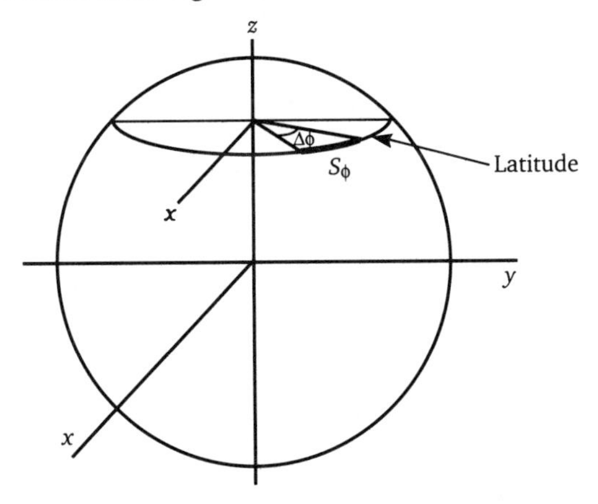

Figure 8.12
Lines of constant θ form the latitudes. Along these lines, φ changes, so the arc length is proportional to Δφ.

This displacement is again an arc length, but it is not

$$S_\phi \neq r\,\Delta\phi.$$

Can you see why? Is there any difference between latitudes and longitudes? Check Figure 8.13, and see whether you can find the difference between latitudes and longitudes.

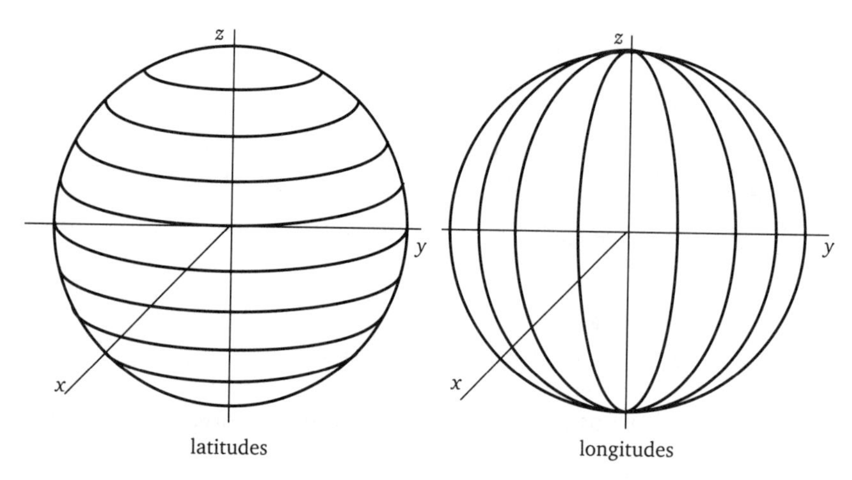

Figure 8.13
What's the difference?

It's the difference between cutting a tomato into wedges or into slices. All the wedges look exactly the same, but the slices vary in radius, depending from which part of the tomato the slice was cut. In general, latitudes are not great arcs because only the equitorial latitude contains the center of the sphere at its center. From Figure 8.14, we see that the radius of a latitude is tempered by a factor of $\sin\theta$.

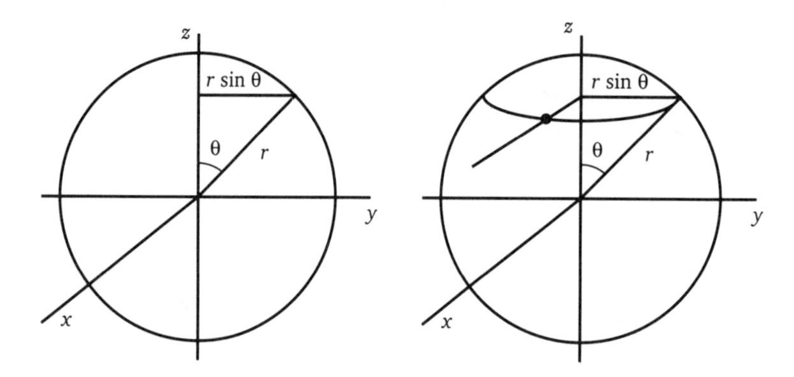

Figure 8.14
The radius of a latitude is tempered by a $\sin\theta$ factor.

The arc length formula for traveling along a latitude is given by

$$S_{\phi} = (r \sin\theta)\,\Delta\phi. \tag{8-27}$$

where θ may be in degrees, but $\Delta\phi$ must be in radians because S_ϕ is an arc length.
Note: as we choose a latitude closer to the pole, the radius goes to zero, just as $\sin\theta$ goes to zero because θ goes to zero.

EXAMPLE 8-3

The radius of the earth is 6.4×10^6m. Find the distance **a)** along a latitude from McDonald Island, Australia (70° E, 55° S) to Tierra del Fuego (70° W, 55° S). **b)** Is the distance found in Part (a) the shortest distance?

Solution:
a) The place to start is to draw a picture like that shown in Figure 8.15. Note that McDonald Is. and Tierra del Fuego are on the same latitude 55°S. The latitudes are measured with respect to the equator. The longitude was set up during the British Empire, so it should be no surprise that the reference longitude (0°), the *prime meridian*, runs through the Royal Observatory in Greenwich, England. McDonald Is. is 70° east of the prime meridian, and Tierra del Fuego is 70° west of the prime meridian. So the total angle $\Delta\phi$, between the two is 140°.

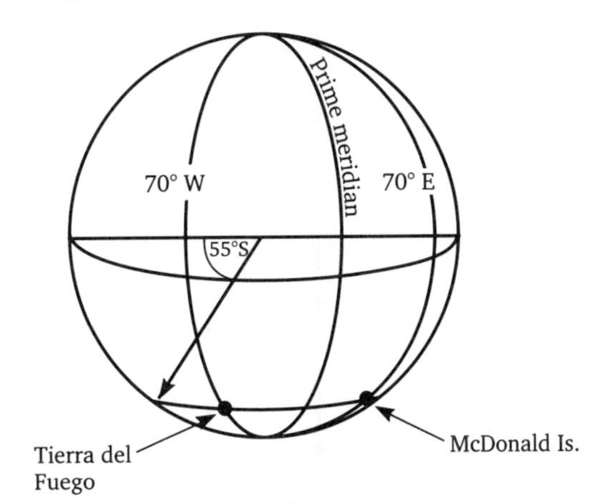

Figure 8.15
McDonald Is. and Tierra del Fuego lie approximately on the same latitude.

Since we are measuring along a latitude

$$S_\phi = (r\sin\theta)\,\Delta\phi. \qquad \text{latitudinal arc length} \qquad (8\text{-}28)$$

where $\Delta\phi$ is the total angle between 70° W and 70° E as shown in Figure 8.16. Taking east as positive and west as negative

$$\Delta\phi = 70° - (-70°) = 140°.$$

Be careful, because the arc length formula requires that $\Delta\phi$ be in radians. So we must write

$$\Delta\phi = 140°\,(\pi/180°) = 2.44$$

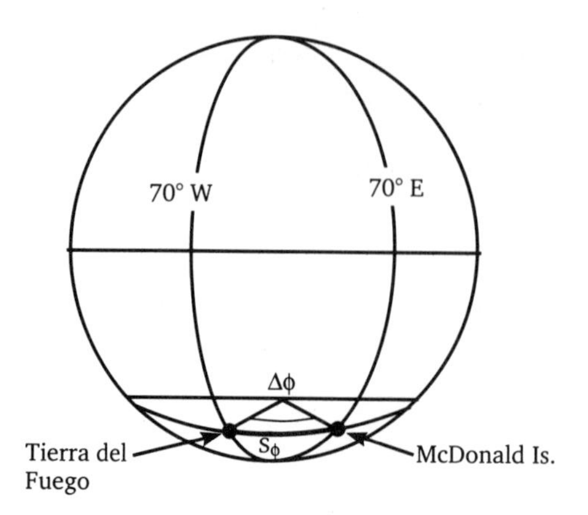

Figure 8.16
The arc length S_ϕ is subtended by the angle $\Delta\phi$.

θ is the angle with the polar axis and not the equator, and thus,

$$\theta = 90° + 55° = 145°.$$

The radius of the latitude is $r \sin 145° = r \sin 35°$. Note that we could also have used $r \sin 35°$ because 145° and 35° are supplementary, so $\sin 145° = \sin 35°$.

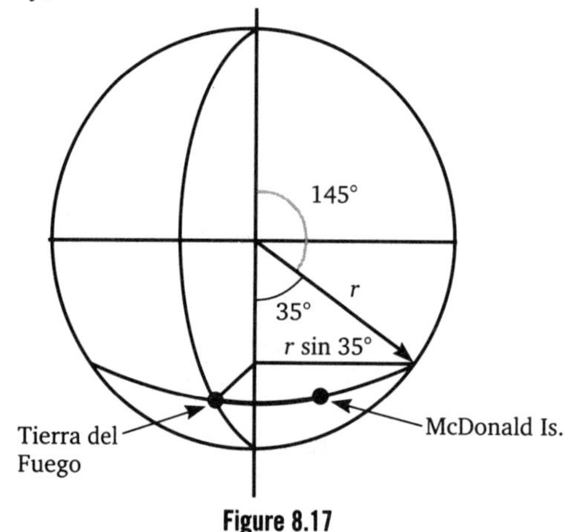

Figure 8.17
The radius of the latitude is $r \sin 125°$.

Putting it all together:

$$S_\phi = r \sin 35° \, \Delta\phi = (6.4 \times 10^6 \text{m})(\sin 35°)(2.44) = 9.0 \times 10^6 \text{ m}$$

or about

$$S_\phi = 5{,}600 \text{ miles}.$$

b) The shortest distance is always given by a great arc whose center lies at the center of the earth. Such an arc lies south of the latitude, as shown in Figure 8.18. Although in the 2-D projection shown, it is very hard to tell the difference between the latitude and the great arc, on a globe it becomes apparent that the great arc connecting Tierra del Fuego and McDonald Is. bends far south of the 55° latitude and crosses over Antarctica.

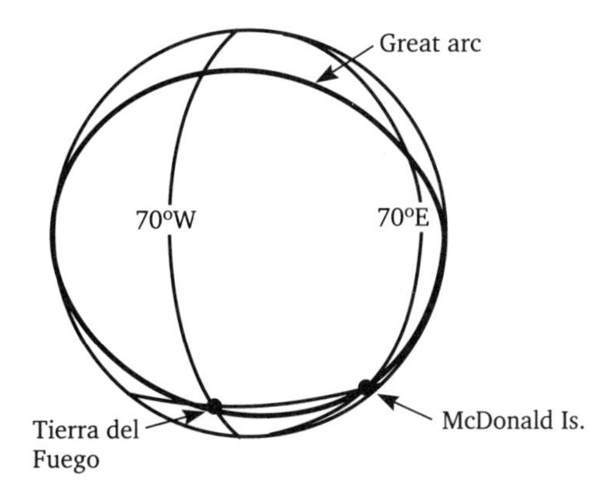

Figure 8.18
A great arc is the shortest distance between two points. The center of curvature of a great arc includes the center of the sphere.

The purpose of this section has been to give you practice playing with some of the aspects of spherical polar coordinates.

8.5 AREA AND VOLUME: ELEMENTS OF SYMMETRY

In this section we will cover a variety of examples to illustrate how various coordinate systems might be used. In each case you should attempt to exploit the natural symmetry of the problem.

EXAMPLE 8-4

A long rod of radius R and length L has one end at 0°C, and the other at 100°C. Find the area through which the heat flows.

Solution:
We can treat the rod as long cylinder. To determine the area through which the heat is flowing, we need to draw a picture of the cylinder and heat flow as in Figure 8.19.

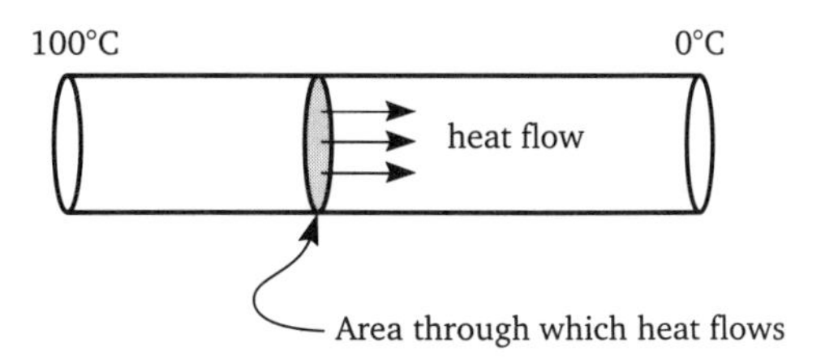

Figure 8.19
Heat flows parallel to the axis of the rod.

Heat will flow from hot to cold, so it will flow parallel to the axis of the rod from the 100°C end to the 0°C end. The heat flows through the cross sectional area,

$$A = \pi R^2.$$

In problems like Example 8-4 it is important to use the cylindrical geometry of the problem, and to visualize the flow so that it is clear what area is involved. For instance, the energy source of stars is nuclear fusion. You can think of a star as a nuclear fusion reactor in which light nuclei are fused together to form heavier nuclei, releasing energy in the process. The fusion process happens deep in the interior of the star where the density and pressure is extremely high. The heat that is generated must be released from the interior by flows out to the surface where it is radiated away. It is important to understand through which area heat is being transported.

EXAMPLE 8-5

The sun radiates 6.27×10^7 watts/m^2 and has a radius of 7×10^8m. Find the total power (P) radiated by the sun (in watts).

Solution:

The sun is basically a sphere that is radiating power from its spherical surface of area $4\pi r^2$ as shown in Figure 8.20.

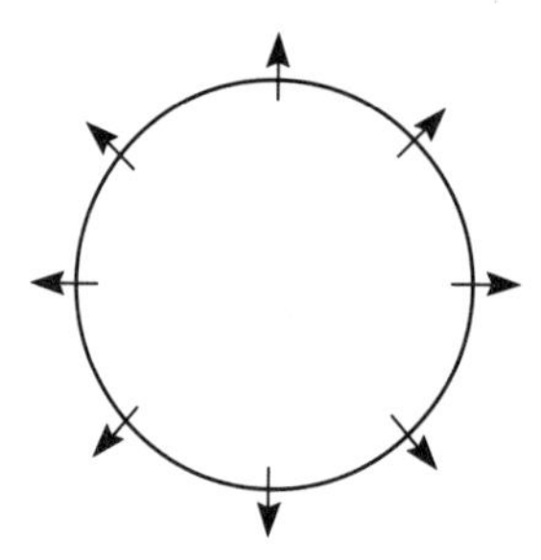

Figure 8.20
In the sun, the energy flows outwards through the $4\pi r^2$ surface area.

Then

$$P = (6 \times 10^7 \text{ watts/m}^2)(4\pi)(7 \times 10^8\text{m})^2 = 3.7 \times 10^{26} \text{ watts}$$

Whenever there are flows, area plays an important role. The flow can be the flow of heat, energy, mass, or any quantity. The symmetry, rectangular, cylindrical or spherical also plays a role.

EXAMPLE 8-6

Air is blown into a spherical balloon at a constant rate of 120 cm³/s. Find the rate at which the radius is changing when the balloon has a radius of 4 cm.

Solution:
We are given the rate at which the volume of the balloon is increasing. In a small amount of time, Δt, the radius will increase a small amount Δr. The base area is $4\pi r^2$ and the height is Δr, so the change in the volume of the balloon is

$$\Delta V = 4\pi r^2 \Delta r$$

as shown in Figure 8.21.

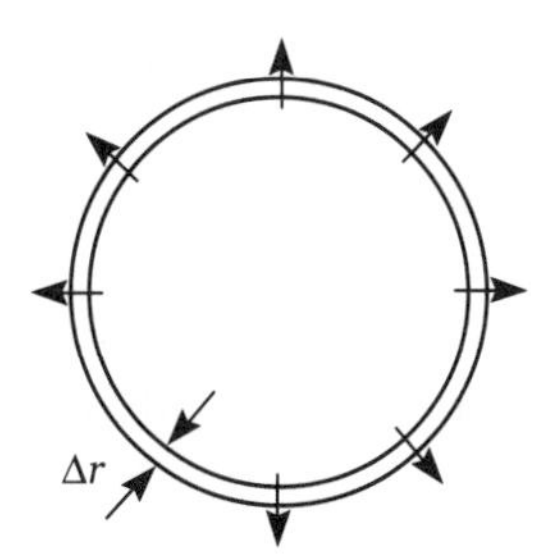

Figure 8.21
An expanding balloon increases its volume by $4\pi r^2 \Delta r$ in a time Δt (base area × height).

Dividing by Δt

$$\frac{\Delta V}{\Delta t} = 4\pi r^2 \frac{\Delta r}{\Delta t}.$$

Note that $4\pi r^2$ is the area through which the flow occurs. Solving for $\Delta r/\Delta t$,

$$\frac{\Delta r}{\Delta t} = \frac{1}{4\pi r^2}\frac{\Delta V}{\Delta t}$$

$$\frac{\Delta r}{\Delta t} = \frac{120\ \text{cm}^3/\text{s}}{4\pi(4\ \text{cm})^2} = 0.6\ \text{cm/s}.$$

Now we are ready to tackle the question of density in more than one dimension. Recall from Chapter 6 that the density relates the mass to the geometry, i.e., it tells us how the mass is distributed in space. If we had mass distributed over a surface, then it would make sense to use the mass per unit area σ, and for a volume, the mass per unit volume ρ.

$$\bar{\sigma} = \frac{M}{A} \qquad \text{surface mass density} \qquad (8\text{-}29)$$

and

$$\bar{\rho} = \frac{M}{V} \qquad \text{volume mass density} \qquad (8\text{-}30)$$

Any symmetries that the mass distribution possesses can be exploited.

EXAMPLE 8-7

A block of length L, width w, and height h has a uniform mass density ρ. Find the total mass of the block.

Solution:
The block has rectangular symmetry and is oriented with one side along the x axis as shown in Figure 8.22.

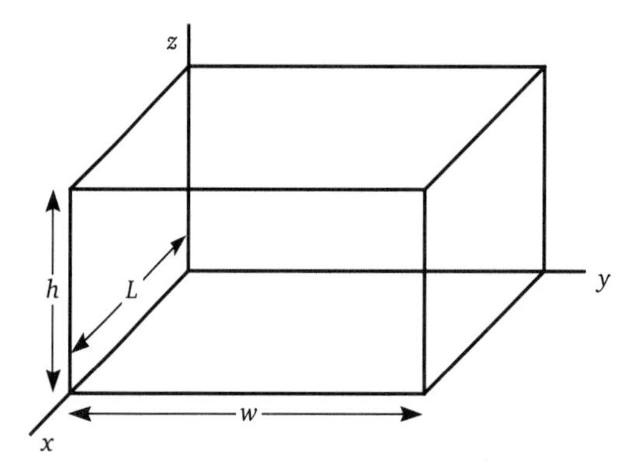

Figure 8.22
A block of uniform density.

Because of the rectangular symmetry, the volume is just

$$V = L \times w \times h.$$

The mass of this slab is

$$m = \rho V = \rho L w h.$$

The choice of the coordinate system used should depend on the symmetry of the object. In Example 8-7 the object had rectangular symmetry, so we used a rectangular coordinate system. In Example 8-8 the object possesses cylindrical symmetry:

EXAMPLE 8-8

A thin hollow cylindrical shell of radius 3 cm and length 17 cm has a uniform mass density of 5 g/cm^2. Find the total mass.

Solution:

A thin hollow cylinder means that the mass is not distributed throughout a volume, but exists approximately as a surface. So, the density should be a density per unit area. Checking the units on 5 g/cm^2, the cm^2 reveals that this is the case. The mass is then

$$m = \frac{m}{A}A = \sigma A$$

Over what surface area is the mass distributed? Draw a picture of a thin hollow cylinder as shown in Figure 8.23.

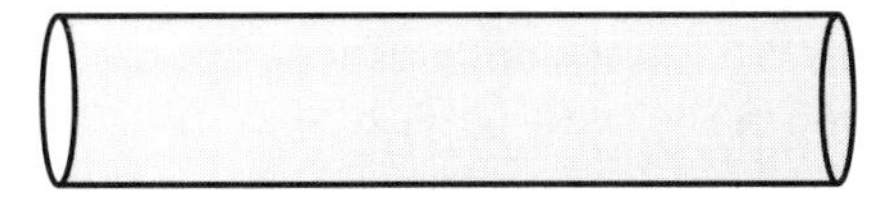

Figure 8.23
Mass distributed over a thin hollow cylindrical shell.

The area of the mass is the curved part of the cylinder which is the $2\pi rL$ surface area.

$$m = \sigma(2\pi rL)$$

$$\text{m} = (5\text{ gm/cm}^2)(2\pi)(3\text{ cm})(17\text{ cm}) = 1.6 \times 10^3\text{ gm} = 1.6\text{ kg}.$$

Symmetry is a very powerful tool that can allow you to do seemingly very difficult problems easily. Even if the symmetry isn't perfect, sometimes you can find one close enough to give you a good approximation. Symmetries simplify problems tremendously. You should be able to recognize them on sight and immediately know the appropriate density and coordinate system to use.

Chapter 8 Problems

[8.1] **a)** Draw $\hat{r}$ and $\hat{\theta}$ on the diagram 8.24 below. **b)** Resolve $\hat{r}$ and $\hat{\theta}$ into Cartesian components for the vector $\vec{r}$ shown below.

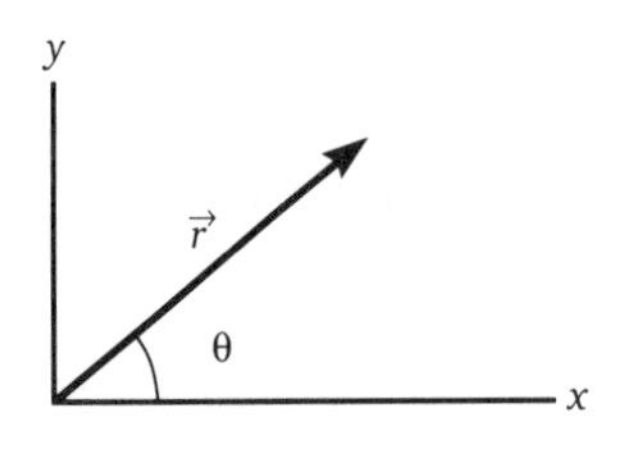

Figure 8.24

[8.2] Prove that $\hat{r}$ and $\hat{\theta}$ are unit vectors.

[8.3] At a given moment an object is at the point (16 meters, 12 meters) and a displacement of 3 m is made directly towards the origin in 5 s. Find the velocity vector.

[8.4] **a)** Write the following points as position vectors using the Cartesian components and unit vectors. For example, $(3, 3) \Rightarrow \vec{r} = 3\hat{\imath} + 3\hat{y}$. **b)** Next draw the position vectors on the graph. **c)** At the tip of the position vector, draw the $\hat{r}$ and $\hat{\theta}$ unit vectors for each. **d)** Finally, resolve $\hat{r}$ and $\hat{\theta}$ into Cartesian components for each.

1) $(1, 1)$ **2)** $(-5, -5)$ **3)** $(-6, 3)$ **4)** $(1, -3)$ **5)** $(-2, -5)$

[8.5] The moon has a period of 27.3 days for one revolution about the earth. If the distance to the moon is 3.85×10^8 m, then **a)** What is the angular speed of the moon? **b)** What is the linear speed of the moon? **c)** Calculate the ratio of the centripetal acceleration of the moon to the acceleration due to gravity at the earth's surface.

[8.6] You wish to build a small turbine of radius 2.0 cm that will rotate at an extremely high rate of 36,000 rpm. **a)** Find the linear velocity of a point on the tip of the turbine blade. **b)** The material you are using can withstand a centripetal acceleration of 4.0×10^5 m/s^2 at this size. Will the turbine withstand being spun this fast, or will it blow apart?

•**[8.7]** A small crumb rests on top of a half cylinder of radius R that is laid flat, as shown in Figure 8.25. Find the smallest horizontal velocity that must be given to the crumb if it is to leave the sphere without sliding down it.

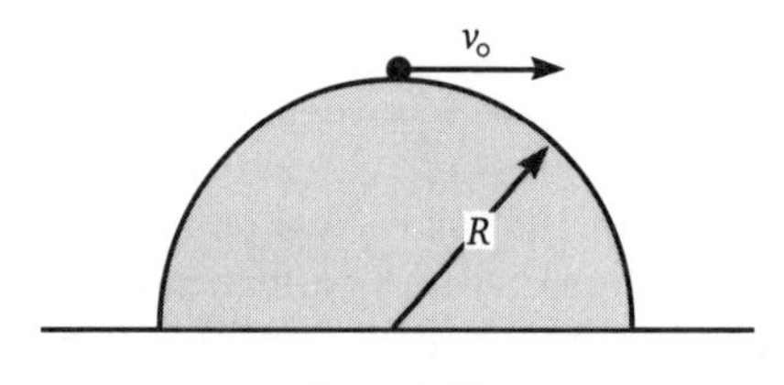

Figure 8.25

[8.8] A roller coaster starts with speed v_0 from point P and travels down the loop-the-loop track, Figure 8.26. At no point does it leave the track. The positive direction is along the track. Label all the points marked by a solid black dot (including P) that have the following characteristics: **a)** speed is maximum, **b)** speed is minimum, **c)** speed is increasing, **d)** speed is decreasing, **e)** centripetal acceleration is greatest, **f)** centripetal acceleration is least, **g)** speed is constant, **h)** net acceleration is zero, **i)** net acceleration is parallel to the velocity, **j)** net acceleration is perpendicular to the velocity.

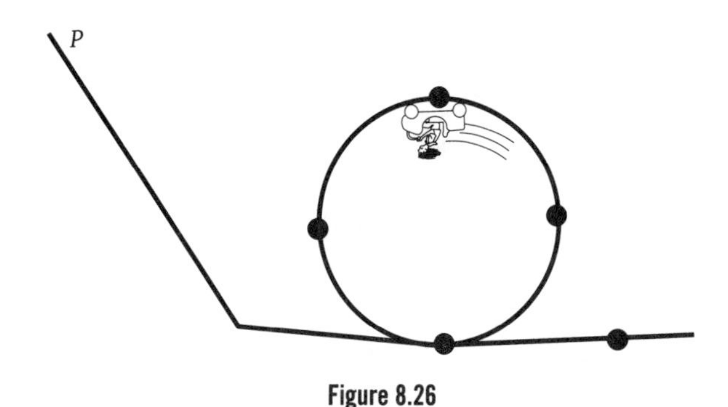

Figure 8.26

[8.9] A point on a cylinder has polar coordinates (4 m, 0.4 radians, 3 m). Find the Cartesian coordinates of this point.

[8.10] Superman flies due north from Guadalajara to Winnipeg. If the radius of the earth is 6.4×10^6 m and the coordinates of the cities are Guadalajara, 20° north latitude, 100° west longitude; Winnipeg, 50° north latitude, 100° west longitude, then find the total distance traveled.

[8.11] Superman flies due west from Winnipeg to Vancouver. If the coordinates of the cities are Winnipeg, 50° north latitude, 100° west longitude; Vancouver, 50° north latitude, 125° west longitude, find the total distance traveled.

• **[8.12]** A sphere of radius R is spinning with constant angular velocity ω about a central axis through the north pole. Describe the velocity of points on the surface of the sphere, i.e., in what direction is the velocity? How does the velocity depend on position? Does it depend on one of the polar angles?

[8.13] The cylindrical bore in a long vertical glass tube sealed at the bottom end has a diameter of 2 mm, and contains 10 cm^3 of alcohol. **a)** Find the height of the alcohol in the tube **b)** If the volume expands with increasing temperature as 0.01 cm^3/°C, then find the height of the alcohol in the tube if the temperature increases by 100°C. Assume the expansion of the glass is negligible.

[8.14] Water flows into a cylindrical tank of height 15 m and radius 10 m at a rate of 3 m^3/s through a fill pipe. At what rate does the level of the water inside the tank increase?

[8.15] A steam pipe of radius r and length L carries hot water and has an insulating jacket which is not perfect. Draw a picture of the pipe, showing the heat flow. Find the area through which the heat flows out of the pipe.

[8.16] A balloon has a mass of 7 g and a radius of 12 cm. Find the surface density of the balloon.

[8.17] A solid sphere of plutonium has a mass of 2 kg. If the density of plutonium is 1.98×10^4 kg/m^3, what is the radius of the sphere?

[8.18] A heating duct in a building has a cross section of 15 cm $\times$ 30 cm and is 100 m long. If the aluminum sheet the duct is made out of has a density of 3 kg/m^2, find the mass of the duct.

[8.19] In the third century B.C., Eratosthenes, head of the library in the Egyptian city of Alexandria, read that a vertical stick planted in the ground in outpost city of Syrene threw no shadow at noon on the twenty-first day of June. Eratosthenes hired a man to pace the distance between Alexandria and Syrene and found it to be about 800 km. If a vertical stick casts a shadow in Alexandria at noon on June 21, such that a line drawn from the tip of the shadow to the tip of the stick makes an angle of 7° with the stick as shown in Figure 8.27, then find the radius of the earth. Assume the rays coming from the sun are parallel.

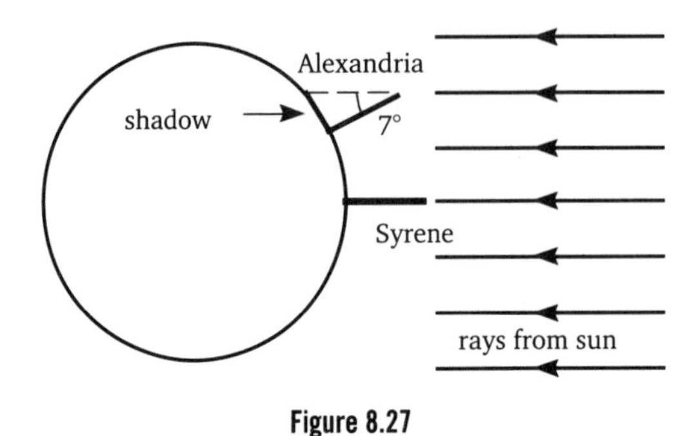

Figure 8.27
A vertical stick casts a 7° shadow in Alexandria at noon on June 21.

•**[8.20]** A particle travels with constant speed in a circular path of radius 5.0 m from the origin. At $t = 0$, it begins at $x = 5.0$ m and $y = 0$ and takes 100.0 s for a complete revolution. **a)** What is the speed of the particle? **b)** For $t = 20.0$ s, draw a picture of the velocity vector, and calculate the x and y components of the velocity vector at that moment.

[8.21] A wheel with radius 70 cm rolls without slipping along a horizontal floor. P is a dot on the outside edge. Initially P is at the point of contact between the floor and the wheel. Find the net displacement of P after the wheel has moved through one half revolution. Specify the magnitude and direction of the displacement vector.

[8.22] This exercise requires the use of a globe which you might find at a library. By looking at a globe, list the countries you would fly over if you flew along the great arc from New York City to Moscow, Russia.

• **[8.23]** If Superwoman flies due east from Ankara, Turkey at 900 mph, **a)** how long will it take her to get to Peking, China. The radius of the earth is 6.4×10^6 m where 1,600 m = 1.0 mi., and the coordinates of the cities are:

Ankara (40° N, 33° E)
Peking (40° N, 116° E).

b) Is Superwoman flying along the shortest path? Explain.

CHAPTER 9
Force

"May the Force . . . naw."

In this chapter, for the first time, we actually introduce some physics. You will be introduced to Newton's laws of motion and asked to find the acceleration of an object that is subject to many forces acting at once. Newton's laws are used to find the cause of motion, and in the cause of motion lies the real meat of physics. As much as is possible, physics is a causal discipline. Physics is the attempt to find the cause of a system's behavior, the reason that an object has the acceleration it has.

9.1 NEWTON'S LAWS AND DYNAMICS

In this section, we begin the formal study of *dynamics,* which is the study of the causes of motion. In particular, we are concerned with what causes acceleration. Dynamics is built on the *Newton's three laws of motion,* in recognition of Isaac Newton, who created the first real theory of motion.

Newton's Three Laws of Motion

1. An object will continue in a state of rest or move along a straight line with uniform constant velocity unless acted upon by a net force.
2. The net force which is the vector sum of all the forces acting on an object equals the mass times acceleration
$$\Sigma\vec{F} = m\vec{a}.$$
3. When a force $\vec{F}$ is placed upon an object, *A*, by a second object, *B*, object *A* will exert an equal but opposite force $-\vec{F}$ on B.

The SI unit for mass is a kilogram. The standard reference kilogram is a platinum-iridium cylinder stored in Sevres, France at the International Bureau of Weights and Measures. This standard cylinder defines the unit of mass, and all mass measurements are made by comparison with copies of this cylinder. The official United States copy is at the National Institute of Standards and Technology in

Washington D.C. It has been taken to Sevres twice since 1889 for comparison with the master cylinder.

Because the official system of units for the United States is the English system (**note:** Great Britain uses the metric system so the English system units are now called *American customary units*), you will often have to convert between the SI and the English system. The English system unit of mass is the *slug*. You must be careful because often the *pound* is used for mass, but the pound is the English unit of force. Using Newton's second law, the relation between mass and force at the earth's surface is

$$F = mg, \tag{9-1}$$

where g is the acceleration due to gravity, and which implies that the relation between a slug and a pound is

$$1 \text{ lb} = (1 \text{ slug})\, g = (1 \text{ slug})(32 \text{ ft/s}^2)$$

At the earth's surface, the slug and the pound are directly proportional—hence, the confusion between the unit of mass and the unit for weight. It is best to keep the distinction clear. It can sometimes cause annoying problems that waste too much time.

The SI unit of force is a Newton. Based on the second law, it can be broken down into fundamental units, as "a kilogram meter per second squared."

The Newton (Unit of Force)

$$1 \text{ N} = 1 \text{ kg} \cdot \text{m/s}^2$$

9.2 THE FIRST LAW

The first law states that the natural state of motion is one of constant uniform motion, which sounds trivial, but its purpose is very important. The first law requires that force and only force can change the velocity; hence, we need not consider anything else to determine the acceleration.

This law was actually quite revolutionary because from the time of Aristotle it was believed that the natural state of motion was one of rest, and indeed, this comes from experimental evidence because everyone knows that if you push a cart it will come to rest if you stop pushing it. Therefore, shouldn't the natural state of motion be one of rest, where a force is required to maintain an object in uniform motion? The answer was determined by Galileo, who arrived at it only after much careful experiment and deliberation. Galileo's experiment is quite interesting and illustrates the manner in which science progresses, so we study it here.

The key, of course, is friction. The reason that everything stops is because of an unbalanced frictional force acting upon the object—but how to prove it, that is the problem! We can't devise an experiment without friction, but we can reduce the effect of friction by using objects that roll. You've probably heard the tale about Galileo dropping the balls off the Leaning Tower of Pisa; however, he probably never did that experiment, or if he did, he found the result not useful because masses falling through a great distance in air reach a terminal velocity due to air resistance. For simple cases, air resistance is proportional to the speed:

$$f = -bv, \tag{9-2}$$

where b is a constant that depends on the size of the object. The faster an object moves, the greater the air friction, so the retarding force due to the air friction increases until at the terminal velocity it balances the force due to gravity. So even dropping balls off a high tower does not get rid of friction. Air friction plays a large role. We can minimize the effect of friction by making the balls fall a short distance, which will prevent the balls from ever attaining a large velocity, and therefore, the air friction will never play an important role. Unfortunately for Galileo, that also created an impossible situation. He did not have an accurate clock, and due to the large gravitational acceleration, he could not accurately get times for objects falling for short time intervals. So how? Why not reduce the acceleration due to gravity by not letting a ball fall through air but rather by letting a mass roll down an inclined plane? As this chapter shows, the acceleration of a mass rolling down an inclined plane depends on the angle of inclination. The steeper the plane, the greater the acceleration, which is known to any child who has ridden a bike down a hill. There is still friction involved in the rolling, but provided that the surfaces are hard, and the balls do not slip, the rolling friction is negligible.

So now we have an experiment that Galileo has hopes of completing. By keeping the velocity low, air friction doesn't play an important role. By reducing the acceleration, the timing becomes less critical. Still, Galileo had to invent his own accurate water clock, made by dripping water at a constant rate into a beaker. By weighing the beaker, Galileo could quite accurately determine the time that had passed.

Using inclined planes, Galileo discovered something that he would have had no hope of discovering had he just dropped the balls from a tower. What if—after rolling down an inclined plane—you let the ball roll back up another inclined plane? How high would it roll? Galileo found that the ball rolled up to approximately the same height as it began from, as illustrated in Figure 9.1, regardless of the inclination of the uphill plane.

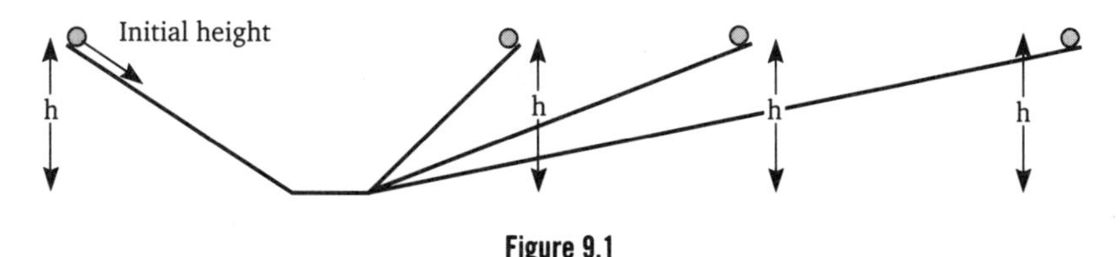

Figure 9.1
The ball returns to the initial height regardless of the angle of inclination.

In actuality, the ball does not return to exactly the same height, due to friction. This is particularly true for steep inclinations where the ball will start to slip because gravity no longer holds it tightly against the board. When the ball starts to slip, it dissipates much of the ball's energy as heat. Also, for very small angles, the ball must roll a great distance to return to its original height, and the additive effects of friction become considerable. What's important is that by using the inclined plane, Galileo found a way to quantify the frictional effects. He was also able to show that the behavior is universal. For example, pendulums behave the same way; i.e., a pendulum will return to the height from which it was released. By comparison, the ball of a pendulum comes up just a little short of its release, so friction plays a minor role with pendulums, too.

We need quite a proof to overthrow 2,000 years of dogma, and Galileo devised an ironclad proof that the natural state of motion is not one of rest but rather one of motion. Suppose that rather than incline the second plane, it is kept flat. How can the ball return to its original height? Of course, it never can. So in the absence of frictional forces, the ball must continue to roll, forever trying to reach its original height!

EXAMPLE 9-1

A hockey puck is tied to a string and whirled in a circle upon the ice. At one point the string breaks. Which of the trajectories shown in Figure 9.2 does the puck take? Please explain.

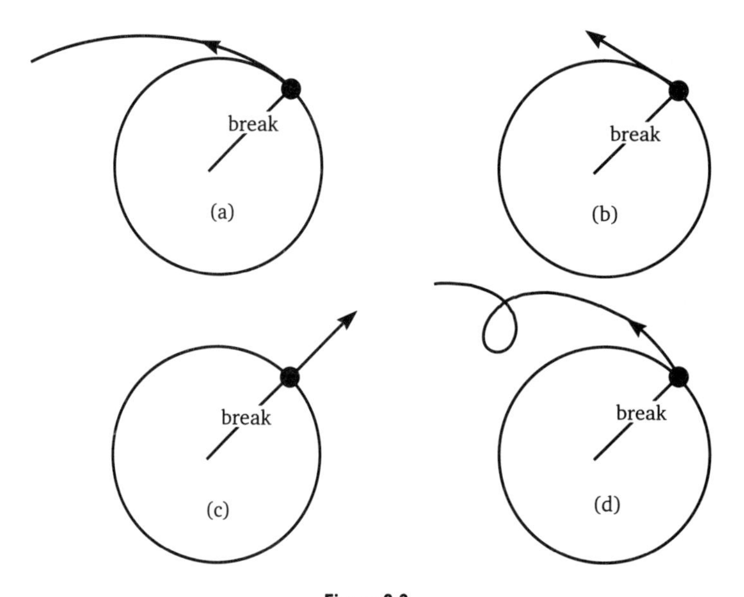

Figure 9.2
A hockey puck is whirled in a circle on the ice when the string breaks.

Solution:
The string is exerting a force on the puck, keeping it moving in a curved path. When the string breaks, that force is removed, so there is no longer a net force on the puck. By Newton's first law, it must move in a straight path. This brings the choices down to two, either (b) or (c). The puck must move in the direction of its velocity because there is no longer any force to change its velocity. Because the velocity is tangential to the circle, the puck must follow choice (b).

9.3 THE SECOND LAW

Although the first law states that only a force can cause an acceleration, it does not tell how it does so. That is the domain of the second law, which states that the sum of the forces acting on a given object equals the mass times the acceleration of the object:

$$\Sigma F = ma. \tag{9-3}$$

For this law to be useful, we need to know the definitions of acceleration, mass, and force independent of this law. In other words,

$F = ma$ does not define force!

To use $F = ma$ to define force would leave us with a definition only valid for accelerating objects. As we will see, it is possible to have balanced forces acting on an object, such that, the object will not accelerate, and these forces would not be covered by such a definition. Newton's second law is not a definition of force because it allows us to find the acceleration. We can then use kinematics to calculate the velocity and the position so that Newton's second law plays a fundamental role in the description of mechanical systems. This means we must define force in a way that is independent of Newton's second law. Unfortunately, there is not one single definition of force but rather many, depending on the source of the force. For instance, a table can hold up a book. The force of the table on the book is normal to the table's surface, and is thus called a normal force (see Figure 9.3).

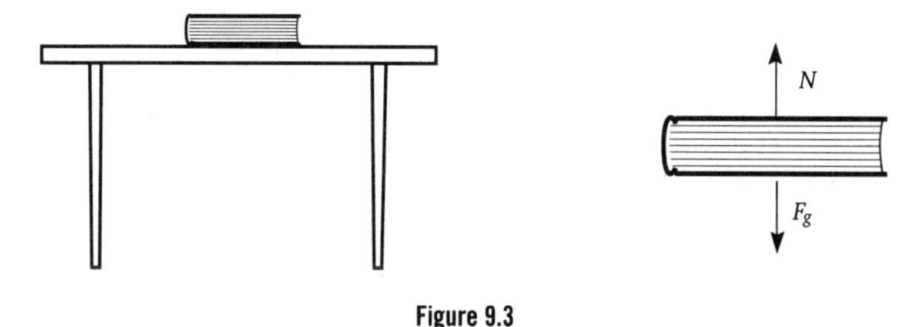

Figure 9.3
A table exerts a force normal to its surface on a book.

The normal force N is necessary to counteract the force due to gravity F_g. We can determine the normal force by first determining the force due to gravity. Suppose that the table were not there and the book were undergoing free-fall. Then the only force acting on the book is that due to gravity (ignoring air friction). Using Newton's law

$$F_g = mg \tag{9-4}$$

where g is the acceleration due to gravity and is a constant near the earth's surface. As long as g is a constant, we can take this as the definition of gravitational force near the earth's surface.

Now apply Newton's laws to the book on the table by first summing the forces on the book. Because the sum of the forces determines the acceleration of an object, it is extremely important to correctly determine all the forces acting on an object. To aid us in focusing our thoughts on this task, we use a *free-body diagram.* To draw a free-body diagram, see the box on the following page.

Free-Body Diagram

1. Isolate one single object.
2. Draw arrows indicating the direction and approximate magnitude of each force acting on the object. Only show arrows for forces acting on the object in the free-body diagram. Do not show forces acting on other objects.
3. Think hard about what forces act on the object. Don't miss any!

From the free-body diagram in Figure 9.3, the table exerts a normal force N up in the positive y direction, and gravity g acts down in the negative y direction. Summing the y forces,

$$\Sigma Fy = N - F_g = N - mg, \tag{9-5}$$

and then applying Newton's second law,

$$N - mg = ma. \tag{9-6}$$

Because the book is resting quietly on the table, its acceleration is zero, such that,

$$N - mg = 0, \tag{9-7}$$

and

$$N = mg. \tag{9-8}$$

Right now, you should be screaming, "Wait! You have just used Newton's laws to define both the force due to gravity and the normal force. We are supposed to define the forces independently." First, Equation (9-8) is not a definition of the normal force. As we will see later, the normal force can vary, depending upon circumstances. The normal force is a passive force: A table will supply whatever force is necessary to constrain the object from moving through the surface, provided the required force does not exceed the limits of the table. Second, experimentation gives mg for the gravitational force. It is an experimental fact that at the earth's surface, the force due to gravity is directly proportional to the inertial mass times a constant. The force due to gravity, as of this writing, is a fundamental force of nature. By "fundamental" we mean that it cannot be derived from anything else and can only be determined by experiment. You will see shortly that the mg form is not generally true and can be derived from a gravitational force due to Newton. In turn, Newton's gravitational force can be derived from a richer theory of gravity, due to what Einstein called "general relativity." In any theory, there will always be at least one force that can only be determined by experiment. This will always be necessary because we will always need to specify which of the infinite mathematical possibilities defines our universe. Currently, there are three fundamental forces* from which all others can be defined:

* There is some debate about the existence of a new limited range force linked to the mass, but the experiments are still very ambiguous.

Fundamental Forces

1. Gravitational force
2. Electroweak force
3. Strong force.

The gravitational force binds our solar system and perhaps our universe together. The electroweak force is responsible for all of chemistry, electricity and magnetism, and the beta decay of nuclei. Electricity and magnetism define the forces between particles carrying an electric charge. The weak force defines the interaction between leptons (electrons) and hadrons (protons and neutrons). We used to think of electricity, magnetism, and the weak force as separate forces, but in 1867, Maxwell showed how electricity and magnetism were manifestations of one electromagnetic force. This was the first unification of forces. Then in 1969, Weinberg, Salaam, and Glashow unified the weak force with the electromagnetic force. Many believe that one day we will understand how all the forces are different manifestations of one force, but that day is still probably distant.

The gravitational force was first defined by Newton and is termed *Newton's Law of Gravitation.* The two masses M and m may be either point masses or spherical masses.

Newton's Law of Gravitation

$$\vec{F} = -\frac{GMm}{r^2}\hat{r}, \tag{9-9}$$

where r is the distance between the centers of the two masses, and G is the *universal gravitational constant. Universal* means that as far as we know, G is the same everywhere in the universe.

$$G = 6.67 \times 10^{-11} N \cdot \mathrm{m}^2/kg^2$$

The unit vector $\hat{r}$ means that the force is a *central force;* that is, it acts along a line joining the two centers of the spherical masses, and the minus sign means that the force is attractive. From Newton's law of gravitation, we can derive the acceleration due to gravity.

EXAMPLE 9-2

Find the acceleration of an apple in free-fall near the earth's surface. Assume there is no air friction.

Solution:

Let the mass of the earth be M, the mass of the apple be m, and r be the radius of the earth. The radius of the apple is negligible compared to that of the earth. The only force on the apple is that of gravity, so applying Newton's laws,

$$\Sigma F = ma$$

$$\frac{GMm}{r^2} = mg.$$

Note that the rate of acceleration of the apple is independent of the apple's mass because it divides out! The acceleration of any object near the earth's surface is independent of mass because the gravitational force law depends on the same mass as the inertial force *ma*!

$$g = \frac{GM}{r^2} = \frac{(6.67 \times 10^{-11}\text{N} \cdot \text{m}^2/\text{kg}^2)(6 \times 10^{24}\text{kg})}{(6.4 \times 10^6\text{m})^2} = 9.8\ \text{m/s}^2 \qquad (9\text{-}10)$$

There are many other forces that we could talk about. All of these other forces, like the force due to gravity near the earth's surface, can be derived from fundamental forces. The following discussion lists a few common forces.

Another force that a surface can exert on an object is a frictional force. Unlike the normal force, the frictional force is always tangent to the surface and always opposes the tendency of two surfaces to slide over one another. We always use f for frictional force. In Figure 9.4, the top block is pulled to the right at velocity v_1 Friction will drag the bottom block to the right because this is the way the bottom block would have to move to avoid sliding against the upper block. According to Newton's third law, the force due to friction on the top block would be equal and opposite to that on the bottom.

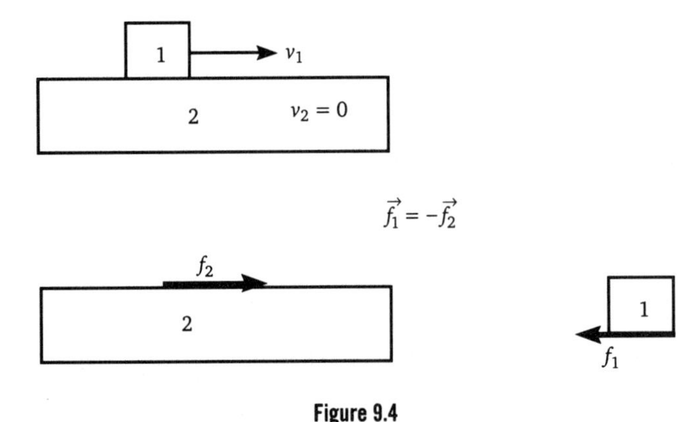

Figure 9.4
Friction acts to slow block 1 down and to speed block 2 up.

The frictional force between two hard surfaces is experimentally found to depend on the roughness of the surface and on the normal force. It does *not* depend on the *surface area.* The dependence upon the normal force is obvious. Rub your hands together. The harder you push your hands together, the greater is the frictional force. Also, note that your hands get warmer, so there must be a conversion of mechanical energy into heat. The fact that it does not depend on surface area seems to contradict common sense. After all, don't the dragsters use big wide tires to increase traction? The restriction here is that the frictional force between two hard surfaces does not depend on the area. When a dragster starts, a portion of the tire melts and forms an *adhesive* layer, similar to glue. Adhesion does depend on the surface area. You must be careful because the distinction between friction and adhesion can sometimes be a rather gray area, because there can be transitions from one to the other during a system's operation. The roughness of the surface is taken into account by defining a *coefficient of friction* μ so that the frictional force can be written

$$f = \mu N = \mu \, mg \qquad (9\text{-}11)$$

Each force has characteristics peculiar to its source. For example, the force that the tension in a string can exert on an object must always lie along its axis, and it must always be a pull, never a push. We call the tension force T. (See Figure 9.5.)

$$\vec{F} = \vec{T} \qquad \text{Tension} \qquad (9\text{-}12)$$

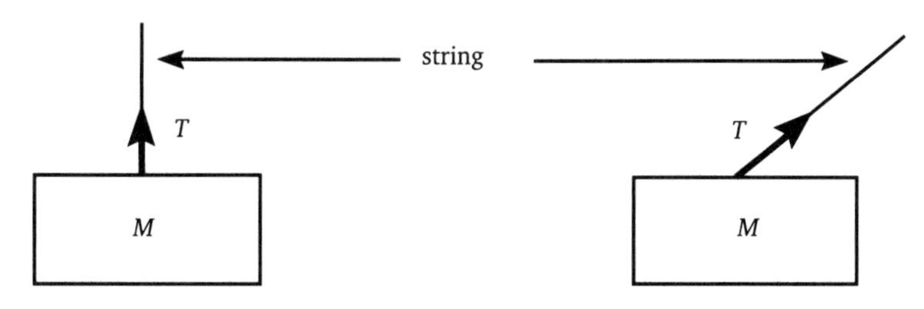

Figure 9.5
The tension force that a string exerts on an object must always lie along the axis of the string.

Because the purpose of the second law is to find the acceleration by summing the forces acting on an object, it is very important to include all the forces. An aid to doing this is to draw a *free-body diagram.* A free-body diagram consists of drawing one object and only one object, and then labeling each force that acts on the object with an arrow. The purpose is to concentrate all your thoughts upon one object and to clearly label all the forces acting on it. Only include forces acting on the one object. Never include forces acting on other objects, or the acceleration and velocity of the object.

EXAMPLE 9-3

Draw a free-body diagram for each of the labeled masses in Figure 9.6, and sum the forces on each object. Assume there is no friction between the string and the pulley, and the mass of the string is negligible. The tension in the string on either side of the pulley will be the same tension.

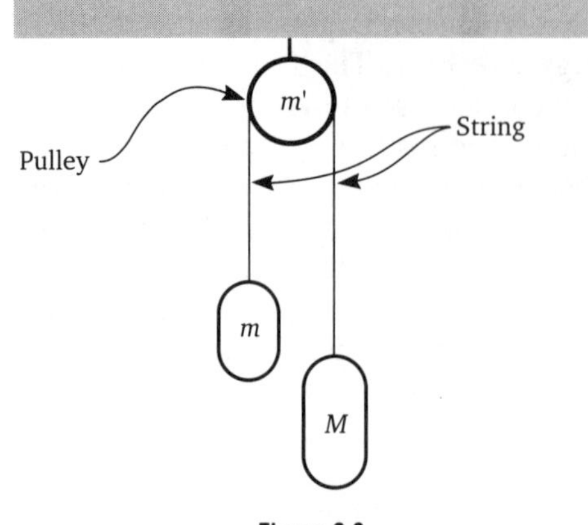

Figure 9.6
Atwood's machine.

Solution:
Because the string slides freely over the pulley and because it has no mass, the tension in the string must be the same on either side of the pulley (see Problem 9.12). It is fairly easy to draw free-body diagrams for the two masses attached to the ends of the string. On each mass the tension acts upward and gravity acts downward.

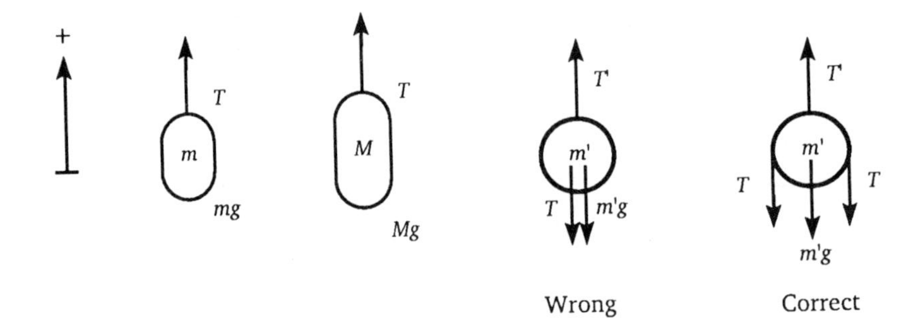

Figure 9.7
Free-body diagrams for the individual masses.

The pulley is more complicated. The first impulse is to say that the hanging masses both pull down upon the pulley so the net force down is $Mg + mg + m'g$, but that is wrong. The M and m masses do not touch the pulley, and so cannot exert an mg type of force on the pulley. The string touches the pulley, and so the string exerts a force on both sides of the pulley. (See Figure 9.7.) This is more that just cosmetic, because the tension in the string is not necessarily $Mg + mg$. The tension is also determined by the acceleration of the masses. If $M \neq m$, the masses will accelerate. This can be seen by summing the forces with the up direction being positive and the down direction negative.

$$\Sigma F_m = T - mg = ma_m$$

and

$$\Sigma F_M = T - Mg = Ma_M$$

Note that $a_M = -a_m$ because if mass m moves up, then mass M must move down by the same amount. Thus, the tension depends on the acceleration of the masses. Summing the forces on the pulley,

$$\Sigma F_p = T' - 2T - m'g = 0.$$

The pulley does not accelerate because T' is a passive force, supplying whatever is necessary to keep the pulley stationary.

You can see from the preceding example the importance of using a free-body diagram and of clearly labeling the forces according to their direct source. It is very important to find all the forces acting on an object. If you leave out any forces, you will not get the correct acceleration. A free-body diagram helps to focus your thoughts clearly on the process, one object at a time.

The force that a spring can exert on an object is defined by Hooke's law

Hooke's Law

$$\vec{F} = -kx\hat{\imath}, \tag{9-13}$$

where x is the distance the object has been displaced from the equilibrium point and k—termed the spring constant—is a measure of the stiffness of the spring. The force that a spring can exert can be either repulsive if the spring has been compressed or attractive if the spring has been stretched (see Figure 9.8.). There is an equilibrium point between the compression and stretch modes.

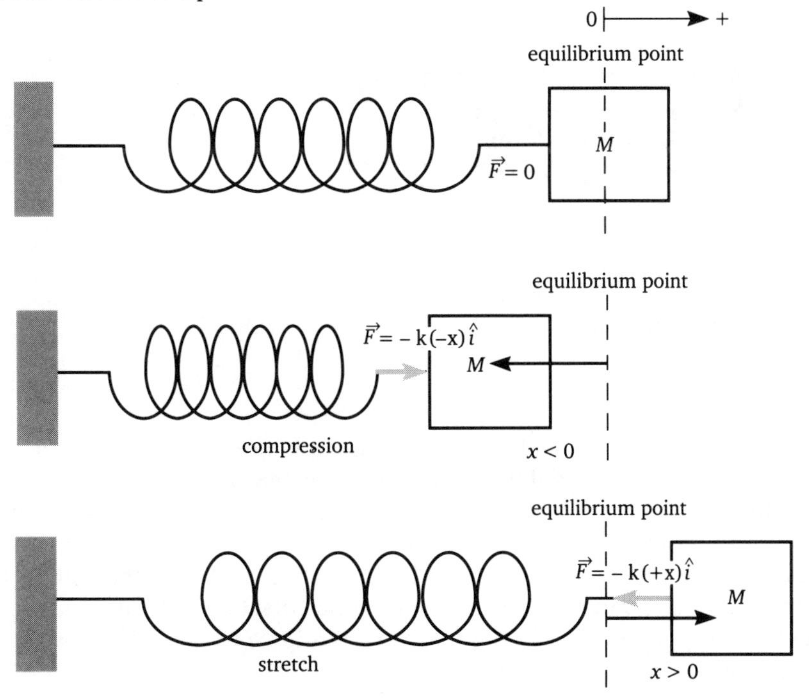

Figure 9.8
The force that a spring exerts can be either attractive or repulsive, depending on whether the spring is stretching or compressing.

9.4 THE THIRD LAW

The third law prevents a system from exerting a net force on itself. Essentially, this is the law that prevents you from picking yourself up by the bootstraps. There are some subtleties in this law, as illustrated by the horse and the cart paradox, described in Example 9-4.

EXAMPLE 9-4

A horse pulls on a cart to take a farmer's hay to the barn. By Newton's third law, if the horse exerts a force F on the cart, then the cart exerts a force $(-F)$ back on the horse. Those two forces sum to zero, so it is impossible for the horse and cart to move! Thus, Newton's third law proves that motion is impossible! Solve this paradox.

Solution:
Again, the first step is to concentrate on the individual objects by *drawing a free-body diagram*. A free-body diagram shows there is no paradox for the cart, because there are only two horizontal forces on the cart, friction opposing the motion and the force exerted by the horse. As long as F is greater than the friction the cart will accelerate (see Figure 9.9). The force the cart exerts against the horse acts on the horse and not on the cart; therefore, it cannot cancel the force the horse exerts against the cart, that is, the third law forces act on different objects and cannot cancel each other.

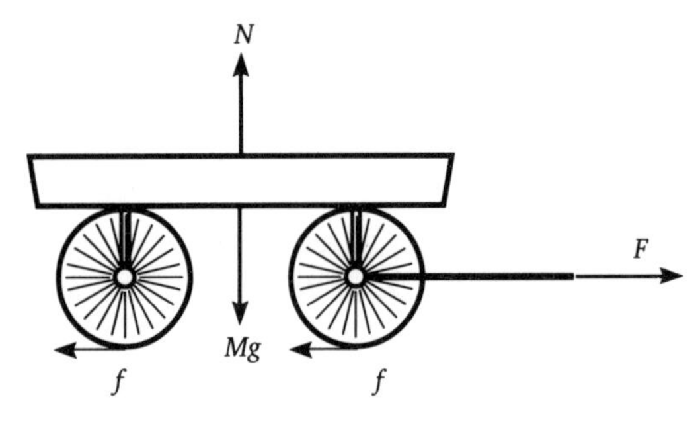

Figure 9.9
There is a net horizontal force acting upon the cart, so it accelerates.

The horse experiences a force opposing his motion but there are additional horizontal forces acting on the horse. The horse also exerts a tangential frictional force upon the ground. There is a tendency for students to draw the frictional forces acting back toward the cart because that is the direction in which the horse pushes against the ground, but this is wrong, and again, a free-body diagram can save you. The horse pushes to the left on the ground, but that force is acting on the ground and *not* on the horse. In a free-body diagram, we only include forces acting on the horse! By Newton's third law, the ground must push back upon the horse. Thus, the ground exerts a force to the right upor the horse causing the horse, to accelerate to the right. (See Figure 9.10.)

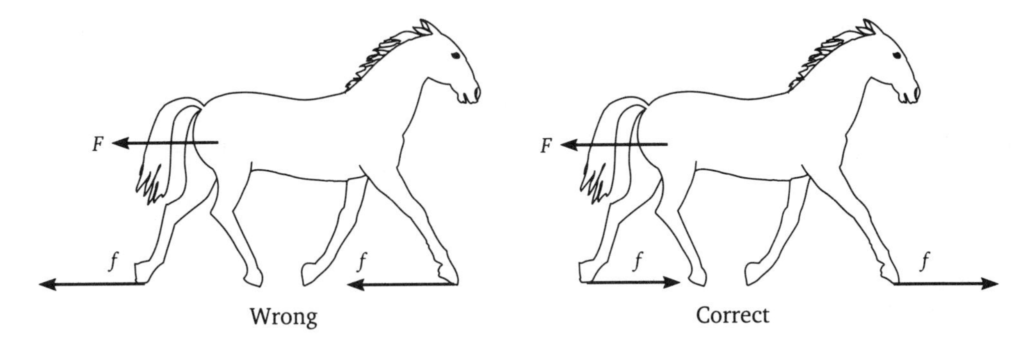

Figure 9.10
The frictional force of the ground on the horse acts to accelerate the horse.

If the force of the ground on the horse acting to the right seems wrong, ask what would happen if the horse were put on smooth ice? It would not be able to accelerate to the right at all because the tangential force of the ground upon the horse would be reduced to nil.

The solution to the horse–cart paradox becomes apparent when we draw a free-body diagram. Newton's third-law forces cannot cancel because they act on different objects. One acts on the horse and the other acts on the cart. This illustrates another important point: Always keep in mind on which object a given force acts.

9.5 MASS, YOUR PERCEPTION OF WEIGHT, AND MOMENTUM

Mass and weight are often confused with one another, but there is a very important distinction. *Weight* is a perception of how heavy an object feels; i.e., the weight of an object is correlated to the force required to lift the object off some surface. The force required to lift an object depends on the circumstances, and for a given object, this can vary from one instant to the next. For example, a person will weigh one sixth as much on the moon as on the earth. This means that the person will be pulled against the surface of the moon one sixth as hard as against the surface of the earth. Thus, weight can vary depending on where you are. Note also that your sensation of weight always involves a normal force. Your weight is the same as the normal force that an object must exert against you to hold you up. Weight is a force! In the example of the book lying upon the table, the weight of the book W is

$$W = mg. \qquad \text{weight} \qquad (9\text{-}14)$$

On the other hand, *mass* is an attribute of a body, independent of anything else. It does not change as a body moves from one place to another. A 7 kg mass has a mass of 7 kilograms, whether it is at the surface of the earth or deep in outer space. The distinction between mass and weight is important because there are effects that depend on the mass and *not* the weight. For example, a roller skate traveling at 5 m/s is much easier to stop than a truck traveling at the exact same 5 m/s. Every moving body possesses a quantity termed *momentum* that is a direct measure of how hard it is to stop the body from moving. Obviously, the faster the body is moving, the harder it will be to stop, so the momentum must be proportional to the velocity. Also, the more massive the body, the harder it will be to stop, so the momentum must also be proportional to the mass.

$$\vec{p} = m\vec{v} \qquad \text{momentum} \qquad (9\text{-}15)$$

To change the momentum, we must place a force on the body. In fact, Newton first stated his second law in terms of the momentum. To quote Newton from an English translation of his book *Principia**:

Definition II

The quantity of motion is the measure of the same, arising from the velocity and quantity of matter conjointly.

Law II

The change of motion is proportional to the motive force impressed; and is made in the direction of the right line in which that force is impressed.

Newton first defines momentum, although he calls it "motion," in Definition II, and then gives his second law in terms of the momentum. Using calculus, the statement is much more compact

$$\vec{p} = m\vec{v} \qquad (9\text{-}16)$$

$$\Sigma\vec{F} = \frac{\Delta\vec{p}}{\Delta t}. \qquad \text{Newton's second law} \qquad (9\text{-}17)$$

What Newton's second law tells us is that even though you are out in space and may be weightless, you are not massless, and to change your momentum requires a force. The following example illustrates some of the complexities involved.

EXAMPLE 9-5

Sky Lab was an orbiting space laboratory used during the early to mid 1970s. The astronauts in Sky Lab made some films of acrobatics in a zero-gravity environment that amazed us earth-boundlings. They did daredevil flips and bounced from wall to wall, and hung in midair. **a)** Determine the acceleration due to gravity in Sky Lab. Assume that Sky Lab's orbit was circular and the height above the surface of the earth was 200 km. **b)** If the acceleration of gravity is nonzero, why did the astronauts in Sky Lab appear weightless in the films they took? **c)** Why—when the astronauts jumped off one wall—did it hurt just as much as on earth when they hit the opposing wall?

Solution:

a) According to Example 9-5, the only force on Sky Lab was the earth's attraction. That attraction kept Sky Lab in a circular orbit of radius $r = R + h$, where R is the earth's radius and h is the height of Sky Lab above the earth's surface. The acceleration was

* Although Newton was English, like all scholarly works of that time period, Newton's *Principia* was written in Latin. This translation was done by Andrew Motte in 1729. Sir Issac Newton, *Principia,* Vol.I, University of California Press, 1966.

$$\Sigma F = ma$$

$$\frac{GMm}{r^2} = ma$$

where m is the mass of Sky Lab and M is the mass of the earth. Given that r is the distance of Sky Lab from the center of the earth $\Rightarrow r = 6{,}400 \text{ km} + 200 \text{ km}$

$$a = \frac{GM}{r^2} = \frac{(6.67 \times 10^{-11}\text{Nm}^2/\text{kg}^2)(6 \times 10^{24}\text{kg}^2)}{(6{,}600 \times 10^3\text{m})^2} =$$

$$a = 9.2 \text{ m/s}^2$$

Thus, the acceleration due to gravity of the astronauts is almost as great as at the surface of the earth! Why then are the astronauts weightless?

b) You only perceive weight if you are being pulled against a surface. Although the astronauts are accelerating, so is Sky Lab! Essentially Sky Lab and the astronauts are accelerating at the same rate, so they fall together. Thus, gravity does not pull an astronaut against Sky Lab, so there is no normal force on the astronaut, and the astronauts do not perceive any weight. Thus, because Sky Lab is falling at the same rate as the astronauts, they feel weightless. c) Although the astronauts are weightless, they are not massless. When an astronaut hits the wall, the momentum changes very suddenly. By Newton's second law, this exerts a large force on the astronaut, just as would happen on earth.

From Example 9-5, we gain an appreciation that weight depends on our surroundings and situation. The astronauts in Sky Lab were being pulled toward the earth almost at the same rate as at the earth's surface, yet they felt weightless. This apparent weightlessness is caused by a lack of a normal force; i.e., gravity wasn't pulling them against a surface, so the astronauts "felt" weightless. Because gravity acts uniformly over every part of our body, we cannot directly feel gravity. What we feel is the effect of gravity. We are pulled by gravity against a surface, normally the floor, and this surface exerts a normal force against us. We interpret the normal force as our weight. This link between the normal force and our weight provides the excitement in many amusement park rides, as the next example illustrates.

EXAMPLE 9-6

In an amusement park ride, people are spun in a large vertical steel drum. When the drum reaches a particular speed, the floor drops from beneath the riders' feet, but the riders do not fall. Why not? In what direction does the wall exert a force upon the riders?

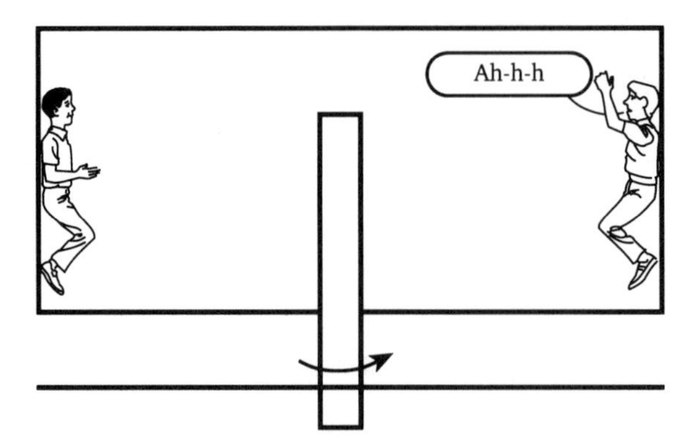

Figure 9.11
The riders are plastered against the side while the floor drops from underneath.

Solution:
The riders do not fall because they are moving in a circular orbit, and the wall must exert a centripetal force pushing them in toward the center of the circle and causing a centripetal acceleration. Thus, the wall exerts a normal force against the people (see Figure 9.11.). This normal force gives rise to a friction force μN that counteracts gravity. Drawing a free-body diagram for one of the people, as shown in Figure 9.12,

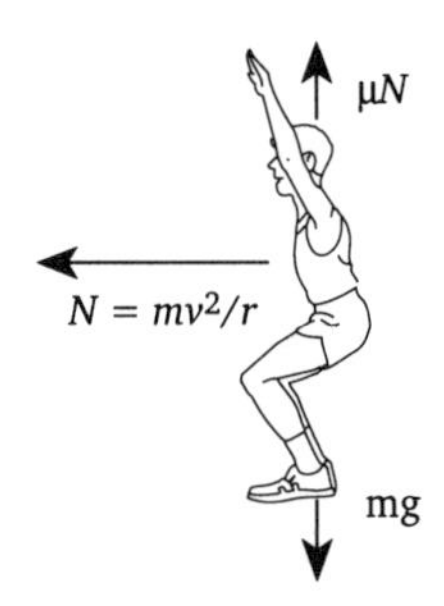

Figure 9.12
Free-body diagram for one of the riders. Remember to draw only the forces acting on the rider and not forces that he rider exerts against the wall.

we see that the wall exerts two forces against the person: a centripetal force in toward the center of the circle, and a tangential frictional force directed up to counteract gravity. The frictional force is significantly less than the normal force because the coefficient of friction is typically less than a half for surfaces that see a lot of wear such as amusement park rides. The net force that the wall exerts upon the person is the vector sum of the normal force and the friction force (see Figure 9.13).

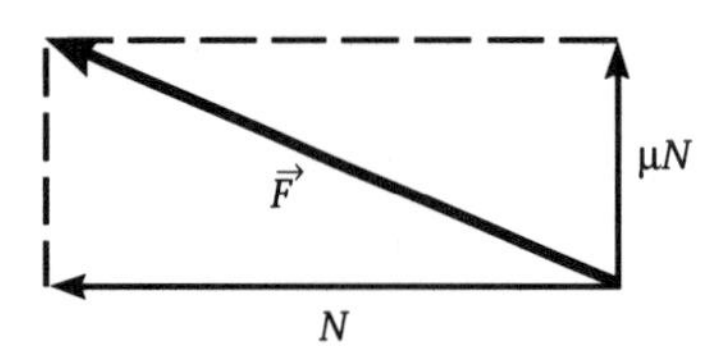

Figure 9.13
The net force that the wall exerts on the person is the vector sum of the normal force and the tangential friction force.

When the ride starts, the riders feels their weight on their feet, but as the speed increases, the riders feel a greater and greater sensation of weight upon their back. When the floor drops, the entire sensation of weight is on the back of the riders, and the thrill begins as the ride defies the riders' common sense reaction that they will fall down.

Chapter 9 Problems

[9.1] You are an engineer for ACME Drilling Co., which is building a new pile-driver drilling machine. The pile driver is a large weight which is accelerated downward where it smashes into a pipe and drives it into the ground. If the pile-driver has a mass of 500 kg and must accelerate at a rate of 25 m/s^2 during the interval before it smashes into the pipe, then what force must the drilling machine exert on the pile-driver during this time?

For Problems 9.2–9.5, draw free body diagrams for each of the labeled masses. Indicate with an arrow each force acting on the object, and label each force; μ means that there is friction between surfaces.

[9.2]

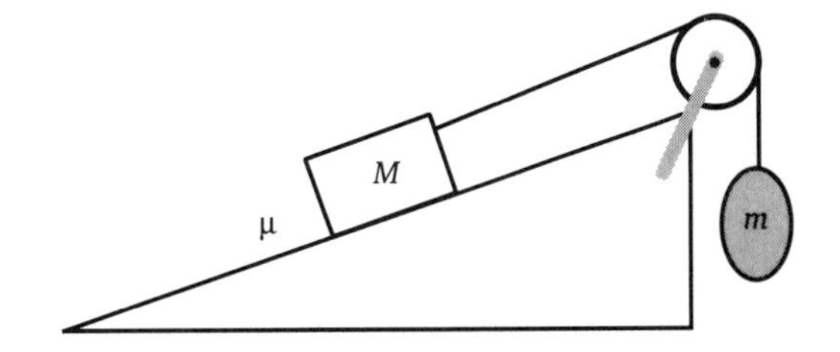

[9.3]

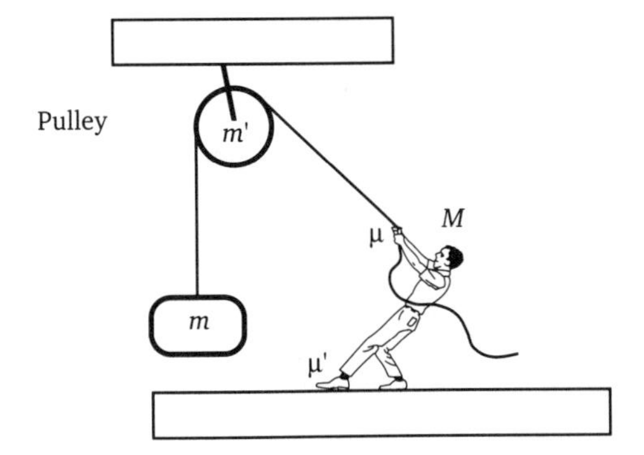

[9.4]

[9.5]

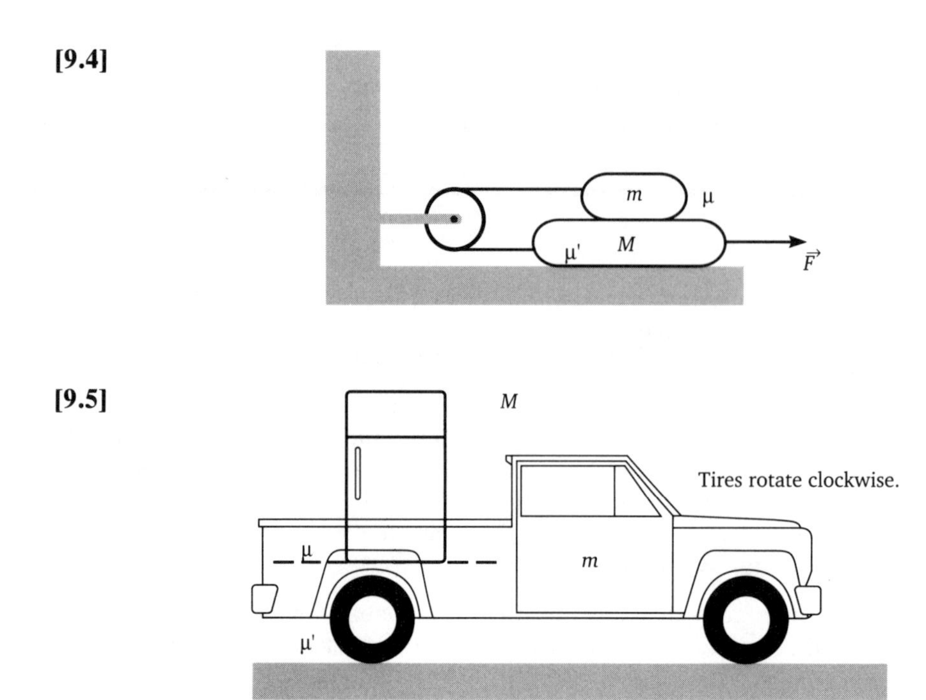

The truck and refrigerator are accelerating at the same rate.

[9.6] Wyle E. Coyote is standing on an overhang at the edge of a cliff. **a)** Draw a free-body diagram for Wyle. Suddenly the overhang on which Wyle is standing breaks free, and both the overhang and Wyle begin to fall freely. **b)** Draw a free-body diagram for Wyle and for the overhang. **c)** From your free-body diagrams, explain why Wyle E. Coyote feels weightless after the cliff overhang breaks free.

•**[9.7]** **a)** Explain where a person on a Ferris wheel would feel heaviest and where the person would feel lightest. Please explain the cause. You may use diagrams to illustrate your point. **b)** If the Ferris wheel is rotating at 1 revolution every 15 seconds and has a radius of 20 m, calculate the force against the Ferris wheel seat that a 70 kg rider exerts at the moment the rider reaches the top of the wheel. **c)** Calculate the force against the seat as the rider reaches the bottom of the wheel.

[9.8] An ice skater is skating at 20 mph. **a)** She falls down onto the ice. Does how hard she hits the ice depend primarily upon her weight or mass? Please explain. **b)** She runs into a tree sticking out of the ice. Does how hard she hits the tree depend primarily upon her weight or her mass? Please explain.

[9.9] An 80 kg person is riding in an elevator at the surface of the earth. How much weight will the person experience when the elevator is accelerating down at 3 m/s^2?

[9.10] A roller coaster starts at rest from point P and travels down the loop-the-loop track, as in Figure 9.14. If the radius of curvature of the roller coaster track at the top of the loop is R, what must the roller coaster's speed be at the top of the loop so that a passenger will just become weightless but will not fall out?

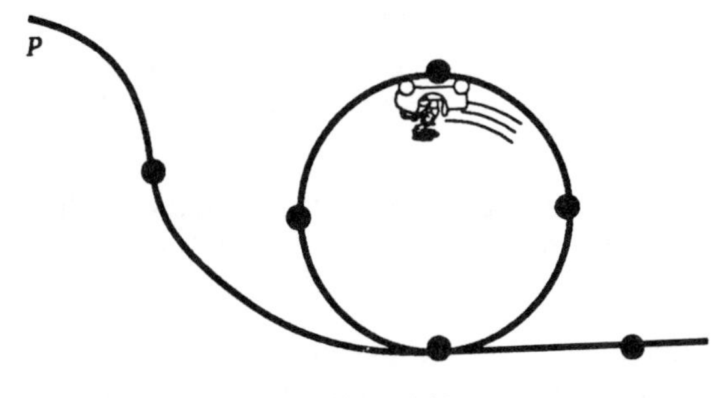

Figure 9.14

[9.11] An airplane is flying in level flight with a constant velocity. Is a person inside the airplane weightless? If so, explain why, if not, explain why not. Draw a free-body diagram for a person sitting in a seat on the airplane.

•**[9.12]** A string of uniform mass per unit length λ and total length L hangs from one end which is tied to a hook in the ceiling. The other end is free. **a)** Find the tension in the string as a function of height h from the bottom of the string. **b)** Find the tension in the string as a function of height if the mass of the string is negligible, but it has a mass M hanging at the bottom. c) What do you conclude about the tension in the massless string for the pulley mass arrangement of Example 9-3?

[9.13] A block is free to slide down a frictionless plane inclined at an angle of θ. **a)** Draw a free-body diagram for the block. **b)** Find the acceleration of the block down the plane in terms of θ and g. **c)** At 45° the plane is halfway to being vertical, but what is the acceleration of the block? Is it $\frac{1}{2}g$? What implications might your result have for skiers and their perception of the steepness of a slope?

•**[9.14]** A car traveling at a high rate of speed, v, comes over the top of a hill whose radius of curvature, r, is small. **a)** Will the occupants of the car tend to feel lighter or heavier than if the car is traveling along a straight path? **b)** Draw a free-body diagram for a person in the car. **c)** Sum the vertical forces on the person to show that the normal force is given by $N = mg - m\frac{v^2}{r}$

d) Under what condition would an occupant of the car become weightless? e) To train astronauts, NASA flies an airplane along a path that produces weightlessness for about 2–3 minutes. What type of path should the airplane follow?

•**[9.15]** At a given moment in time, a pendulum of length L makes an angle θ with the vertical and is traveling with a tangential speed of v, as shown in Figure 9.15. At this moment in time, **a)** Draw a free-body diagram for the pendulum. **b)** Construct equations for the sum of the tangential forces and the sum of the radial forces. **c)** Find the tangential and radial accelerations of the pendulum at the given moment in time.

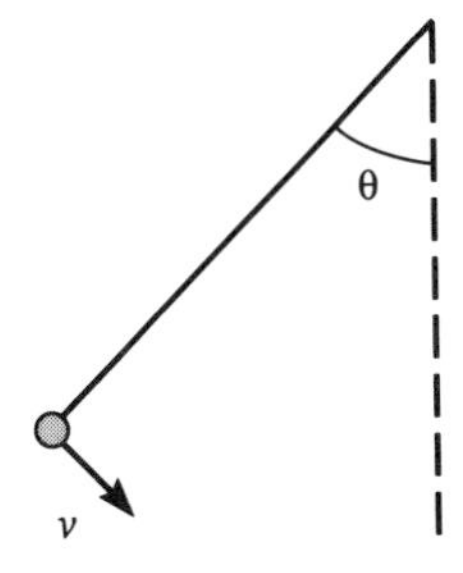

Figure 9.15
A simple pendulum at a given instant in time.

•**[9.16]** At a given moment, a pendulum hangs vertically but is swinging through its equilibrium point with a tangential speed v. **a)** Draw a free-body diagram for the pendulum at the given moment. **b)** Sum the vertical forces. Is the tension in the string greater or less than if the pendulum were not swinging but just hanging vertically?

•**[9.17]** Wyle E. Coyote inadvertently steps off a cliff and starts to fall through the air. If his mass is m and the frictional force is $\vec{f} = -b\vec{v}$, then find the Coyote's terminal speed. Start by drawing a free-body diagram.

•**[9.18]** The truck in problem 9.5 accelerates forward with a uniform acceleration. If the coefficient of friction between the refrigerator and the truck bed is 0.6, then at what acceleration will the refrigerator just start to slip?

CHAPTER 10
Vector Projections

(Live long and prosper.)

In Chapter 7, we set about constructing a vector algebra, and we set the formal rules for vector addition and subtraction, trying all the while to have as many properties of scalar algebra carry over as possible. In this chapter, we discuss other operations with vectors, such as changing the length of a vector, the projection of one vector parallel to and perpendicular to another vector, and we will give examples of how each is useful. In formulas in which one vector quantity multiplies another, you must generally specify how the vectors relate to one another. This involves projecting one vector relative to another. Because there are many situations in physics that require the relationship between two vectors, the projection of vectors is a very important and often used topic, but it can be very confusing. For instance, work involves the projection of the force vector along the displacement vector (see Section 10.4), and the force of a magnetic field on a moving charged object depends on the projection of the velocity perpendicular to the magnetic field that is the reason charged particles tend to spiral around magnetic field lines. This chapter provides you with some practice in this difficult subject of vector projections.

10.1 MULTIPLICATION OF A VECTOR BY A SCALAR

To multiply a vector $\vec{A}$ by a scalar n we can simply use the distributive property. In two dimensions

$$n\vec{A} = nA_x\hat{i} + nA_y\hat{j} \tag{10-1}$$

Note that if n is positive, the product only changes the magnitude of the vector by a factor of n

$$\text{n}\ |\vec{A}| = \sqrt{(nA_x)^2+(nA_y)^2} \tag{10-2}$$

and not the direction.

$$\theta = \tan^{-1}\frac{nA_y}{nA_x} = \tan^{-1}\frac{A_y}{A_x}. \tag{10-3}$$

This multiplication changes the length of the vector and is thus referred to as a change of scale. If n is negative, it flips the direction of A by 180° as well as changing the scale as represented in Figure 10.1.

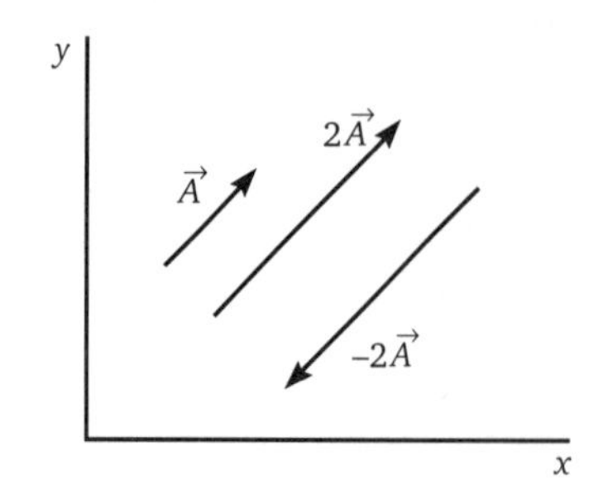

Figure 10.1
Multiplying a vector by a positive scalar changes the length but not the direction. Multiplying the vector by a negative scalar changes the length and flips the direction.

10.2 PROJECTIONS PARALLEL TO A VECTOR

Often when considering two distinct vectors, the relationship between the direction of the vectors is important. In this spirit we define the projection of the vector $\vec{B}$ along $\vec{A}$ to be the component of $\vec{B}$ that lies parallel to $\vec{A}$, $B_{\parallel}$. To find the projection of $\vec{B}$ parallel to $\vec{A}$, first place the tails of the two vectors together, and next drop a perpendicular onto $\vec{A}$ from the tip of $\vec{B}$ as shown in Figure 10.2. Letting θ be the angle between the two vectors, $B_{\parallel} = B \cos \theta$.

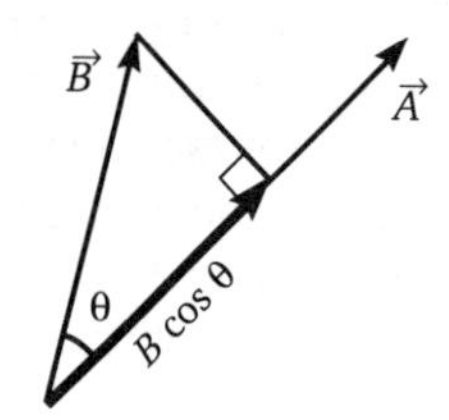

Figure 10.2
The projection of $\vec{B}$ parallel to $\vec{A}$ is $B \cos \theta$.

EXAMPLE 10-1

Find $B_{\parallel}$ for the two vectors

$$\vec{A} = 3\,\hat{i} + 5\,\hat{j}$$
$$\vec{B} = 2\hat{i} + \hat{j}.$$

Solution:
First find the angle between the two vectors. The angle $\vec{A}$ makes with the x-axis is

$$\theta_A = \tan^{-1}\tfrac{5}{3} = 59°$$

and for $\vec{B}$

$$\theta_B = \tan^{-1}\tfrac{1}{2} = 26.6°.$$

The difference is the angle between them

$$\theta = \theta_A - \theta_B = 32.4°$$

$$B_{\parallel} = B \cos \theta = \left(\sqrt{2^2 + 1^2}\right) \cos 32.4°$$

$$B_{\parallel} = 1.9$$

When multiplying two vectors, we must decide which components are to multiply which. There are two schemes that are particularly useful. One scheme involves finding the projection of one vector along another, and then multiplying the first vector with this parallel projection of the other, i.e.,

$$\vec{A} \cdot \vec{B} = AB_{\parallel} \tag{10-4}$$

which is termed the *dot product* or *scalar product.* The dot $(\cdot)$ is used to represent this product and the resultant is a scalar, i.e., it is just a number. This multiplication of vectors occurs many places in physics, sometimes so subtly that it goes unnoticed. For instance, in the equation for kinetic energy $K = \frac{1}{2}mv^2$, the velocity is a vector. v^2 means multiplying the vector v by itself or taking the projection of the vector v along itself. Because the angle between v and v is 0, we have

$$v_{\parallel} = v \cos 0 = v. \tag{10-5}$$

The multiplication of a vector by itself, however, is not always trivial. For instance, consider the case of two arbitrary vectors $\vec{A}$ and $\vec{B}$ that have an angle θ between them when their tails are placed together as shown in Figure 10.3. If we consider the vector $\vec{C}$ formed by their subtraction

$$\vec{C} = \vec{A} - \vec{B}, \tag{10-6}$$

then the length of $\vec{C}$ depends on how much of $\vec{B}$ is parallel to $\vec{A}$. The more that $\vec{B}$ is parallel to $\vec{A}$, the less is the length of $\vec{C}$, as shown by comparing the large θ in Figure 10.3(a) with the smaller θ shown in 10.3(b).

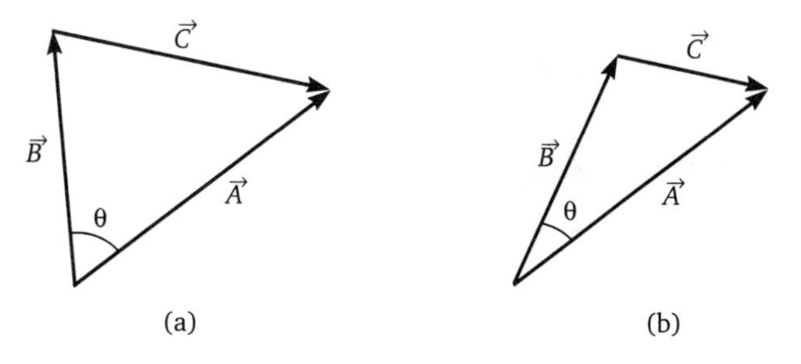

Figure 10.3
The vector $\vec{B}$ has a greater projection parallel to $\vec{A}$ in (b) than in (a)

This effect we would expect from the law of cosines, $C^2 = A^2 + B^2 - 2\,AB \cos \theta$, but from where does the law of cosines come? We can easily tackle this trigonometry problem in the following example.

EXAMPLE 10-2

A vector $\vec{C}$ is given by $\vec{C} = \vec{A} - \vec{B}$. Use the dot product to show $C^2 = A^2 + B^2 - 2\,AB\cos\theta$.

Solution:

The vectors are shown in Figure 10.3. In particular pay attention to the given angle θ which is the angle between $\vec{A}$ and $\vec{B}$. C^2 is the projection of C along itself so that

$$C^2 = \vec{C} \cdot \vec{C};$$

however, $\vec{C}$ is comprised of two other vectors

$$C^2 = (\vec{A} - \vec{B}) \cdot (\vec{A} - \vec{B})$$

where we are taking the projection of $(\vec{A} - \vec{B})$ parallel to itself. In multiplying the terms, A is parallel to itself making A^2 trivial and likewise for B, but we must use the projection of B along A for the cross terms.

$$C^2 = A^2 - 2\,\vec{A} \cdot \vec{B} + B^2$$

$$C^2 = A^2 - 2\,AB_{\parallel} + B^2$$

Using $B_{\parallel} = B\cos\theta$,

$$C^2 = A^2 + B^2 - 2\,AB\cos\theta.$$

This result is the law of cosines.

The Dot Product

$$\vec{A} \cdot \vec{B} = AB_{\parallel} = AB\cos\theta \tag{10-7}$$

The dot product has the simplest rule: multipy the length of A by the parallel projection of B. However, the dot product is symmetric in that it also contains the projection of $\vec{A}$ along $\vec{B}$ because the amount of $\vec{A}$ parallel to $\vec{B}$ is

$$A_{\parallel} = |\vec{A}|\cos\theta \tag{10-8}$$

as shown in Figure 10.4.

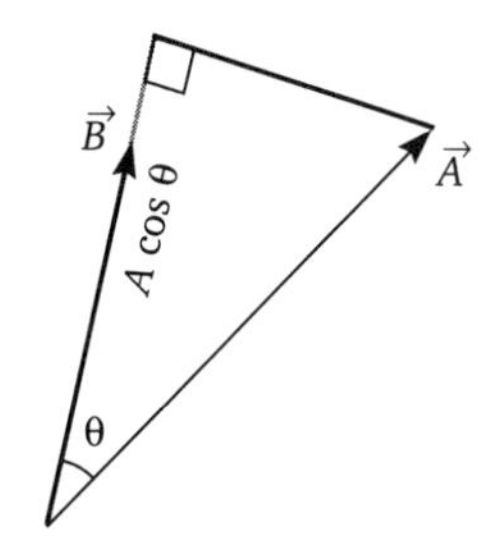

Figure 10.4
The projection of *A* along *B*.

Because

$$A(B \cos \theta) = (A \cos \theta)B, \tag{10-9}$$

$$AB_{\parallel} = A_{\parallel}B \tag{10-10}$$

The dot product contains both the projections of $\vec{A}$ along $\vec{B}$ and of $\vec{B}$ along $\vec{A}$. We can use this symmetric property of the dot product to our advantage in Section 10.4 when we discuss the work done by a force.

10.3 PROJECTIONS PERPENDICULAR TO A VECTOR

In the last section, we found a vector product that uses the projection of one vector onto another. Another useful scheme for multiplying vectors is to multiply one vector by the perpendicular projection of the other vector. The projection of B perpendicular to A, $B_{\perp}$ is shown in Figure 10.5 where $B \sin \theta$ is the projection of $\vec{B}$ perpendicular to $\vec{A}$.

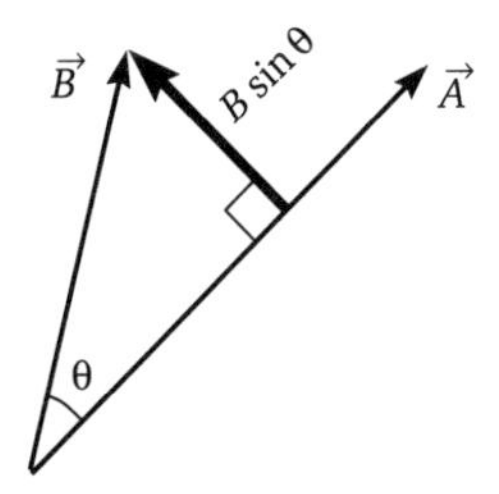

Figure 10.5
The projection of *B* perpendicular to *A*.

We will call this product the *cross product* or *vector product* and use the cross-multiplication symbol, $\vec{A} \times \vec{B}$, to designate it. The resultant is a vector. The magnitude of this product will be given by

$$|\vec{A} \times \vec{B}| = AB \sin \theta \tag{10-11}$$

where $B \sin \theta$ is the projection of B perpendicular to A, and θ is the smallest angle between the vectors.

The Cross Product

$$|\vec{A} \times \vec{B}| = AB_{\perp} = AB \sin \theta \tag{10-12}$$

The cross product is more complicated than the dot product because there is also a direction defined by the right-hand rule, as shown in Figure 10.6.

The Right-Hand Rule

Lay the fingers of your right hand along the first vector in the product (for $\vec{A} \times \vec{B}$, along $\vec{A}$). Orient your hand such that the palm faces the second vector in the product, and your fingers can cross the first vector into the second. Your thumb will point in the direction of the cross product.

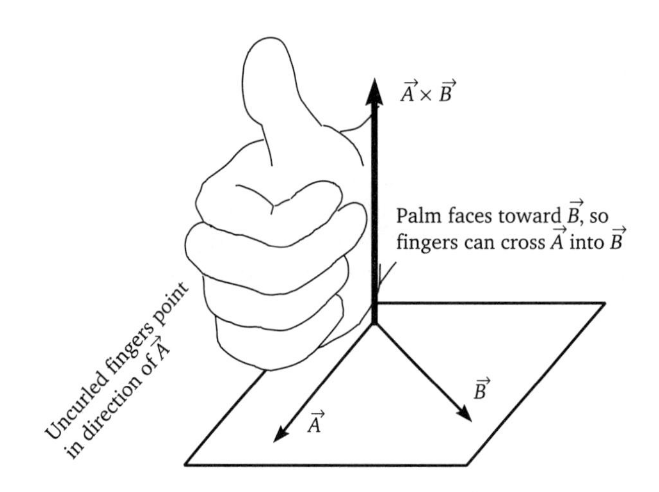

Figure 10.6
The right-hand rule for crossing two vectors.
$\vec{A} \times \vec{B}$ is perpendicular to the plane containing *A*, and *B*,.

The right-hand rule preserves the sense of rotation. If $\vec{A}$ is crossed into $\vec{B}$ as in Figure 10.7(a), the angle is counter-clockwise and positive, and the thumb points up in the positive *z*-direction. This crossing of $\vec{A}$ into $\vec{B}$ is like a rotation of $\vec{A}$ into $\vec{B}$. However, if $\vec{B}$ is crossed into $\vec{A}$ as in Figure 10.7(b), then the angle is negative, and the thumb points down. So we get a sense of rotation of $\vec{B}$ into $\vec{A}$ that is opposite that of $\vec{A}$ into $\vec{B}$. This sense of rotation comes from the anti-commutation property of the cross product,

$$\vec{A} \times \vec{B} = -\vec{B} \times \vec{A}. \tag{10-13}$$

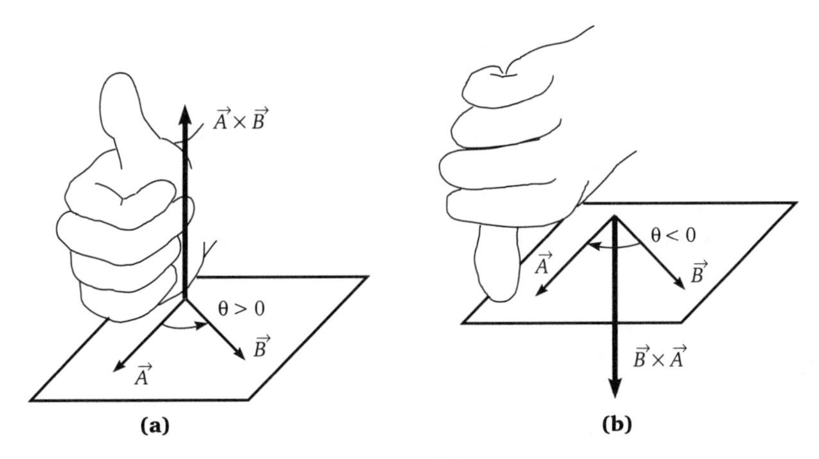

Figure 10.7
The cross product has a sense of rotation because it anticommutes.

The Cartesian coordinate system is a right-handed system in that the projection of $\hat{j}$ perpendicular to $\hat{i}$ has a magnitude of 1, producing a unit vector, and the direction is in the $\hat{k}$ direction. $\hat{i} \times \hat{j}$ forms a unit vector, $\hat{k}$, perpendicular to both $\hat{i}$ and $\hat{j}$:

$$\hat{i} \times \hat{j} = \hat{k}. \tag{10-14}$$

The major result of the right-hand rule is that the cross product between any two vectors always lays perpendicular to the plane containing those two vectors.

EXAMPLE 10-3

Vector $\vec{B}$ has a length of 3 m, makes an angle of 36.9° with the positive x-axis, and vector $\vec{A}$ has a length of 6 m and lies on the y-axis. Find the magnitude and direction of $\vec{A} \times \vec{B}$.

Solution:

First, we must see where the two vectors lie, so draw a picture of the vectors as in Figure 10.8. It will help us to avoid mistakes. $\vec{A}$ lies along the $\hat{j}$ axis and $\vec{B}$ makes an angle of 36.9° with the x-axis.

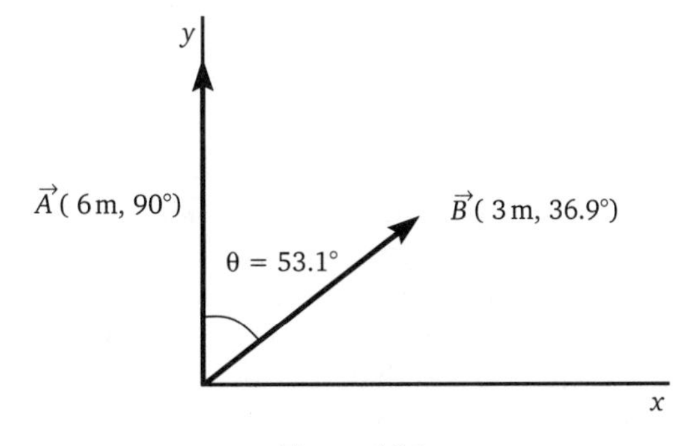

Figure 10.8

Use $|\vec{A} \times \vec{B}| = AB \sin\theta$ to get the magnitude, and then use the right-hand rule to get the direction. If $\vec{A}$ is along the $\hat{j}$ axis, and $\vec{B}$ makes an angle of 36.9° with the x-axis, then the angle between the vectors is $\theta = 90° - 36.9° = 53.1°$.

$$|\vec{A} \times \vec{B}| = (6\text{ m})(3\text{ m}) \sin 53.1° = 14.4\text{ m}^2 \tag{10-15}$$

We next need to find the direction. $\hat{i} \times \hat{j} = \hat{k}$ so $\hat{k}$ must be coming out of the page (fingers along $\hat{i}$, palm facing $\hat{j}$, so your fingers cross $\hat{i}$ into $\hat{j}$, and your thumb will point out of the page.) $\vec{A}$ cross $\vec{B}$ is opposite to $\hat{i}$ cross $\hat{j}$, so $\vec{A}$ cross $\vec{B}$ points into the page in the $(-z)$-direction.

10.4 WORK AND KINETIC ENERGY

The ideas of the parallel and perpendicular projections of vectors are central to our description of vector fields, and indeed, lie at the heart of the theory. We should, thus, spend a few moments elaborating upon these ideas. A very important quantity in mechanics is kinetic energy, K, given by

$$K = \tfrac{1}{2}mv^2 \tag{10-16}$$

where m is mass and v is the speed. As mentioned before, normally, you will just plug the speed in for v and not think about the vector nature, but there are certain times when you will need to refer to v as a vector. Then, you will need to treat v^2 as the square of a vector, i.e., take the projection of v with itself. In fact, the vector nature of velocity and displacement is central to understanding kinetic energy and results in a very important theorem known as the work-energy theorem.

Work is a word you are already familiar with in some respects, but the definition in physics is not quite the same as the common usage. The basic idea behind work is that if you must place a net force on an object to displace the object, then you do work on the object. The work done is the net force times the net displacement.

$$W = F\Delta r \tag{10-17}$$

Of course, it's not this simple. As you may have guessed, both the force and the displacement are vectors. It is only the component of the force along the displacement that does work. If the force does not cause the object to move, then there is no work done. The overlap between the two vectors is given by

$$W = F_{\parallel}\Delta r \tag{10-18}$$

This may at first seem strange since if I asked you to hold this book at arm's length for ten minutes, you become exceedingly tired and would think you were doing a lot of work; however, you would be doing no work on the book by our definition because the book is not moving. In order to do work, something must move. In holding the book, your muscles do work. The muscles in your arm are composed of cells that can contract only for an instant, and then they must relax. These striated muscles are perfect for moving, but not for maintaining a steady force. At any instant, only some will be contracted so that the average effect of all the muscles acting over time is a steady force. After a while, waste chemicals build up in the cells, and they don't contract as effectively. As the firing of the muscles becomes erratic, it becomes harder to maintain a steady force, and your arm may start to shake. The point is that the muscles are moving, changing their shape, and hence, doing work, but work is not being done on the book. It is stationary. We could just as well replace you by a table that would expend no energy. The work done by the muscles is dissipated as heat causing you to sweat.

The unit of work is the Joule, named after James Prescott Joule who did many experiments relating mechanical energy and heat.

$$[W] = (\text{Newton}) \cdot (\text{meter}) = \text{Joule}$$

EXAMPLE 10-4

A worker walks at constant velocity up an inclined plane while carrying a 5.0 kg crate. The plane is inclined at 36.9° from the horizontal and has a length of 10.0 m. How much work does the worker do?

Solution:

To find the work, we need to find the force the worker exerts on the crate. To analyze forces, always start with a free-body diagram as shown in Figure 10.9. Since the crate is not accelerating, the forces must sum to zero. Gravity pulls down, so the worker must counter with a force straight up.

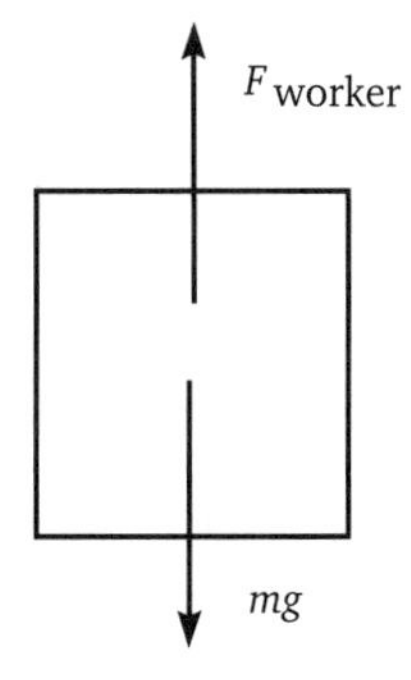

Figure 10.9
Free-body diagram for the crate.

The force the worker exerts is given by

$$\Sigma F_y = F_w - mg = 0$$

$$F_w = mg$$

This force points straight up. The dot product commutes so it doesn't matter if we take the projection of the force along the displacement or the projection of the displacement along the force. The projection of the displacement along the force is just the vertical height the worker lifts the crate, determined in Figure 10.10.

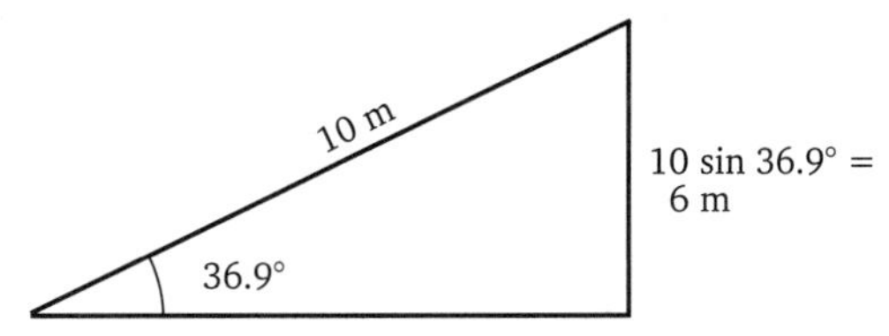

Figure 10.10
The worker lifts the crate through a vertical height of 6 m.

The work done is then

$$W = F_y \, \Delta y = mg \, \Delta y$$

$$W = (5 \text{ kg})(9.8 \text{ m/s}^2)(6 \text{ m}) = 294 \text{ J}$$

The work–energy approach is equivalent to using Newton's laws. In Chapter 9 we saw that the net force determined the acceleration. Once found, the acceleration can be used to find the velocity and position. An alternative method is to use work and energy to find the velocity. This method is based on a very powerful theorem called the "work–energy" theorem. To understand this theorem, we start with Newton's second law.

$$\Sigma\vec{F} = m\frac{\Delta\vec{v}}{\Delta t} \tag{10-19}$$

The net work done by these forces over a displacement $\Delta\vec{r}$ is

$$W = \Sigma\vec{F}\cdot\Delta\vec{r} \tag{10-20}$$

From Newton's second law,

$$W = m\frac{\Delta\vec{v}}{\Delta t}\cdot\Delta\vec{r} \tag{10-21}$$

The displacement $\Delta\vec{r}$ happens during the same time that the velocity changes, so we can move the Δt over to the $\Delta\vec{r}$:

$$W = m\,\Delta\vec{v}\frac{\Delta\vec{r}}{\Delta t} = m\,\Delta\vec{v}\cdot\vec{v} \tag{10-22}$$

where $\Delta\vec{r}/\Delta t$ is the velocity $\vec{v}$. The next step is subtle. $\Delta\vec{v}$ is simple because it is just the change in velocity $\Delta\vec{v} = \vec{v} - \vec{v}_o$, but the $\vec{v}$ is not constant. It starts with $\vec{v}_o$ and finishes with $\vec{v}$, so what $\vec{v}$ should we use? We confronted this problem before in Chapter 2 in the equation $x = x_o + vt$. Then we decided that v was the average velocity. Here too, $\vec{v}$ must represent the average velocity $\vec{v} = \frac{1}{2}(\vec{v} + \vec{v}_o)$

$$W = m(\vec{v} - \vec{v}_o)\cdot\tfrac{1}{2}(\vec{v} + \vec{v}_o) \tag{10-23}$$

This time there are no cross terms between v and v_o, so we are simply left with

$$W = \tfrac{1}{2}mv^2 - \tfrac{1}{2}mv_o^{\,2} \tag{10-24}$$

The $\frac{1}{2}mv^2$ is called the *kinetic energy*. This is an amazing result. What has happened is that we have been able to relate the work to the change in kinetic energy that does not directly involve the forces at all. This is the work energy theorem. The work done by the net force equals the change in the kinetic energy. The net force is the sum of *all* the forces acting on an object, $\Sigma\vec{F}$.

$$W_{net} = \Delta K. \tag{10-25}$$

EXAMPLE 10-5

Find the speed of a ball after it has freely fallen through a height h if it starts from rest.

Solution:

Because we are looking for the speed, we should try the work energy theorem first. To apply this theorem we need the net force on the ball, so we can calculate $W_{net} = \Sigma\vec{F} \cdot \Delta\vec{r}$. For a ball in free-fall, the only force on it is due to gravity

$$\vec{F} = mg\,\hat{j}$$

Next, we need the displacement. It occurs in the negative $\hat{j}$ direction

$$\Delta\vec{r} = -\,h\hat{j}$$

$$W_{net} = mgh$$

By the work–energy theorem,

$$mgh = \tfrac{1}{2}mv^2$$

Solving for v

$$v = \sqrt{2gh}.$$

10.5 DIRECTED SURFACE AREA AND FLUX

Normally, we think of the area of a rectangle as the length times the width. We can generalize this to any parallelogram by forming its sides from the two vectors $\vec{L}$ and $\vec{W}$ (length and width) as shown in Figure 10.11.

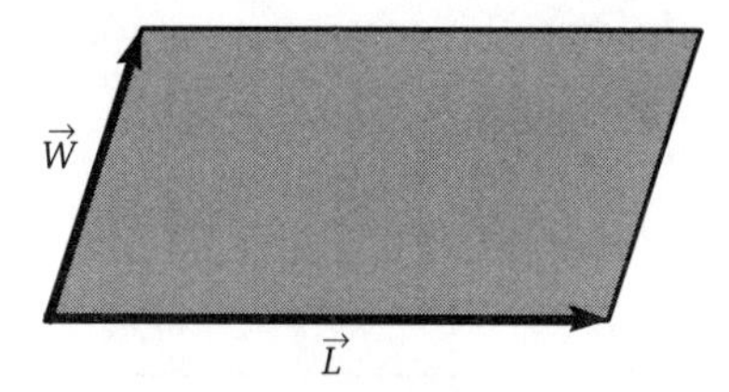

Figure 10.11
A parallelogram formed by two vectors $\vec{L}$ and $\vec{W}$.

To find the area of this parallelogram we need to multiply $\vec{L}$ and $\vec{W}$, but how should we do this? If we use the dot product, we will not obtain an area, because if $\vec{L}$ and $\vec{W}$ are parallel, they confine zero area. However, the dot product is not zero, and instead, yields its maximum value.

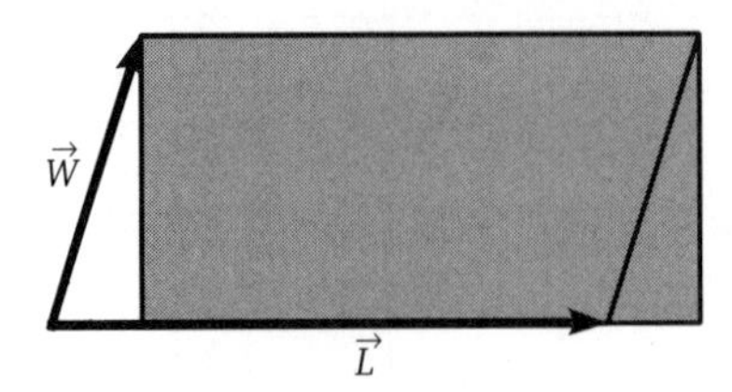

Figure 10.12
The area of the shaded rectangle is the same as the area of the parallelogram.

From the geometrical construction in Figure 10.12, the area of the shaded rectangle is the same as the area of the parallelogram, because the area of the shaded triangle is the same as that of the unshaded triangle. Thus, we find that it is the amount of $\vec{W}$ which is perpendicular to $\vec{L}$ that contributes to the area, and the area of the parallelogram is given by the cross product

$$A = LW_{\perp} \tag{10-26}$$

but the cross product also has a direction. What can this direction possibly mean?

$$\vec{A} = \vec{L} \times \vec{W}. \tag{10-27}$$

Note that this vector area is perpendicular to the plane containing $\vec{L}$ and $\vec{W}$, and thus, is perpendicular to the original surface! Certainly, it seems absurd to give a direction to an area; however, it is just this direction that makes the vector area useful.

For example, suppose we have a situation where the orientation of the parallelogram is important. Perhaps there is fluid flowing through the surface. The amount of fluid that flows through the parallelogram depends upon the orientation of the surface relative to the flow. If the plane of the surface is parallel to the flow, obviously, no fluid will flow through the surface.

On the other hand, if the plane of the parallelogram is perpendicular to the flow, we will see the maximum amount of fluid flowing through the surface per unit time. The *flux* is the amount of fluid flowing per unit time through the surface. Figure 10.13(a) has zero flux, and 10.13(b) has the maximum amount of flux

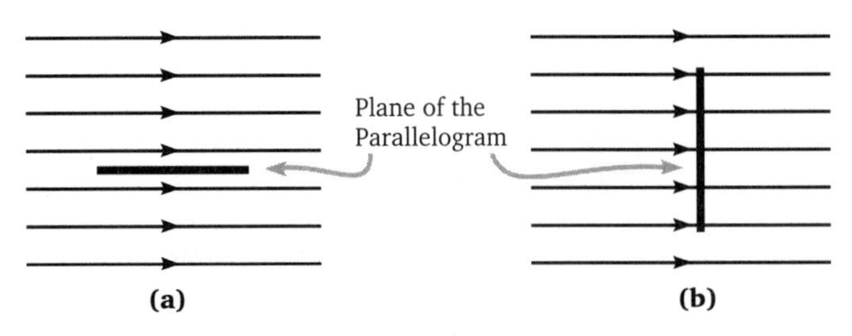

Figure 10.13
The flux through the parallelogram in (a) is zero, and the flux through the parallelogram in (b) is the maximum possible for this flow.

possible for that surface. Intuitively, the flux should depend on the speed of the flow and the area of the surface. Consider a surface of area A shown in Figure 10.14.

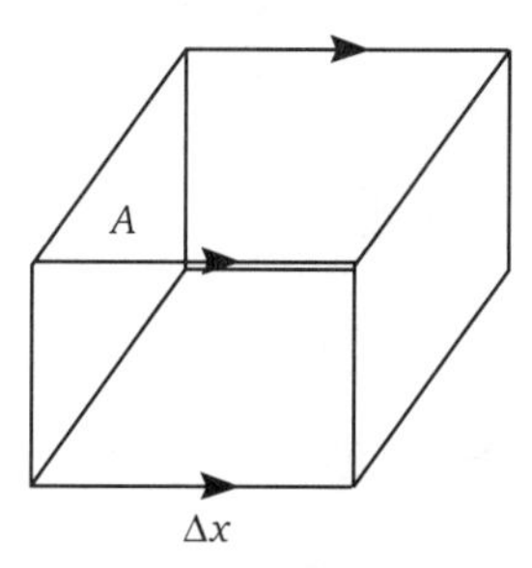

Figure 10.14
The fluid will sweep out a volume $A\,\Delta x$ in time Δt.

In a time Δt, the fluid flowing through the surface A will have moved a distance Δx and has a volume of $\Delta V = A\Delta x$. We can multiply and divide by the volume $A\Delta x$ to write the flux as:

$$\left.\frac{\Delta m}{\Delta t}\right|_{\text{passing through } A} = \frac{\Delta m}{A\Delta x}\frac{\Delta x}{\Delta t}A \tag{10-28}$$

The first factor is the density. If the fluid flows a distance Δx in a time Δt, then the second factor must be the velocity.

$$\left.\frac{\Delta m}{\Delta t}\right|_{A} = \rho v A \tag{10-29}$$

We have found, the denser the fluid, the greater the flux; the faster the flow, the greater the flux; and the larger the surface, the greater the flux.

If we tip the surface relative to the flow, then only the part of the surface that is perpendicular to the flow will contribute to the flux, as shown in Figure 10.15. The projection of the surface perpendicular to the flow has area $A_{\perp} = A\cos\theta$. The flux becomes

$$\left.\frac{\Delta m}{\Delta t}\right|_{A} = \rho v A\cos\theta \tag{10-30}$$

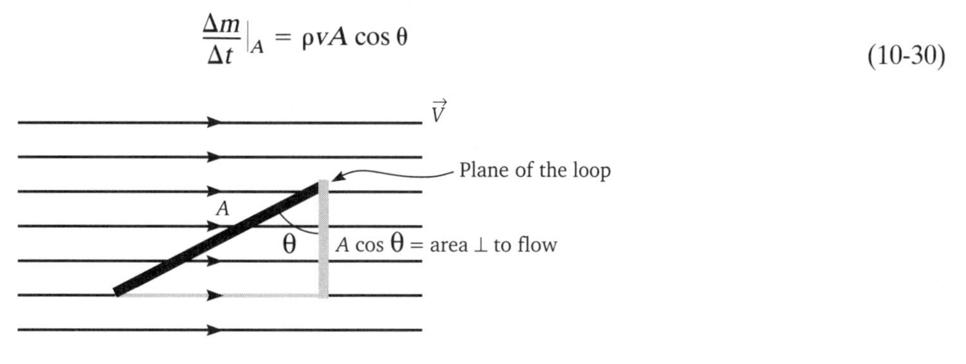

Figure 10.15

v is actually a vector, so we need to describe the orientation of A to v. We can use the cross product to describe the orientation of A relative to the flow in a coordinate-free manner, i.e., without reference to a particular coordinate system. Then the orientation of A is given by the vector normal to the surface

$$\vec{A} = \vec{L} \times \vec{W} = LW\hat{n} \tag{10-31}$$

where $\hat{n}$ is the unit normal to the surface as shown in Figure 10.16.

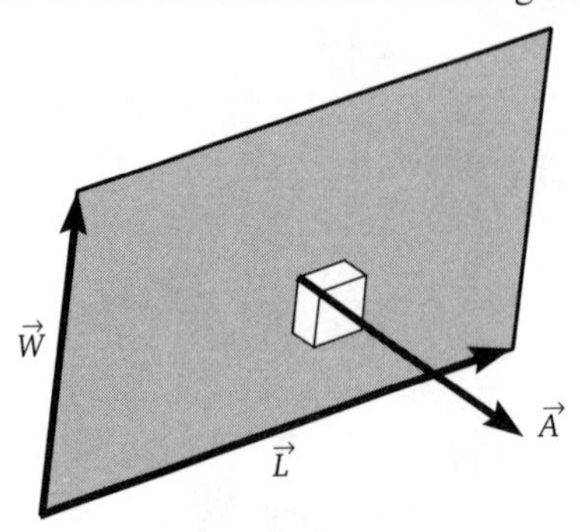

Figure 10.16
The area vector uniquely determines the orientation of the plane no matter how we rotate the plane.

Since the vector $\vec{A}$ is perpendicular to the surface, the maximum flux occurs when $\vec{A}$ is parallel to the velocity of the flow as shown in Figure 10.13(b), i.e., the flux will be a maximum when the area vector for the surface overlaps the velocity vector for the flow (see Figure 10.17). When $\vec{A}$ is parallel to $\vec{v}$, the surface will be perpendicular to the flow. Mathematically we use the dot product to state this:

$$\vec{v} \bullet \vec{A} = vA \cos \theta \tag{10-32}$$

$$\frac{\Delta m}{\Delta t}\Big|_A = \rho\vec{v} \cdot \vec{A} \tag{10-33}$$

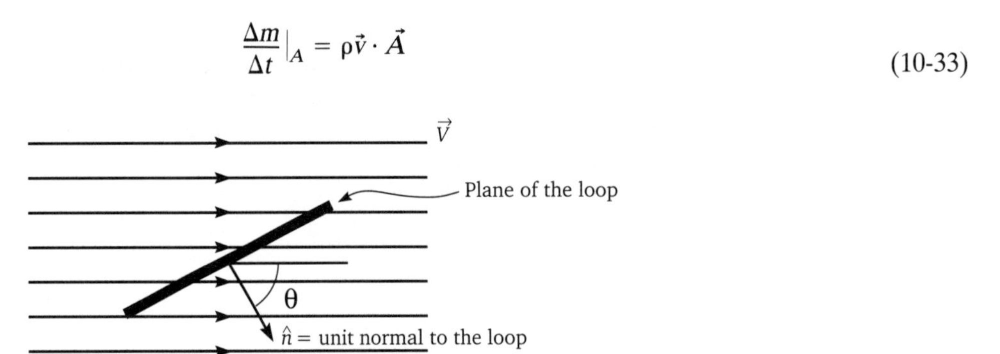

Figure 10.17
The overlap of the area $\vec{A}$ with the velocity $\vec{v}$ gives the flux through the loop.

EXAMPLE 10-6

A pipe necks down to a smaller pipe of half the radius as shown in Figure 10.18. If the speed of the fluid is v in the first length of the pipe, then find the speed of the fluid in the second part of the pipe. Assume the density of the fluid remains constant.

Solution:
With flux, you must visualize the surface through which the flow passes, otherwise you will go nowhere. The flow is along the axis of the pipe, so the flow is through the πr^2 cross-sectional area of the pipe.

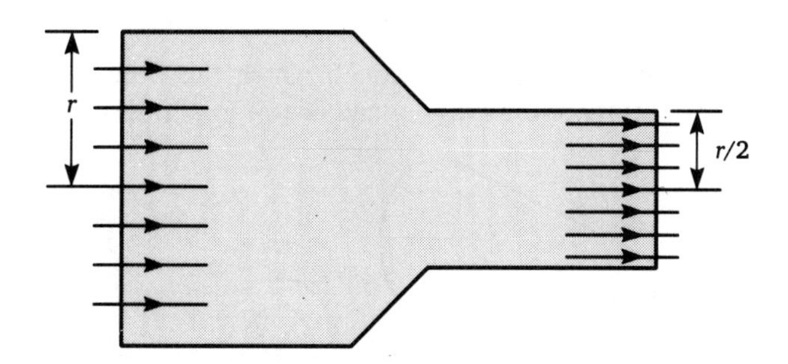

Figure 10.18
A pipe that necks down to half the original radius.

The flux follows from Equation (10-33)

$$\frac{\Delta m}{\Delta t}\Big|_A = \rho\, vA$$

Because we cannot create or destroy fluid in the pipe, what flux we have through the large pipe must also pass through the small pipe, i.e., if 5 kg/s is flowing out of the large pipe, then 5 kg/s better be flowing through the small pipe. In order to have the flux the same, the speed must increase in the small pipe.

$$\rho v_1 A_1 = \rho v_2 A_2 \tag{10-34}$$

$$v_2 = \frac{A_1}{A_2} v_1$$

Because $A = \pi r^2$,

$$v_2 = \frac{r_1^{\,2}}{r_2^{\,2}} v_1$$

$$v_2 = \frac{r^2}{(r/2)^2} v = 4v.$$

Thus, we have indeed found that the speed must increase by a factor of 4 in the smaller pipe.

EXAMPLE 10-7

A garden soaker hose is 30 m long and has an inside radius of 8 mm and an outer radius of 12 mm. A soaker hose, made of porous material, allows the water to slowly flow radially outwards along its entire length. If the velocity of the flow is 1 mm/s at the inner surface, **a)** What is the flux through the inner surface? **b)** What is the flux through the outer surface? **c)** What is the speed of the flow at the outer surface?

Solution:
Again, you must first visualize the surface through which the flow passes. The flow is radially outward, so we can draw a picture of the cylinder and draw radial flow lines as in Figure 10.19(a).

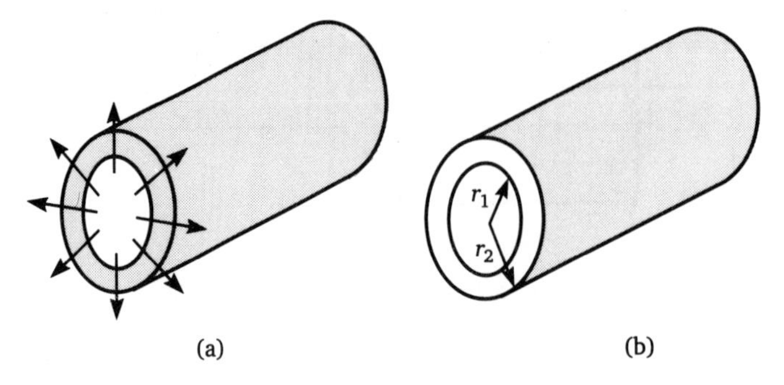

Figure 10.19
(a) The flow is radially outward. (b) The surface perpendicular to the flow has an area of $2\pi rL$.

The surface that is perpendicular to the flow is the $2\pi rL$ shaded cylindrical surface shown in Figure 10.19(b). It is not the πr^2 cross-sectional area. When using flux, it is very important to see the surface that is perpendicular to the flow. Visualizing the surface can save you from making major mistakes.

$$A = 2\pi rL.$$

The flux is given by Equation (10-33)

$$\frac{\Delta m}{\Delta t}\bigg|_A = \rho\, vA$$

where the unit normal to the surface is always parallel to the velocity so

$$\vec{v} \cdot \vec{A} = |\vec{v}|\,|\vec{A}| \cos 0° = v(2\pi rL)$$

and the flux is

$$\frac{\Delta m}{\Delta t}\bigg|_A = \rho\, v 2\pi rL$$

a) At the inner radius, the flux is

$$\frac{\Delta m}{\Delta t}\bigg|_{r_1} = (1000 \text{ kg/m}^3)(0.001 \text{ m/s})(2\pi)(0.008 \text{ m})(30 \text{ m}) = 1.5 \text{ kg/s}$$

b) Just as in Example 10-6, the water cannot be created or destroyed, so the amount flowing through the inner surface must be equal to the amount flowing through the outer surface, i.e., if 1.5 kg/s flows through the surface at r_1, then 1.5 kg/s must flow through the surface at r_2.

$$\frac{\Delta m}{\Delta t}\bigg|_{r_1} = \frac{\Delta m}{\Delta t}\bigg|_{r_2} = 1.5 \text{ kg/s}$$

c) Because we know the flux through the outer surface,

$$1.5 \text{ kg/s} = \rho\, v_2 2\pi r_2 L$$

we can solve for v_2

$$v_2 = \frac{1.5 \text{ kg/s}}{(1000 \text{ kg/m}^3)2\pi(0.012 \text{ m})(30 \text{ m})} = 6.6 \times 10^{-4} \text{ m/s} = 0.66 \text{ mm/s}$$

Therefore, because the 1.5 kg/s is passing through a larger surface area at r_2 than at r_1 the flow slows down. This slowing down is the inverse of what happened in Example 10-6.

10.6 THE FINAL WORD (REALLY)

In this book we have studied many topics that have sharpened your skills, beginning with Fermi Questions and basic problem solving techniques and ending with the sophisticated topics of vector products. The word "skills" is important. We have not covered the vast content of science knowledge, but we have concentrated on developing the skills that will allow you to successfully begin that task. Some of these skills you may need immediately in your physics course, and others you may not need so soon. Even so, this is just the beginning. There are skills you will acquire in the future that are not even hinted at here. This is not a bad thing, rather it is part of the journey. Along any journey, you grow and change, emerging from the journey a different person than you were when you began. But at the end there will always be more that you don't understand. This is a natural part of the process of life. It is a wonder in itself. Do not think that you are alone in these feelings; the final word I leave you with is Newton's.

I do not know what I may appear to the world; but to myself I seem to have been only like a boy playing on the seashore, and diverting myself in now and then finding a smoother pebble or a prettier shell than ordinary, while the great ocean of truth lay undiscovered before me.

Chapter 10 Problems

[10.1] For the two vectors

$$\vec{A} = 3\hat{\imath} + \hat{\jmath}$$
$$\vec{B} = -2\hat{\imath} + 2\hat{\jmath}$$

a) Find the magnitude and direction of $3\vec{A}$. **b)** Find $\vec{A} \cdot \vec{B}$. **c)** Find the magnitude and direction of $\vec{A} \times \vec{B}$.

[10.2] For the two vectors

$$\vec{A} = 3\hat{\imath} = 5\hat{k}$$
$$\vec{B} = 2\hat{\imath} - 5\hat{k}$$

a) Find the angle between the two vectors $\vec{A}$ and $\vec{B}$. **b)** Find a unit vector which is perpendicular to the two vectors $\vec{A}$ and $\vec{B}$.

[10.3] Prove for any two arbitrary vectors $\vec{A}$ and $\vec{B}$: $\vec{A} \bullet (\vec{A} \times \vec{B}) = 0$.

[10.4] Prove for any arbitrary vector $\vec{A}$: $\vec{A} \bullet \vec{A} = |\vec{A}|^2$ and $\vec{A} \times \vec{A} = 0$.

[10.5] Find the magnitude and direction of the area formed by the two vectors

$$\vec{l} = 4\hat{i} + 2\hat{j}$$
$$\vec{w} = 3\hat{i} - \hat{j}.$$

[10.6] Find the angle between the two vectors $\vec{A}$ and $\vec{B}$ if $\vec{A} \cdot \vec{B} = -AB$.

[10.7] The potential energy U of an electric dipole, whose moment is $\vec{p}$ in an applied electric field $\vec{E}$, is given by the relation

$$U = -\vec{p} \cdot \vec{E}$$

Describe the orientation of $\vec{p}$ relative to $\vec{E}$ that produces the minimum potential energy (To work this problem you need to know nothing about the physics behind p and E).

[10.8] A constant force of 50 N, directed above the horizontal, causes a horizontal displacement of an object of 3.0 m. If 100 J of work is done, find the angle that the force makes with the horizontal.

[10.9] Write a table of all the possible combinations of dot products among the Cartesian unit vectors $\hat{i}$, $\hat{j}$, and $\hat{k}$. For example, $\hat{i} \bullet \hat{k}$.

[10.10] Write a table of all the possible combinations of cross products among the Cartesian unit vectors $\hat{i}$, $\hat{j}$, and $\hat{k}$. For example, $\hat{i} \times \hat{k}$.

[10.11] A girl pulls a sled of mass 5.0 kg with a constant force of 40.0 N, but at an angle of 40° with respect to the horizontal. Find the work she does on the sled in pulling it 20 m. (See Figure 10.20.)

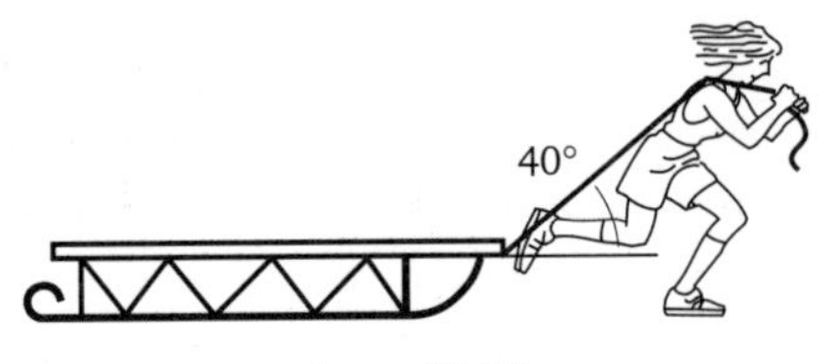

Figure 10.20

[10.12] A flux of 5 kg/s flows into a square pipe of side length l. Find the flux leaving the gray surface of each of the three pipes shown in Figure 10.21.

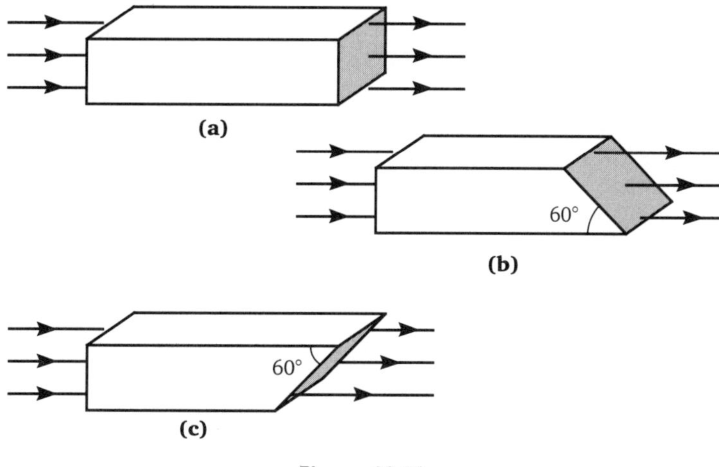

Figure 10.21

[10.13] Prove that the angle between the plane of the surface and its projection perpendicular to the flow, shown in Figure 10.15, is the same angle as that between the normal to the surface and the flow direction, shown in Figure 10.17.

• **[10.14]** Consider a spherical source of fluid. The fluid flows radially out from this source in all directions. **a)** Find a formula for the area of a surface, such that, at each point on the surface, the flow is perpendicular to the surface. **b)** Derive a formula for the mass flux through this surface. Hint: See Example 10-7. This example had cylindrical symmetry. What type of symmetry is present in the present problem?

• **[10.15]** The rays of light reaching the earth from the sun are almost parallel. Find the area of the projection of the earth's surface that is perpendicular to the rays of light from the sun. Only use the portion of the surface in the sun's light. The radius of the earth is 6.4×10^6 m.

• [10.16] A very long string vibrates, creating sound waves that travel radially away from the axis of the string. **a)** what geometric shape will the wavefronts assume? **b)** Find the area that is perpendicular to the flow of sound emitted from a section of the length L of the string. Assume the surface is a distance r from the axis of the string and surrounds the section L.

• [10.17] A 3.0 kg block starts from rest and slides 5.0 m down a frictionless plane inclined at 36.9° above the horizontal. **a)** How much work has been done on the block by gravity? **b)** How much work has been done on the block by the surface of the inclined plane? **c)** What is the final velocity of the block?

APPENDIX A
Homework and Exams

A.1 Writing Solutions

Obviously, if you are building a bridge you must get the correct answers to your calculations or the bridge will fall down, possibly injuring people. However, the problems assigned to you in a physics course serve a different purpose than to just get the right answer. You are learning a logic system for solving problems, therefore it is necessary to see the logic you used in arriving at your answer, and you can expect your papers to be graded on that logic. To grade the logic, the grader has to understand your solution to a problem. Therefore, if you scribble down a partial solution, without writing complete equations, you probably will not receive a very good grade on that problem. Remember that on any graded assignment, you are trying to convince the grader that you know how to solve the problem. Always write your solutions in a neat logical fashion with each major step written down and a brief sentence of explanation between major steps. Use complete equations and not abbreviations.

Figure A.1 shows examples of two students' solutions to a quiz problem.

The pilot of an airplane traveling 160 km/hr wants to drop supplies to flood victims isolated on a patch of dry land 190 m below. How far, i.e., what distance, ahead of the patch of land should the supplies be released by the plane in order to hit the patch?

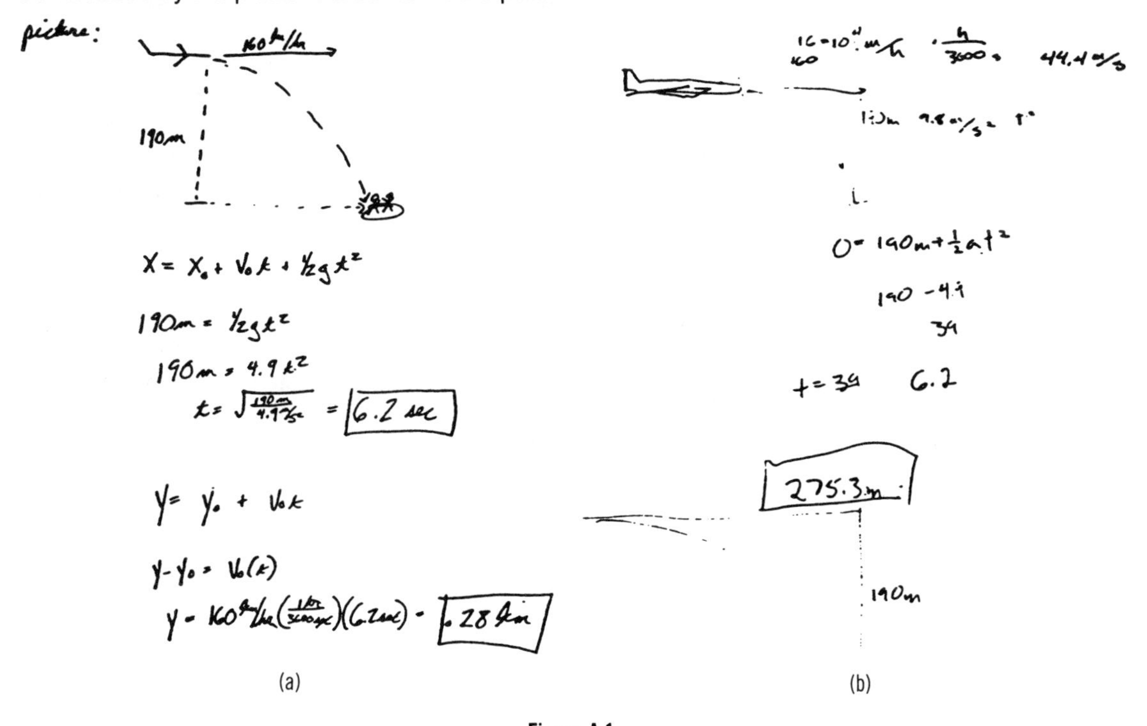

Figure A.1

Two solutions to a quiz problem. (a) has a well-developed solution and (b) does not.

Solution (a) shows a clear logical solution with the major steps written. Note there is a carefully drawn picture with the given information marked on it. The solution starts with a general equation from kinematics, which is then applied to both the vertical and then horizontal components of the position. In stark contrast to the clear solution in (a) is Figure A.1 (b) in which no general equation ever appears. The first written equation has only numbers in it. After this first equation, you cannot follow what the person is doing because there are no equations. No partial credit can be given past the first equation because only the time (6.2), without units, appears 3 lines down, and then magically, the answer appears in the middle of the page. The solution in (b) earns only a few points partial credit because it is not an acceptable solution.

Another reason for clearly showing your work in complete equations is that, what if you made a mistake? If you do not record your steps the grader cannot give you partial credit. In order to receive as much partial credit on an exam as possible, you should put down all the information you know about the problem and do so in an organized fashion so that the grader can clearly see what you know. Otherwise, you will be unlikely to receive points; the grader cannot give you credit for work that you have not written down on the page.

A.2 Studying for Exams

Preparing for an exam takes time. You should start preparing a week before the exam, so that you are not studying long into the night right before the exam. Know precisely what material the exam will cover. You should spend half your time understanding the theory covered in your lecture notes and the book, and half your time working problems. The problems are a very important part of preparing for an exam. Go over the homework assignments, and know these problems cold. However, the homework problems usually represent the minimum requirements to pass the course. If you wish to do more than just pass, you should be working additional unassigned problems. Sources for extra problems are the odd problems in most textbooks. They have answers, so you can check your work. Another good source of problems is the worked examples in other textbooks that you can find in the library. First read an example's question and try to solve it before looking at the solution. Problems that you have not seen before are a good measure for judging whether or not you are prepared for the exam. If you can work a fair percentage of new problems without help, then you are ready for the exam. If you are having a great deal of difficulty working new problems, then you are not ready.

Often students will try to memorize the solution to each and every assigned problem. Although it may succeed once in a while, the exam will involve thinging: **NOT MEMORIZATION.** It is unlikely that the problems on the exam will be the same as on the assignments. This has two major implications. One, it is not enough to just work problems. You must also understand the theory. The theory and problem-solving skills are complementary. To solve problems effectively, you will need a firm understanding of the principles, as well as an understanding of how to apply them. Two, you should get a good night's sleep before the exam and come to the exam well rested and prepared to think. **CRAMMING WILL NOT HELP**. If you have spent a week diligently preparing for an exam, then the night before the exam, you will need only a light review. Try to do something relaxing; something you enjoy. Reward yourself for studying so hard. If this process sounds like a training regimen for an athlete, that is very nearly correct. You should be coming into an exam with all your faculties at their peak and your skills sharp.

An effective way to study is to start by reviewing the theory and as you do so, construct a review sheet that consists of the major principles, such as Newton's laws, conservation of energy, or Bernoulli's principle, and some sample equations for each. Then, use the review sheet while you work problems. As you work the problems, you may find that you have forgotten important information. Add any forgotten information immediately. By the end of the week, you will have an effective study sheet that you can use to review just before the exam. It will also help you to organize your review for the final exam. Another technique is to study with friends. If you can explain a topic so another person can understand it, then you understand it.

You will do better if you stay relaxed and thinking clearly. Remember that you can solve the problems on the exam by using the techniques you have learned in the course. Get your thoughts down on the paper in an organized manner. If you have problems with stress during an exam, see a counselor about learning stress relaxation techniques. They can help. Above all, you will do much better if you start studying early.

APPENDIX B
Solutions and Answers to Selected Odd Numbered Problems

CHAPTER 1

[1.1] **a)** $8x = 3$, $x = \frac{3}{8}$. **b)** $4 + 1 = 3x - 2x$, $x = 5$. **c)** $7x = \frac{7}{3} + \frac{3 \cdot 2}{3}$, $7x = \frac{13}{3}$, $x = \frac{13}{21}$ **d)** $1 + 3x = 5x$, $1 = 2x$, $x = \frac{1}{2}$.

[1.3] **a)** Clear the denominators, multiply by the LCD, $9ax$: $9a + 3a = x$, $x = 12a$. **b)** The LCD is x: $a - bx = cx$, collect terms: $a = bx + cx$, $a = (b + c)x$. Divide out the $(b + c)$: $x = \frac{a}{b + c}$. **c)** The denominator can be factored: $\frac{a(1 + \sqrt{x})}{(\sqrt{x} - 1)(\sqrt{x} + 1)} = b$, $\frac{a}{(\sqrt{x} - 1)} = b$. Clear the denominator: $a = b(\sqrt{x} - 1)$. Divide by b: $\sqrt{x} - 1 = \frac{a}{b}$. Isolate the square root: $\sqrt{x} = \left(1 + \frac{a}{b}\right)$. Square: $x = \left(1 + \frac{a}{b}\right)^2$.

[1.5] In Davis, California (we'll use Davis as an example) there are about 50,000 people, and Davis, being a bicycle city, has about 1 bike/person $\Rightarrow$ 50,000 bicycles. There may be fewer in your town. The average mass of a bicycle is about 10 kg. Then the total mass is approximately (50,000)(10 kg) = 500,000 kg. (500 Metric Tons!)

[1.7] **a)** Assume about 5–10 hairs/cm^2. Treat the arm as a cylinder with mean radius about 4 cm, length of 50 cm, and about half the cylinder has hair. Then, (5 hairs/cm^2)$\left(\frac{1}{2} 2\pi rL\right)$ = (5 hairs/cm^2) ($\pi \cdot$ 4 cm $\cdot$ 50 cm) = 3,000 hairs. **b)** Assume that there are about 100 hairs/cm^2. If we treat the head as a sphere with a 21 cm circumference, the radius is $r = C/2\pi$ = 8 cm. The area of the head is then $4\pi r^2$ = 800 cm^2. Only about half the head is covered with hair so the total hair area is 400 cm^2. The number of hairs is then:

$$(100 \text{ hairs/cm}^2)(400 \text{ cm}^2) = 40{,}000 \text{ hairs.}$$

[1.9] Average people are about 30″ in circumference at the waist which using $C = 2\pi r$ gives r = 12 cm for an average radius. The average height is 5′ 10″ which is about 1.8 m, and gives a volume of about:

$$V = (\pi r^2)h = (\pi)(.12 \text{ m})^2(1.8 \text{ m}) = 0.08 \text{ m}^3.$$ Assuming an average mass of 70 kg $\Rightarrow$

$$\rho = (70 \text{ kg})/(0.08 \text{ m}^3) = 900 \text{ kg/m}^3.$$

We could arrive at this value along a different route. An object will float if it is less dense than water. Since we float with roughly 9/10 of our body below the water, then we must be about 9/10 as dense as water, or

$$\rho = \frac{9}{10}(1000\ \text{kg/m}^3) = 900\ \text{kg/m}^3.$$

[1.11] Answer, 10^6.

[1.13] Assume the standard classroom is 8 m × 10 m × 3 m = 240 m^3. If we knew how many moles of air were in the classroom, we'd know how many molecules were in it. One mole of water weighs 18 grams and occupies 18 ml = 18 cm^3. On average a gas is about 1000 times less dense than a liquid so one mole of gas occupies about 20 ml × 1000 = 20 liters (at STP a mole of any gas occupies 22.4 liters). There are 1000 liters in a cubic meter so we have (240 m^3)(10^3 liter/m^3)/(20 liter/mole) = 10^4 moles. Then the number of molecules is:

$$(6 \times 10^{23}\ \text{molecules/mole})(10^4\ \text{moles}) = 6 \times 10^{27}\ \text{molecules}.$$

[1.15] Answers: **a)** 1/2 mole = 3×10^{23} molecules **b)** 10^{24} molecules.

[1.17] There are $(2\ \text{paper clips/gm}) \times \dfrac{1000\ \text{gm}}{2.2\ \text{lb}} = 1000\ \text{paper clips/lb}.$

[1.19] Water is fairly dense, especially compared to the vapor form. Ice has a density of 9/10 that of water so we would expect that a water molecule would have about the same linear dimensions as the spacing between molecules. ⇒ 1000 kg/m^3. Note that even if the molecules are twice as far apart as their size, that gives only a factor $(2^3) = 8$ error in our estimate. We are still in the ball park.

[1.21] Answer, 4 and 4.

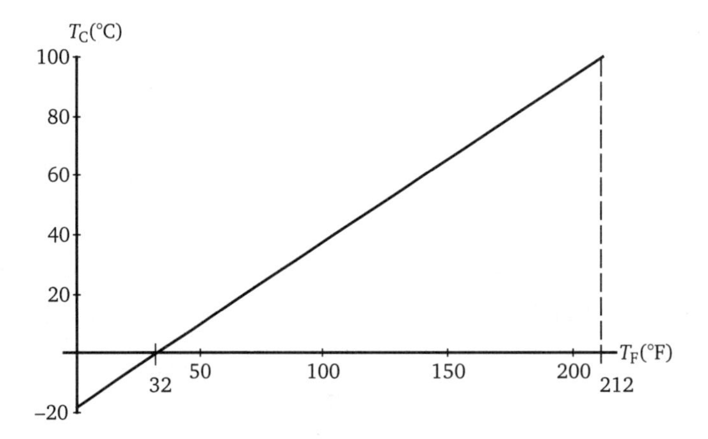

Figure B.1
T_C vs T_F

[1.23] **a)** Since our formula must work for any temperature, we must let a variable represent the temperature. Let T_F represent the temperature in degrees Fahrenheit, and T_C represent the temperature in degrees Celsius. When asked to find a formula, it sometimes helps to plot the function as in Figure B.1 where connecting (212°F, 100°C) and (32°F, 0°C) with a straight line produces the graph shown. The slope is $\dfrac{100°\text{C}}{(212 - 32°\text{F})} = \dfrac{5}{9}$ C°/F°. Let the intercept be b, and the formula becomes

$T_C = \left(\frac{5}{9}\right) T_F + b$. To find the y-intercept, $b = -\left(\frac{5}{9}\right) 32 = -\left(\frac{160}{9}\right)$. So the equation becomes $T_C = \left(\frac{5}{9}\right) T_F - \left(\frac{160}{9}\right)$ **b)** $T_C = \left(\frac{5}{9}\right)$ (98.6° F) − 160/9 = 37° C.

CHAPTER 2

[2.1] Answer, (d)

[2.3] **i)** At points B and E, the slope is zero so the velocity is zero; thus the position isn't changing. **ii)** B and E—same as part (i). **iii)** At point F the positive curvature means the velocity is increasing, i.e., a positive acceleration. **iv)** At point A the velocity is positive (positive slope) but the curvature is negative. **v)** At point C the slope is negative so the position is decreasing, and the curvature is negative so the acceleration is negative. **vi)** At point D the path curves up from the tangent line drawn at D so the acceleration is positive; thus, the velocity is increasing. **vii)** At point A it's decelerating, but moving forwards. **viii)** At point C it's accelerating and moving backward. **ix)** At points D, E, and F there is positive curvature (smiley face). **x)** At points A, B, C there is negative curvature (frowny face).

[2.5] Answer, 47.5 mi/h.

[2.7] $x = x_0 - v_0 t - \frac{1}{2} at^2$ is a frowning parabola as shown in Figure B.2 because the curvature, $\left(-\frac{1}{2}a\right)$, is negative. At the origin the graph goes through the point x_0, and since the initial velocity is $-v_0$, the slope of the graph at the origin is $-v_0$, i.e., negative, so the x axis must intercept the curve where it is a decreasing function.

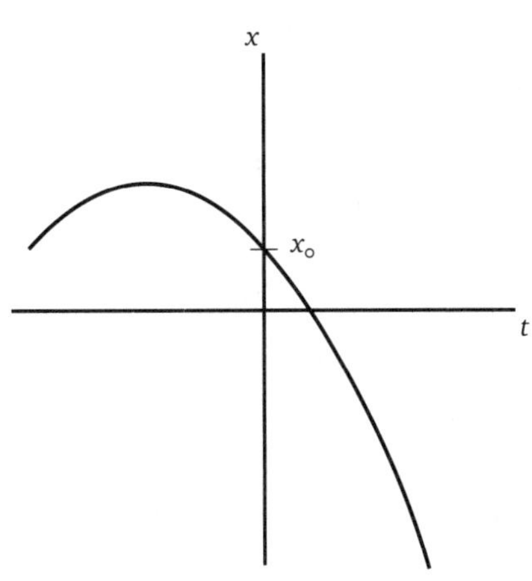

Figure B.2
Graph of $x = x_0 - v_0 t - \frac{1}{2} at^2$

[2.9] **a)** Let t be the time he hits the ground. Then the total distance he falls is $d = \frac{1}{2} gt^2$. During the first $t - 1$ seconds, he fell half the distance:

$$\frac{1}{2} d = \frac{1}{2} g(t - 1)^2. \text{ So } \frac{1}{2}\left(\frac{1}{2} gt^2\right) = \frac{1}{2} g(t - 1)^2.$$

Clear g: $\frac{1}{2}t^2 = (t-1)^2$. Clear $\frac{1}{2}$: $t^2 = 2(t^2 - 2t + 1)$

$$t^2 - 4t + 2 = 0. \text{ Using the quadratic formula:}$$

$$t = \frac{4 \pm \sqrt{16-8}}{2} = (2 \pm \sqrt{2}) \text{ s}$$

The minus root solution can be eliminated because $2 - \sqrt{2} < 1$ s, and thus, violates the conditions of the problem (he falls last half of distance in 1s).

$$t = (2 + \sqrt{2}) \text{ s}$$

b) Substitute t into the equation for $y = \frac{1}{2}g(2 + \sqrt{2})^2 = 57.1$ m.

[2.11] **a)** The radioactivity is greatest at the maximum in the curve, which is at 3 weeks. **b)** Between zero and 3 weeks, the radioactivity is increasing. **c)** Between three weeks and 10 weeks the radioactivity is decreasing. **d)** The radioactivity is increasing the fastest where the slope is the greatest, which is about 2 weeks. **e)** Again, this is where the slope is most negative, at about 5 weeks. **f)** At the maximum of the curve the change in radioactivity is zero because the slope is zero, which is at 3 weeks. **g)** About 1.25×10^{-10} Curies/g. **h)** About 0.75×10^{-10}Curies/g.

[2.13] **a)** At rest means the object maintains a constant position, so the slope of the position vs time curve is zero. Therefore, the object in D maintains a constant position. **b)** For uniform acceleration the velocity curve is a straight line and the position curve is a parabola A and F satisfy these (C would also, it gives a zero acceleration) **c)** A & F have constant non zero acceleration. C & E have zero acceleration. **d)** A has uniform acceleration producing a position graph like in F. B has an acceleration that increases in time so the position curve has increasing curvature, and F is trivially not a straight line. **e)** B, because the velocity would be parabolic.

[2.15] **a)** To find the maximum height reached, we need to use $y = y_o + v_o t - \frac{1}{2}gt^2$, where t is the time the ball reaches the maximum height. Because we want to find the highest point reached above the top of the building, we should place the origin to the coordinate system at the top of the building. Before proceeding it's a good idea to check the unknowns. t is unknown and so is y. There are two unknowns and only one equation, so we need to find another equation. Do we know any thing about the speed at the highest point? That's the place it has momentarily stopped so $v = 0$. Use this value in the velocity equation $v = v_o + \boldsymbol{at}$ where $\boldsymbol{a} = -g$ giving $v = v_o - gt$. Substituting $v = 0$, $0 = v_o - gt$. Solve for t: $t = \frac{v_o}{g}$. Substitute this time into the y equation $y_m = 0 + v_o \frac{v_o}{g} - \frac{1}{2}g\frac{v_o^2}{g^2}$, $y_m = +\frac{v_o^2}{2g} = \frac{(8.7 \text{ m/s})^2}{9.8 \text{ m/s}^2} = 7.75$ m. **b)** The package is in the air for 4.5 s so we can just use $v = v_o - gt$. $v = 8.7 \text{ m/s} - (9.8 \text{ m/s}^2)(4.5 \text{ s}) = -35.4$ m/s which means the package is headed down with a great speed just before it hits!

CHAPTER 3

[3.1] The easiest way to show this is to work out a couple of years and see what happens. Since 100% of M is $1 \times M$, then 1% of M is $\frac{1}{100}M = (0.01)M$, and $6\frac{1}{2}$% is $(0.065)M$. After the first year there will be, $M_1 = 5000 + (0.065)5000 = 5000(1 + 0.065)^1$. This amount now becomes the base amount for the second year. $M_2 = M_1(1 + 0.065) = 5000(1 + 0.065)(1 + 0.065)$ so that $M_2 = 5000(1 + 0.065)^2$. Well, if this is the result after 2 years, then by deductive reasoning after t years $M = 5000(1 + 0.065)^t$. **b)** Use the formula. If the money doubles, $M = 10{,}000 = 5{,}000(1.065)^t$. Dividing by 5,000 $\Rightarrow (1.065)^t = 2$. To solve this for t we need to undo the exponential so we use the log to the base 10. Taking the log of both sides, $\log(1.065)^t = \log(2)$, and $t \log(1.065) = \log(2)$. Solve for t: $t = \dfrac{\log 2}{\log 1.065} = 11$ years. To triple, $t = \dfrac{\log 3}{\log 1.065} = 17$ years.

[3.5] Here you can use the definition of the log and exponential to quickly evaluate each expression. **a)** $\log 10^x = \log 0.001$, $\log 10^x = \log 10^{-3}$, $x = -3$. **b)** $\log_5 5^{x-2} = \log_5 625$, $\log_5 5^{x-2} = \log_5 5^4$, $x - 2 = 4 \Rightarrow x = 6$. **c)** $\log_2 2^{3x} = \log_2 16$, $3x = \log_2 2^4$, $3x = 4$, $x = \frac{4}{3}$. **d)** $6^{\log_6 x} = 6^3$, $x = 6^3 = 216$. **e)** $\log_7 7^{-3} = x$, $x = -3$. **f)** $x^{\log_x 64} = x^{3/2}$, $64 = x^{3/2}$, $s^6 = x^{3/2}$, $2^2 = x^{1/2}$, $x = 2^4 = 16$. **g)** $\log_4 \left(\dfrac{x}{4x + 6}\right) = -2$, $\dfrac{x}{4x + 6} = 4^{-2}$. Divide by 16 and multiply by $4x + 6$, $16x = 4x + 6$, $12x = 6$, $x = \frac{1}{2}$.

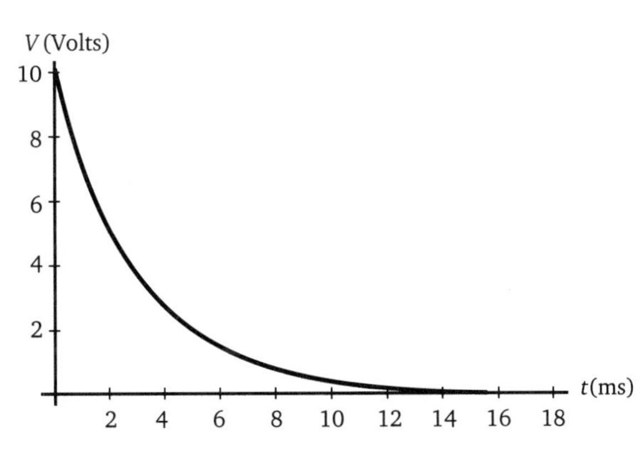

Figure B.3
Graph of $V = 10\, e^{-t/0.003}$.

[3.7] **a)** The exponential decay obeys the form $V = V_o e^{-t/\tau}$ where τ is the decay constant. Comparing to our case we find $\tau = 0.003$ s $= 3$ ms (ms = millisecond). **b)** To make a good looking graph, we want about 6 decay constants, so we mark the t-axis from 0 to 18 ms as shown in Figure B.3. **c)** $V_o = 10$ V (V = Volts). **d)** Substitute 9 in the V and solve for t. $9 = 10\, e^{-t/0.003}$, $0.9 = e^{-t/0.003}$. Take the ln of both sides,

$$\ln 0.9 = -\frac{t}{0.003} \Rightarrow t = -0.003 \ln 0.9,\ t = 3.2 \times 10^{-4}\ \text{s} = 0.32\ \text{ms}.$$

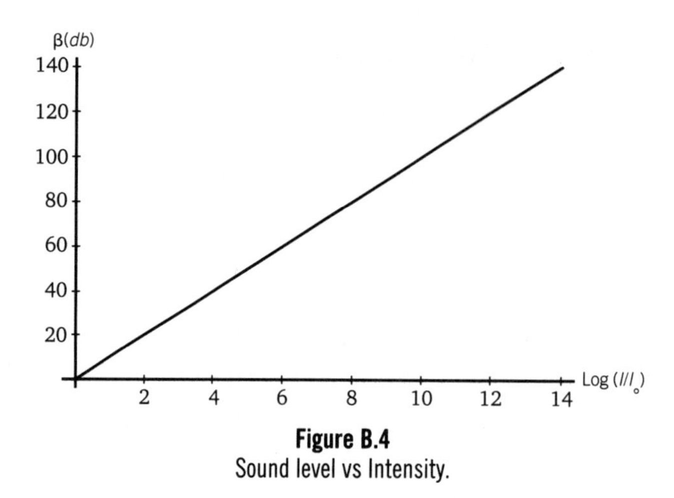

Figure B.4
Sound level vs Intensity.

[3.9] **a)** The db scale is marked linearly. Since the intensity scale spans many orders of magnitude, it should be marked using a $\log_{10}$ scale with each major tick mark on the scale representing one or two orders of magnitude as in Figure B.4. **b)** The graph is a straight line on the semi-log plot so that A in $\beta = A\,f\left(\frac{I}{I_o}\right)$ will be the slope of the graph. $\Rightarrow A = \frac{140db - 0}{14} = 10\,db$, and $f = \log\left(\frac{I}{I_o}\right)$.

$\Rightarrow \beta = (10\,db)\log\left(\frac{I}{I_o}\right)$.

[3.11] Answer, **b)** $F = z^2$

[3.13] You are being asked for the number of cycles per second which is the frequency. $f = \frac{\omega}{2\pi}$ where $v = r\,\omega \Rightarrow f = \frac{v}{2\pi r} = \frac{44\text{ m/s}}{2\pi(7\text{ m})} = 1\,\frac{\text{cycle}}{\text{s}}$. Or one could take the approach that the time for one cycle is $T = \frac{2\pi r}{v}$ and since $f = \frac{1}{T} \Rightarrow f = \frac{v}{2\pi r}$.

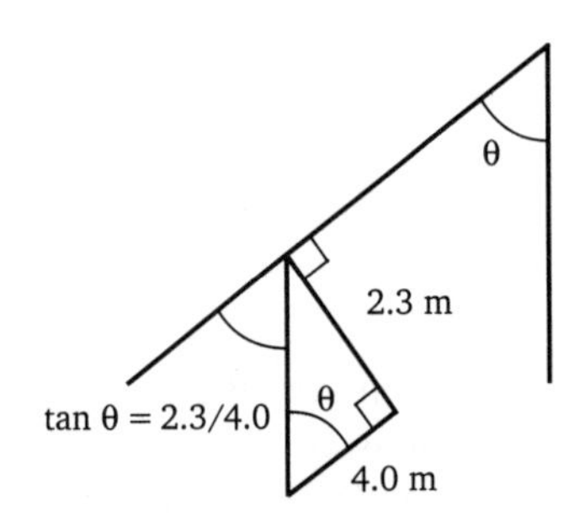

Figure B.5

[3.15] The first step is to relate θ to the given information. Using parallel lines and the bisector, θ is the angle shown in Figure B.5. The tangent gives

$$\tan\theta = \frac{2.3}{4.0} \Rightarrow \theta = \tan^{-1}\left(\frac{2.3}{4.0}\right) = 30°.$$

[3.17] Answers: **a)** 45° **b)** 270° or $3\pi/2$ **c)** undefined.

[3.19] a) The period is the time to complete one revolution $T = 17.9$ s. **b)** $f = \frac{1}{T} = \frac{1}{17.9\text{s/cycle}} =$ 0.056 cycles/s = 0.56 Hz **c)** $\omega = 2\pi f = 0.35$ /s (that's radians per second!) **d)** angular speed = angular frequency = ω = 0.35/s. **e)** $v = r\,\omega$ = (20 m)(0.35/s) = 7.0 m/s.

[3.21] What size must the moon be to perfectly eclipse the sun? Looking at the moon's orbit, the moon must subtend the same angle as the sun when viewed from the earth as shown in Figure B.6. This means the two subtend the same angle. Approximating the diameters as arc lengths

$$\theta = S/R, \text{ where } \theta_S = \theta_m.\ \frac{S}{R} = \frac{S_m}{R_m} \Rightarrow$$

$$R = \frac{S}{S_m} R_m \quad R = \frac{1.4 \times 10^9 \text{ m}}{3.5 \times 10^6 \text{ m}} (3.8 \times 10^8 \text{ m}) = 400\,(3.8 \times 10^8 \text{ m}) = 1.5 \times 10^{11} \text{ m}.$$

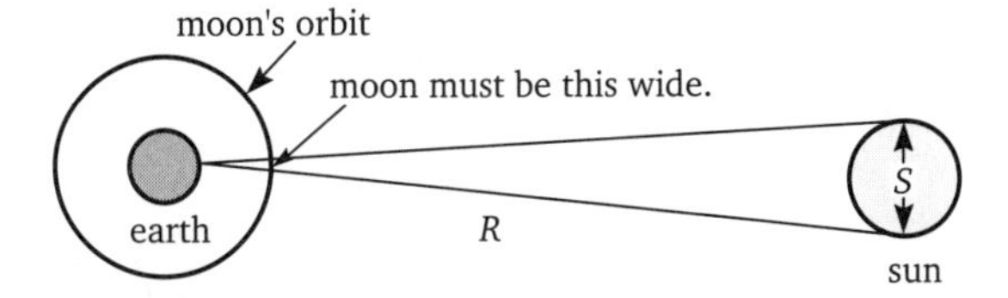

Figure B.6

CHAPTER 4

[4.1] If the wave is increasing in height, then the buoy must be on the leading edge of the wave. The crest has yet to pass the buoy, as shown in Figure B.7.
Using the chain rule $\frac{\Delta y}{\Delta t} = \frac{\Delta y}{\Delta x}\frac{\Delta x}{\Delta t}$ where $v = \frac{\Delta x}{\Delta t}$. Solving for the slope, $\frac{\Delta y}{\Delta x} = \frac{1}{v}\frac{\Delta y}{\Delta t} = \frac{1.3 \text{ m/s}}{12 \text{ m/s}} =$ 0.11.

[4.3] Let P be pressure. Using the chain rule, $\frac{\Delta P}{\Delta t} = \frac{\Delta P}{\Delta h}\frac{\Delta h}{\Delta t} = \frac{10000 \text{ N/m}^2}{1 \text{ m}}(4 \text{ m/s})\ F = 40{,}000$ N/(m$^2 \cdot$ s).

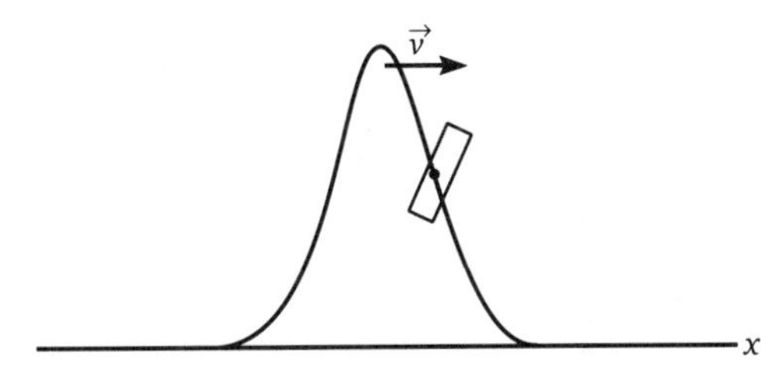

Figure B.7
A buoy on the leading edge of a wave.

[4.5] **a)** Use the chain rule to expand the $\Delta N/\Delta t$: $\frac{\Delta N}{\Delta t} = \frac{\Delta N}{\Delta x}\frac{\Delta x}{\Delta t} = \lambda\, v$ where λ is the linear density and v is the speed. Solving for the speed, $v = \frac{1}{\lambda}\frac{\Delta N}{\Delta t} = \frac{1}{2.5 \times 10^6 \text{ electrons/m}} \cdot 5 \times 10^{14} \text{ electrons/s} = 2 \times 10^8$ m/s. **b)** Let Q be charge. Then the current is $\frac{\Delta Q}{\Delta t}$. The rate at which charge comes out of the accelerator depends on the number that comes out per unit time multiplied by the charge each carries. Let e be charge on an electron. Then $\frac{\Delta Q}{\Delta t} = \frac{\Delta N}{\Delta t} \cdot e = 5 \times 10^{14}$ electrons/s $\cdot\ 1.6 \times 10^{-19}$ C/electron; $\frac{\Delta Q}{\Delta t} = 8 \times 10^{-5}\text{A} = 80\ \mu\text{A}$.

[4.7] First plot the function as shown in Figure B.8. In the region near $x = 2$, the function is at the end of a frowny face so the curvature is negative.

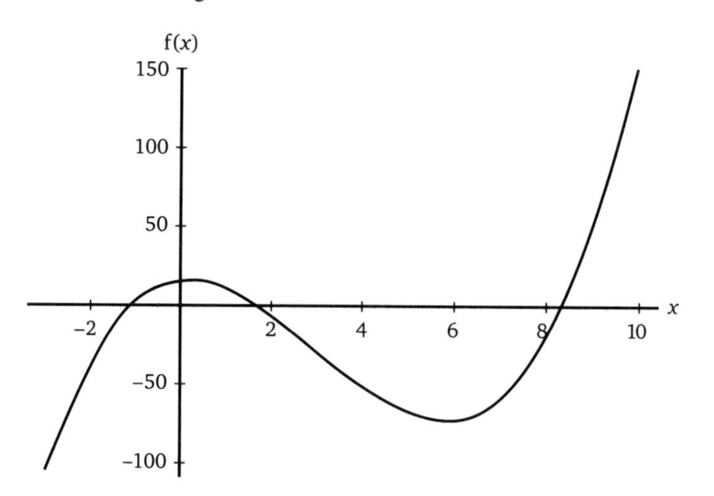

Figure B.8
A plot of $x^3 - 9x^2 + 3x + 14$.

[4.9] **a)** Positive curvature (smiley face) from $0 \to 2, 4 \to 6, 10 \to 12$. **b)** Negative curvature (frowny face) from $3 \to 4, 7 \to 9$. **c)** $f(x)$ has no curvature where the graph is straight, i.e., from $2 \to 3, 6 \to 7, 9 \to 10$.

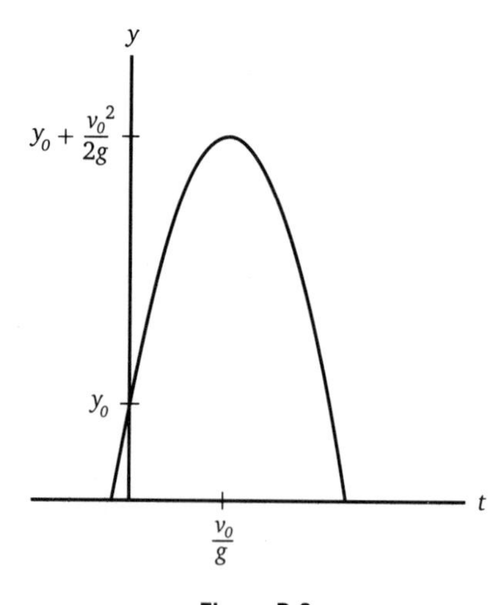

Figure B.9

[4.11] The position is given by $y = y_o + v_o t - \frac{1}{2}gt^2$ as shown in Figure B.9. The extrema occur at the top of the trajectory where the velocity is zero, and y obviously has a maximum.

$$v = v_o - gt = 0 \Rightarrow t = \frac{v_o}{g}.$$

Plugging into the equation for y: $y = y_o + v_o\frac{v_o}{g} - \frac{1}{2}g\frac{v_o^2}{g^2} \Rightarrow y = y_o + \frac{1}{2}\frac{v_o^2}{g}$. The acceleration $a = -g$ and is negative, so the y vs t is a frowny face.

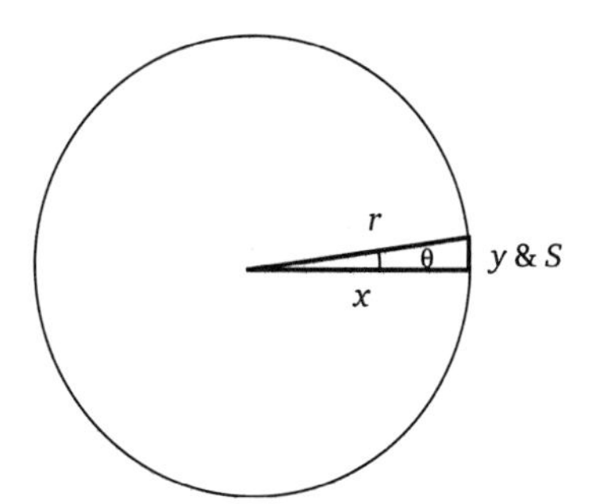

Figure B.10

[4.13] **a)** $\sin 5° = 0.08716$ and $\tan 5° = 0.08749$. Using $\tan 5°$ rather than $\sin 5°$ gives only a 0.4% error. For small angles (<20°) the two are interchangeable. **b)** For the triangle shown in Figure B.10, $\sin\theta = y/r$ and $\tan\theta = y/x$, but $x \simeq r$, so $\sin\theta \simeq \tan\theta$, which is true whenever θ is small, i.e., when $x >> y$.

[4.15] **a)** $\sin\theta \simeq \theta = 20°\frac{\pi}{180°} \simeq \frac{3}{9} = 1/3 = 0.333.$

b) From a calculator, $\sin 20° = 0.3420$. The percent error is

$$\text{error} = \frac{0.342 - 0.333}{0.342} \times 100\% = 2.6\%.$$

[4.17] First construct an expansion parameter, $\frac{d}{x} << 1$. $f(x) = \frac{1}{x^5(1 - d/x)^5} = \frac{1}{x^5}(1 - d/x)^{-5}$. Now approximate using the binomial expansion through the linear term: $f(x) \simeq \frac{1}{x^5}\left(1 - (-5)\frac{d}{x}\right)$.

$$f(x) \simeq \frac{1}{x^5} + \frac{5d}{x^6}.$$

[4.19] Answer, $1 - \frac{x}{d}$.

CHAPTER 5

[5.1] The large fluctuations in results are due to a poor experimental design. Better results would have been obtained by using a long tape measure. A 12" ruler is ill-suited for measuring a float. Since the measurements vary due to random fluctuations of where the ending points and starting points of the individual measurements of the 12" ruler are made, the general scatter in the data is due to **random** errors.

Note that due to the experimental design, the precision of the experiment is low, i.e., even a very careful person is unlikely to make a good measurement.

Note that one data point, 35' 03", is so far from the others that it is probably due to an **illegitimate** error on the part of the student. Because this data point is very suspect, it should not be averaged with the others. Experimental data should always be determined by **random errors never by illegitimate or systematic errors.**

[5.3] The relative error is the standard deviation divided by the mean

$$e_{\text{rel}} = \frac{s}{\bar{x}} = \frac{2''}{397''} = 0.005.$$

[5.5] Using Equation (5-16), $\Delta V = \frac{4\pi(r + \Delta r)^3}{3} - \frac{4\pi r^3}{3}$,

$$\Delta V = \frac{4\pi(4.52 + 0.03)^3}{3} - \frac{4\pi(4.52)^3}{3} = 395 \text{ cm}^3 - 387 \text{ cm}^3 = 8 \text{ cm}^3.$$

$$e_v = (100\%)\frac{\Delta V}{V} = (100\%)\left(\frac{8 \text{ cm}^3}{387 \text{ cm}^3}\right) = 2\%.$$

[5.9] **a)** 1,1; 1,2; 2,1; 1,3; 3,1; 1,4; 4,1; 1,5; 5,1; 1,6; 6,1; 2,2; 2,3; 3,2; 2,4; 4,2; 2,5; 5,2; 2,6; 6,2; 3,3; 3,4; 4,3; 3,5; 5,3; 3,6; 6,3; 4,4; 4,5; 5,4; 4,6; 6,4; 5,5; 5,6; 6,5; 6,6. **b)** There is only one possible way of rolling a double 4 out of the thirty six outcomes. The probability is 1/36. **c)** There are six doubles: 6/36 = 1/6 **d)** 30/36 = 5/6 **e)** You can roll a 7 six different ways ⇒ 6/36 = 1/6. An eleven can be rolled 2 ways ⇒ 2/36. The combined probability is (6 + 2)/36 = 2/9.

[5.11] **a)** The best estimate of the period will be the mean; however, note that the 0.9 s value is a blunder and should not be included. $\overline{T} = \frac{4(1.5)+3(1.6)+1.7+1.4}{9} = 1.54$ s **b)** The uncertainty is determined by the standard deviation:

$$\Delta T = s = \sqrt{\frac{4(0.4)^2+3(0.6)^2+(1.6)^2+(1.4)^2}{9}} = 0.088 = 0.09 \text{ s}$$

c) Measure the total time for ten periods then divide by 10.

CHAPTER 6

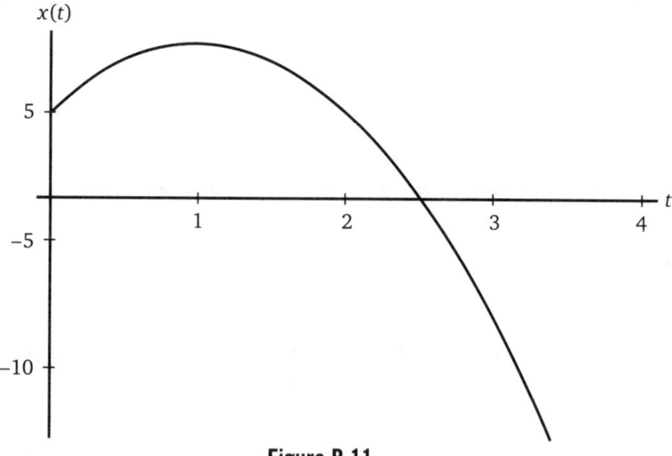

Figure B.11
A plot of $x = 5 \text{ m} + (8 \text{ m/s})\, t - (4 \text{ m/s}^2)t^2$.

[6.1] **a)** the t-intercept is given by $0 = 5 + 8t - 4t^2$. Using the quadratic equation,

$$t = \frac{-8 \pm \sqrt{8^2 - 4(-4)(5)}}{2(-4)} = \frac{-8 \pm 12}{-8} = 1 \pm \frac{3}{2}, t = 2.5 \text{ s}.$$

The maximum value of x is where the slope is zero, i.e., where $v = 0$ (see part **(b)**), at $t = 1$ s. At the maximum value, $x = 5$ m $+ 8$ m $- 4$ m $= 9$ m. **b)** $x/t = 3\text{s} = 5 \text{ m} + (8 \text{ m/s})(3 \text{ s}) - (4 \text{ m/s}^2)(3 \text{ s})^2 =$ 17 m. Note that this has the form $x = x_o + v_o t + \frac{1}{2}at^2$, where $x_o = 5$ m, $v_o = 8$ m/s, and $a = -8 \text{ m/s}^2$. The velocity must be $v = v_o + at$, so that, $v = (8 \text{ m/s}) - (8 \text{ m/s}^2)t$ where $v/t = 3\text{s} = (8 \text{ m/s}) - (8 \text{ m/s}^2)(3 \text{ s}) = -16$ m/s. The acceleration is constant at (-8 m/s^2). **c)** The velocity is the slope of x vs t, so v is positive for $0 \leq t < 1$. **d)** The acceleration is a constant $\boldsymbol{a} = -8 \text{ m/s}^2$ so it is negative as can be seen from the graph shown in Figure B.11 which has negative curvature. **e)** It is still $\boldsymbol{a} = -8 \text{ m/s}^2$, thus negative.

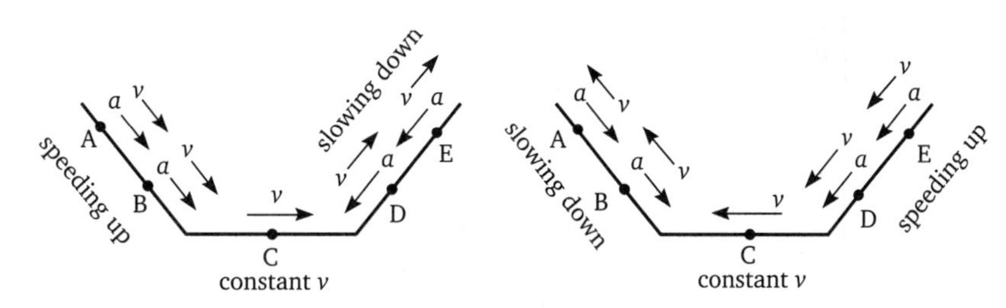

Figure B.12

[6.3] See Figure B.12.

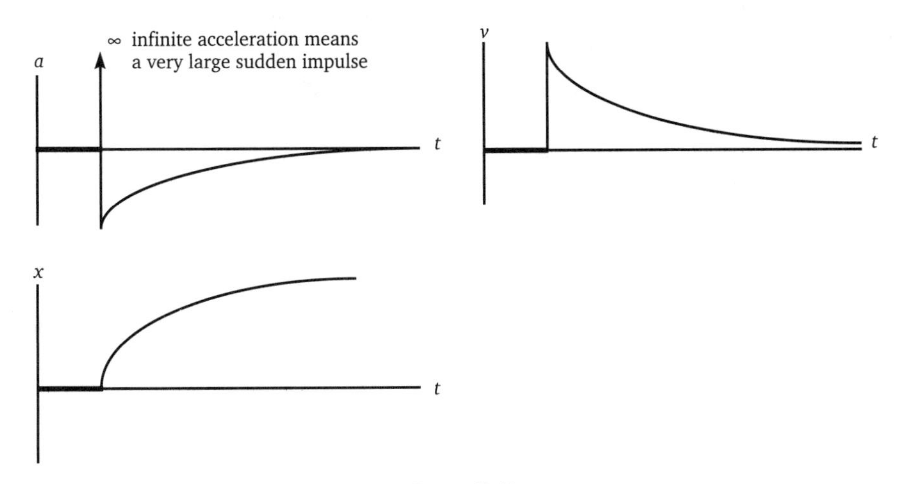

Figure B.13

[6.5] From the velocity graph in Figure B.13, we see the velocity increase dramatically at the step, indicating a large sudden acceleration. It then decreases to zero, so the acceleration, which is the slope, must be negative, i.e., the velocity curve might decrease as $1/t$ in which case, the acceleration is $-(1/t^2)$. The position graph is an increasing graph since the velocity is always positive, but it becomes flat as the velocity decreases to zero. From the acceleration, the position has negative curvature (frowny face).

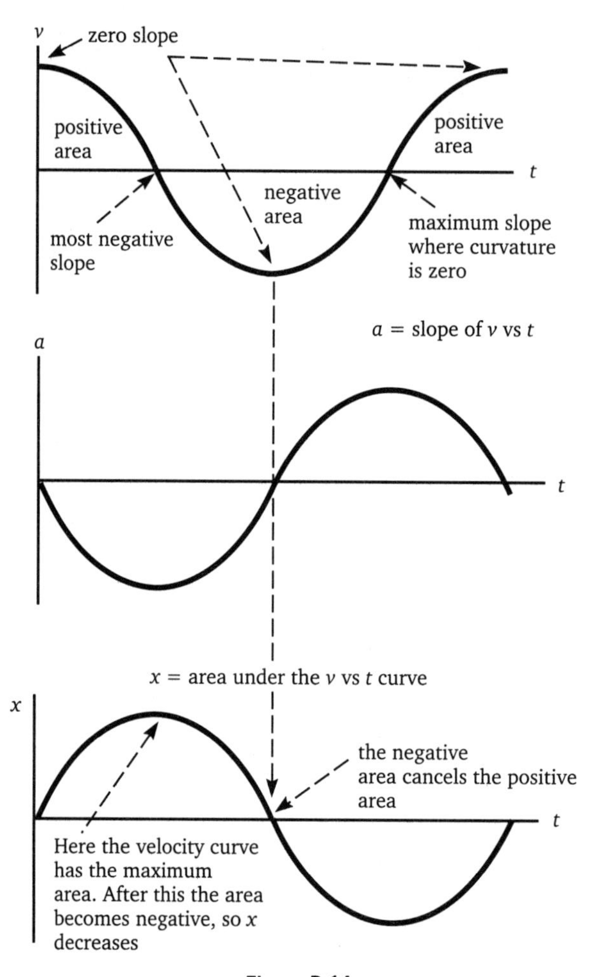

Figure B.14

[6.7] Figure B.14 suggests that if $v = \cos t$, then $a = -\sin t$, and $x = \sin t$. Such motion might be displayed by a pendulum or a mass on a spring.

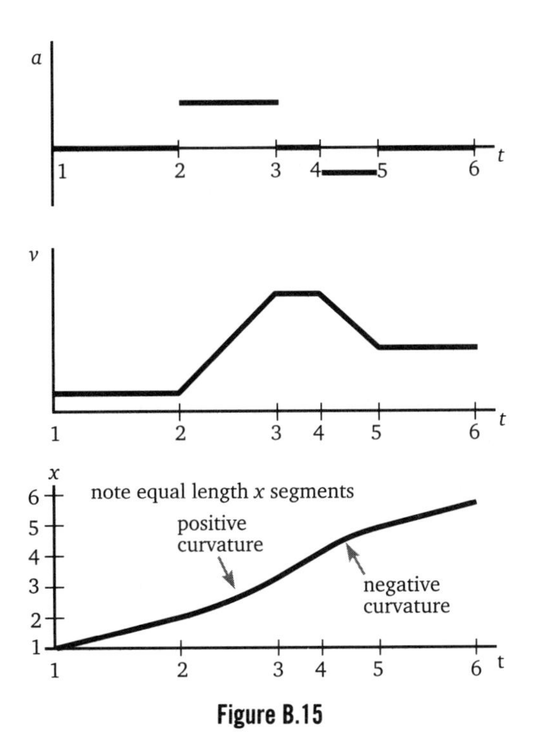

Figure B.15

[6.9] In general, first find the relative time intervals, then the acceleration, velocity, and position, in that order. From 1 to 2 the bead will be moving the slowest because it is highest up and hasn't begun to accelerate yet. It will then speed up as it falls from 2 to 3, and move at a constant speed from 3 to 4 where it will be moving fastest; therefore, 3→4 will require the least time. From points 4 to 5, it will slow down, but will not be going as slow as at point 2 so this segment will be considerably shorter in time than 2→3. Finally, from 5→6 which is halfway between 2 and 3 it will be moving with about the average speed it took to fall from 2 to 3, and will thus take about the same amount of time.

It will take the most amount of time to cover the first segment. See Figure B.15.

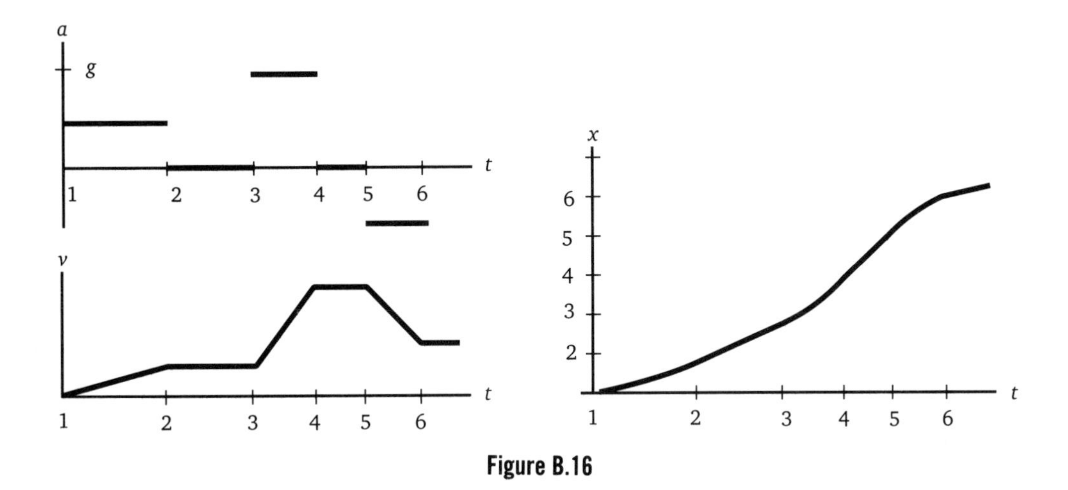

Figure B.16

[6.11] We first figure the times where the events occur. On the segments where the bead is highest on the track, it will be moving the slowest (think of a slick slide—you move slowest at the top and fastest at the bottom), and hence, will take the longest time. The longest time interval is $1 \rightarrow 2$. Where the bead is lowest it will be moving the fastest, and thus will have short time intervals. Shortest time interval is $4 \rightarrow 5$ as shown in Figure B.16.

On the position vs time curve where the acceleration is positive the graph curves up and where the acceleration is negative there is negative curvature (frowny face). The bead never moves back towards the origin and hence the position is an increasing function. It does not decrease.

[6.13] Because the density varies with x, we must sum the area under the density vs x curve. This area is the total mass, **b)** $m = \frac{1}{2}(73 \text{ g/m} - 43 \text{ g/m})L + 43 \text{ g/m} = \frac{1}{2}(73 \text{ g/m} + 43 \text{ g/m})L = (58 \text{ g/m}) \cdot L$. Because we don't know L, this is as far as the mass can be computed. **a)** The average density is $\overline{\lambda} = \frac{\text{total mass}}{L} = \frac{1}{2}(73 \text{ g/m} + 43 \text{ g/m}) = 58 \text{ g/m}$.

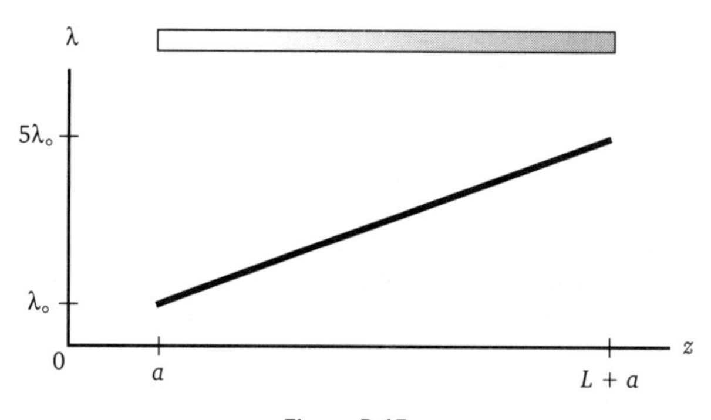

Figure B.17
Density vs Position.

[6.15] Because the density varies with position, start with a graph of the density vs position as shown in Figure B.17. **a)** The average density is $\overline{\lambda} = \frac{1}{2}(5\lambda_o + \lambda_o) = 3\lambda_o$. **b)** The density is a straight line that has slope $\frac{5\lambda_o - \lambda_o}{(L+a)-a} = \frac{4\lambda_o}{L}$, so the equation is $\lambda = \frac{4\lambda_o}{L}z + b$ where b is the y-intercept. At $z = a$, $\lambda = \lambda_o$. Substituting these values into our equation: $\lambda_o = \frac{4\lambda_o}{L}a + b$. Solve for b: $b = \lambda_o - \frac{4\lambda_o}{L}a$ and the equation is $\lambda = \frac{4\lambda_o}{L}z + \lambda_o - \frac{4\lambda_o}{L}a$. Let's check the result. At $z = a$, $\lambda = \frac{4\lambda_o}{L}a + \lambda_o - \frac{4\lambda_o}{L}a = \lambda_o$. This point is fine. Next check $L + a$ where the density should be $5\lambda_o$: $\lambda = \frac{4\lambda_o}{L}(L + a) + \lambda_o - \frac{4\lambda_o}{L}a = 4\lambda_o + \lambda_o$, so this point works too.

[6.17] **a)** The average density is $\overline{\lambda} = \frac{1}{2}(3\lambda_o + \lambda_o) = 2\lambda_o$. **b)** Because the density is linear, we want to use the equation for a straight line, although the mass is distributed along the arc of a circle so the density varies with θ. The equation will have the form $\lambda = m\theta + b$. The slope is $\frac{3\lambda_o - \lambda_o}{\frac{\pi}{3}R} = \frac{6\lambda_o}{\pi R}$. The y-intercept is trivial

$b = \lambda_\circ$. The equation is then $\lambda = \frac{6\lambda_\circ}{\pi R}\theta + \lambda_\circ$. Check the result: at $\theta = 0, \lambda = 0 + \lambda_\circ$, ok, and $\theta = \pi/3, \lambda = \frac{6\lambda_\circ}{\pi R}$ $\frac{\pi}{3} + \lambda_\circ = 2\lambda_\circ + \lambda_\circ$, ok.

[6.21] Answer, $\simeq 15.5$ N · s.

CHAPTER 7

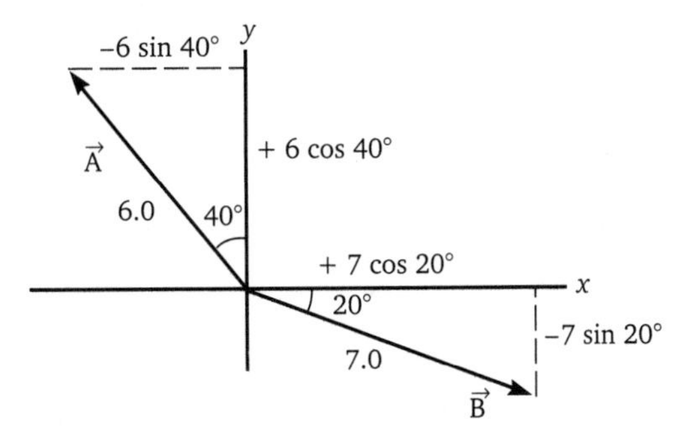

Figure B.18

[7.1] First drop perpendiculars onto the axis from which the given angle is measured as shown in Figure B.18.

$$A_x = -6 \sin 40° + -3.9, A_y = +6 \cos 40° = 4.6,$$
$$B_x = +7 \cos 20° = 6.6, \text{ and } B_y = -7 \sin 20° = -2.4.$$

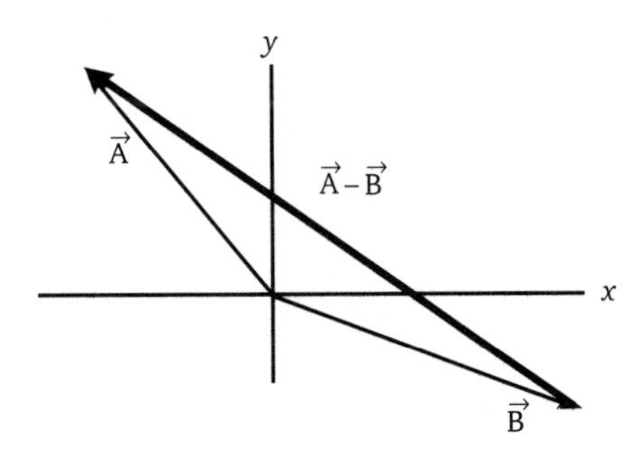

Figure B.19

[7.3] Always draw a picture of the subtraction, so you can estimate the size and direction of the resultant vector. See Figure B.19.

$$\vec{A} = -3.9\,\hat{i} + 4.6\hat{j}$$
$$-\vec{B} = -6.6\,\hat{i} + 2.4\hat{j}$$
$$\vec{A} - \vec{B} = -10.5\hat{i} + 7.0\hat{j}$$

$$|\vec{A} - \vec{B}| = \sqrt{10.5^2 + 7^2} = 12.6$$

$$\theta = \tan^{-1}\frac{7.0}{-10.5} = -33.7° \text{ up from the negative } x \text{ axis}$$

[7.5] First break the vectors into components: $A_x = -6\cos 50° = -3.9$, $A_y = +6\sin 50° = 4.6$, $B_x = 7\cos 30° = 6.1$, $B_y = 3.5$, $C_x = 3\cos 60°$, and $C_y = -3\sin 60° = -2.6$. Then adding :

$$\vec{A} = -3.9\,\hat{i} + 4.6\,\hat{j}$$
$$-\vec{B} = -6.1\,\hat{i} - 3.5\,\hat{j}$$
$$3\vec{C} = 3(1.5)\,\hat{i} - 3(2.6)\hat{j}$$
$$\vec{D} = \vec{A} - \vec{B} + 3\vec{C} = -5.5\,\hat{i} - 6.7\,\hat{j}$$

$$|\vec{D}| = \sqrt{5.5^2+6.7^2} = 8.7,\ \theta = \tan^{-1}\frac{-6.7}{-5.5} = 50.6° \text{ down from the negative } x \text{ axis.}$$

[7.7] Answer, **a)** 26.8 mi/h, $-63.4°$ (south of east). **b)** $v_x = 12$ mi/h, $v_y = -24$ mi/h.

[7.9] **a)** $\vec{A} = \vec{B} - \vec{C}$, **b)** $\vec{A} = \vec{C} - \vec{B}$, **c)** $\vec{A} + \vec{C} = \vec{B} \Rightarrow \vec{A} = \vec{B} - \vec{C}$, **d)** $\vec{A} = \vec{B} + \vec{C}$, **e)** $\vec{A} + \vec{B} = -\vec{C} \Rightarrow \vec{A} + \vec{B} + \vec{C} = 0$.

[7.11] $\vec{v} = \vec{v}' + \vec{V}_{rel}$ **a)** $\vec{v}' = +2{,}200$ km/hr $\hat{i}$ and $\vec{V}_{rel} = +850$ km/hr $\hat{i}$. $\Rightarrow \vec{v} = +3050$ km/hr $\hat{i} \Rightarrow |\vec{v}| = 3050$ km/hr. **b)** $\vec{v}' = -2200$ km/hr $\hat{i}$, and $\vec{V}_{rel} = +850$ km/hr $\hat{i}$. $\Rightarrow \vec{v} = -1350$ km/hr $\hat{i} \Rightarrow |\vec{v}| =$ 1350 km/hr. **c)** Here the velocities are in different directions so it is very important to use a vector addition. $\vec{v}' = 2200$ km/hr $\hat{j}$ and $\vec{V}_{rel} = 850$ km/hr $\hat{i}$. $|\vec{v}| = \sqrt{2200^2 + 850^2} = 2360$ km/hr.

[7.13] We need to determine how the vectors fit together so we begin by looking at the frames of reference. The velocity of the plane can be measured relative to an observer floating with the air, say in a balloon or relative to the ground. According to our procedure the vectors must fit together as $\vec{v}_{p\to g} = \vec{v}_{p\to a} + \vec{v}_{a\to g}$. So first draw $\vec{v}_{p\to a}$. It is 200 mph due east. Next add $\vec{v}_{a\to g}$ to the head. Finally complete the sum with $\vec{v}_{p\to g}$ as depicted in Figure B.20.

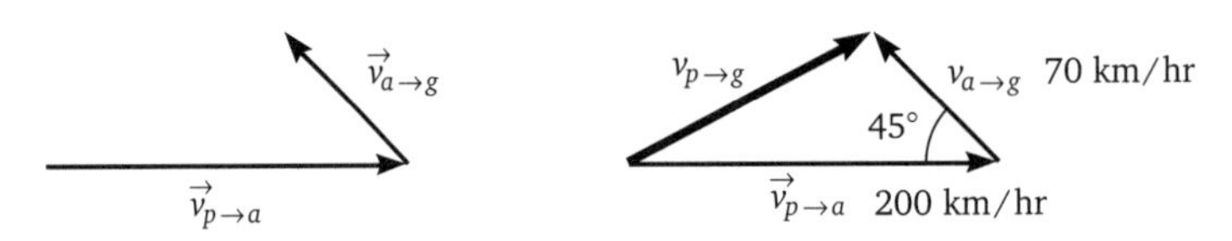

Figure B.20

From here its a matter of doing the vector sum. Break the vectors into components and sum.

$$\vec{v}_{a\to g} = -70\cos 45°\,\hat{i} + 70\sin 45°\,\hat{j} = -49.5\,\hat{i} + 49.5\,\hat{j}$$

$$\begin{aligned}\vec{v}_{p\to a} &= 200\,\hat{i} + 0\,\hat{j}\\ \vec{v}_{a\to g} &= -49.5\,\hat{i} + 49.5\,\hat{j}\\ \hline \vec{v}_{p\to g} &= 150.5\,\hat{i} + 49.5\,\hat{j}\end{aligned}$$

$$|\vec{v}_{p\to g}| = \sqrt{150.5^2 + 49.5^2} = 158 \text{ km/hr.}$$

$$\theta = \tan^{-1} 49.5/150.5 = 18.2°.$$

[7.15] Answer, 15.2 mi/h at 1.7° north of east.

[7.17] We are given the direction of the velocity of the ship relative to the ground and the direction of the wind. We do not know the direction of the ship relative to the water, and this is what we are solving for. Start by drawing $\vec{v}_{s\to g}$. Relativity tells us that $\vec{v}_{s\to g} = \vec{v}_{s\to w} + \vec{v}_{w\to g}$, so we can draw $\vec{v}_{w\to g}$ such that it looks like it is the final vector in the addition for form $\vec{v}_{s\to g}$ as shown in Figure B.21.

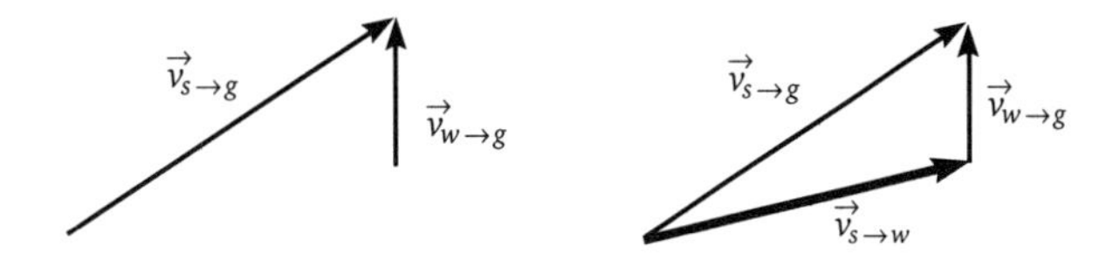

Figure B.21

Because we only know the direction of $\vec{v}_{s\to g}$, the length of $\vec{v}_{s\to w}$, we cannot resolve them into components. A plan of attack might be to figure the angles from the geometry as illustrated in Figure B.22.

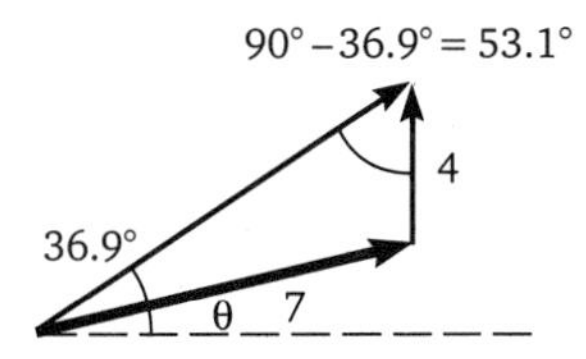

Figure B.22

Using the law of sines, $\frac{\sin(36.9° - \theta)}{4} = \frac{\sin 53.1°}{7}$, and solving for θ $\sin(36.9° - \theta) = \frac{4}{7}\sin 53.1°$.

$\theta = 36.9° - \sin^{-1}(\frac{4}{7}\sin 53.1°) = 9.7°$ north of east.

CHAPTER 8

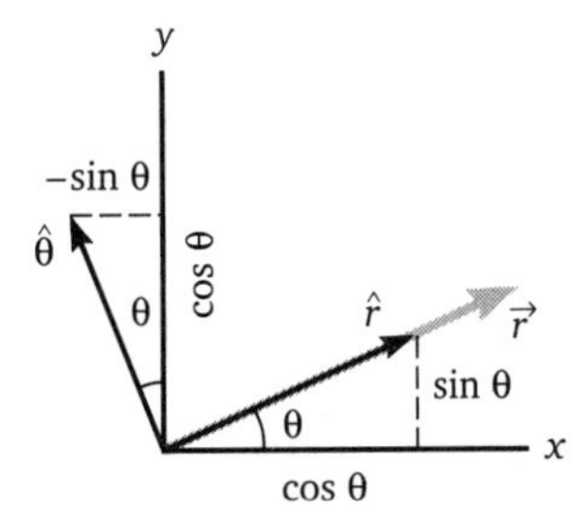

Figure B.23

[8.1] As shown in Figure B.23, first drop perpendiculars from the unit vectors onto the coordinate axes. The length of each unit vector is one by definition. $\hat{r} = \cos\theta\,\hat{i} + \sin\theta\,\hat{j}$, $\hat{\theta} = -\sin\theta\,\hat{i} + \cos\theta\,\hat{j}$.

[8.3] A displacement made directly towards the origin is a radial displacement, so $\Delta\vec{r} = -3\text{m}\,\hat{r}$. The velocity is radially inwards,

$$\vec{v} = -\frac{3\text{m}}{5\text{s}}\hat{r} = -0.6\text{ m/s}\,\hat{r}$$

For the direction we need to find θ because $\hat{r} = \cos\theta\,\hat{i} + \sin\theta\,\hat{j}$. This direction can be found from the position vector $\theta = \tan^{-1}\frac{12}{16} = 36.9°$.

[8.5] **a)** $\omega = \frac{2\pi}{T}$ where T is in seconds. $\Rightarrow \omega = \frac{2\pi}{(23.7\text{ days})(24\text{ hr/day})(3600\text{ s/hr})} = \frac{2\pi}{2.05 \times 10^6\text{ s}}$ $\omega = 3.07 \times 10^{-6}/\text{s}$. **b)** $v = r\omega = (3.85 \times 10^8\text{ m})(3.07 \times 10^{-6}/\text{s}) \Rightarrow v = 1.18 \times 10^3\text{ m/s}$. **c)** $a = r\omega^2$, $a/g = r\omega^2/g \Rightarrow = \frac{a}{g} = \frac{(3.85 \times 10^8\text{ m})(3.07 \times 10^{-6}/\text{s})^2}{9.8\text{ m/s}^2} = 3.7 \times 10^{-4}$. Note that the linear speed of the moon is very large being over a 1000 m/s, but the angular speed is very small. Why? The radius of the moon's orbit is very large. Note that centripetal acceleration of the moon is caused by the attractive pull of the earth, but that from result **(c)** this acceleration of the moon is much less than the gravitational acceleration the earth produces on an object at the surface.

[8.7] Note that gravity is causing an acceleration down towards the center of the circle. We want the limiting case where the crumb at first just follows the arc of the cylinder. To get the crumb to travel in a circle there must be a centripetal acceleration $a = \frac{v_\circ^2}{R}$. Gravity is the only force that could cause this acceleration, so if $\frac{v_\circ^2}{R}$ is less than g, the crumb will be pulled against the surface, and if $\frac{v_\circ^2}{R}$ is greater than g, gravity will not be strong enough to supply the centripetal acceleration, and the crumb will not follow the surface and will fly off. The limiting case is when $\frac{v_\circ^2}{R} = g \Rightarrow v_\circ = \sqrt{Rg}$.

[8.9] $x = r\cos\theta = (4\text{ m})\cos 0.4 = 3.7$ m (don't forget to put your calculator in radian mode). $y = r\sin\theta = (4\text{ m})\sin 0.4 = 1.6$ m. $z = 3$ m.

[8.11] Because superman is flying due west, he is traveling along a latitude, the radius for this arc is $r\sin\theta$. $\Rightarrow \Delta S = r\sin\theta\,\Delta\phi$ where $\theta = 90° - 50° = 40°$, and $\Delta\phi = 125° - 100°$ and $\Delta\phi$ must be in radians $\Delta\phi = 25°\frac{\pi}{180°} = 0.44$ (remember radians is unitless, you don't write it).

$$\Delta S = (6.4 \times 10^6\text{ m})\sin 40°\,(0.44). \quad \Delta S = 1.8 \times 10^6\text{ m.} = 1{,}800\text{ km.}$$

[8.13] **a)** $V = \pi r^2 h$, $h = \frac{V}{\pi r^2} = \frac{10\text{ cm}^3}{\pi(0.2\text{ cm})^2} = 79.6$ cm. **b)** $\Delta h = \frac{1}{\pi r^2}\Delta V$. We are given $\frac{\Delta V}{\Delta T} = 0.01°\text{ cm}^3/°\text{C}$. Expanding $\Delta h = \frac{1}{\pi r^2}\frac{\Delta V}{\Delta T}\Delta T = \frac{0.01°\text{ cm}^3/°\text{C}}{\pi(0.2\text{ cm})^2}100°\text{C} = 8.0$ cm, so the new height is

$h' = h + \Delta h = 87.6$ cm.

[8.15] Answer, **b)** $2\pi rL$.

[8.17] Answer, 2.9 cm.

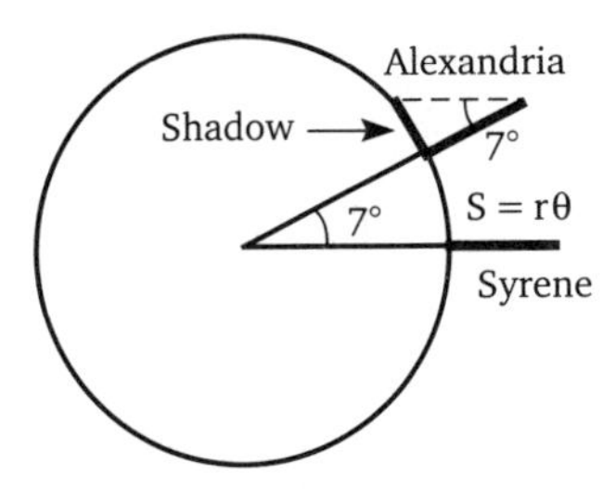

Figure B.24

[8.19] Becasue the shadow subtends an arc of 7°, Alexandria must be 7° north of Syrene, which itself must be on the equator if its stick throws no shadow. Represent the earth with a circle and draw the sticks at Alexandria and Syrene showing the radius of the earth, as shown in Figure B.24. Using the arc length formula for traveling along a longitude: $S = r\theta \Rightarrow$

$$r = \frac{S}{\theta} = \frac{800\text{ km}}{7° \times \frac{\pi}{180°}} = 6{,}500\text{ km}$$

[8.21] Answer, 2.6 m at 32.5° above horizontal.

[8.23] Answer, **a)** 7,100 km. **b)** No.

CHAPTER 9

[9.1] Since the object must accelerate at a rate much larger than gravity, the force of the machine must be in the same direction as gravity. Taking the down direction as positive, and summing forces:

$$F_{\text{pile}} = F_g = ma \Rightarrow F_{\text{pile}} = ma - F_g \Rightarrow F_{\text{pile}} = (500\text{ kg})(25\text{ m/s}^2 - 9.8\text{ m/s}^2) = 7600\text{ N}.$$

[9.5] The hardest part about this problem is figuring out the frictional forces. As the truck accelerates, the mass M will tend to slip backwards. Friction does not oppose motion, rather it opposes the tendency of M to slide backwards. Thus, the friction force must act forwards on M, and hence friction is the only force that causes M to accelerate forwards as shown in Figure B.25. If we were to remove this force, M would not move forward at all. By Newton's third law, the frictional force f_2 must act the other way on the truck. Next, the tires on the truck push backwards on the ground (to the left) so the ground must push forwards on the tires. This force is again a frictional force.

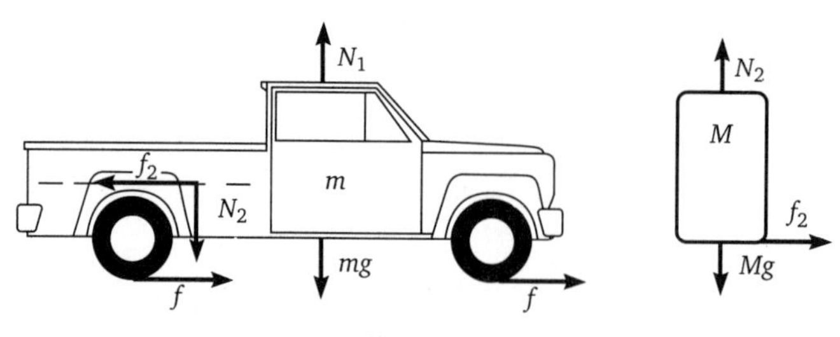

Figure B.25

[9.7] **a)** A person on a Ferris wheel always must experience a net force towards the center of the wheel. This force is necessary to have the rider move in a circular path, and is called the centripetal force. From the free–body diagram shown in Figure B.26, we see that at the top the centripetal force is supplied by gravity, but at the bottom the normal force must supply both the centripetal force and counter-act gravity. Thus, the normal force at the bottom must be larger than at the top, and the person feels heaviest at the bottom and lightest at the top.

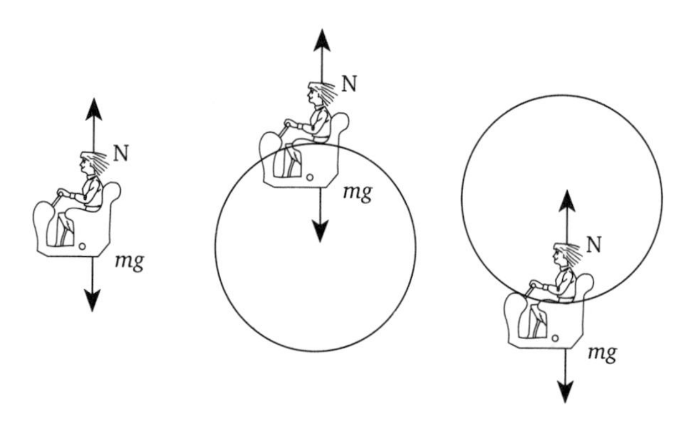

Figure B.26

b) From the free-body diagram at the top $\Sigma Fy = N - mg = -ma_c$ because the centripetal acceleration is down in the same direction as mg. $\Rightarrow N = mg - mr\omega^2 = 686\text{ N} - 246\text{ N} = 440\ N$.

c) $\Sigma F_y = N - mg = +ma_c$ because the centripetal acceleration is now in the same direction as $N \Rightarrow$

$$N = mg + mr\omega^2 = 686\text{ N} + 246\text{ N} = 932\text{ N}.$$

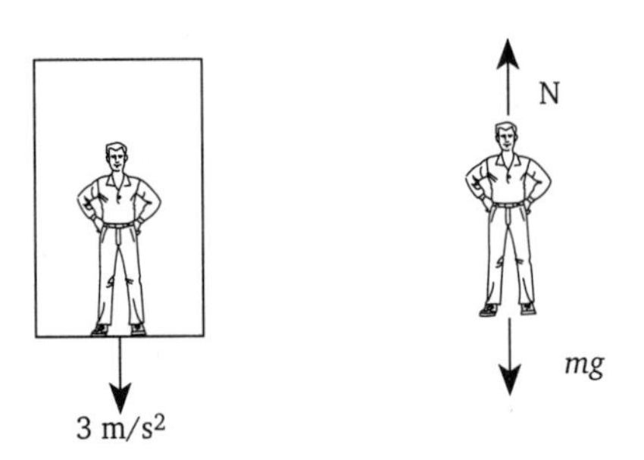

Figure B.27

[9.9] The person's weight will be determined by the normal force exerted upon him/her. First draw a free-body diagram, see Figure B.27. Summing forces: $\Sigma F_y = N - mg$. Now we've defined the force. Use Newton's 2nd law, note the acceleration is negative $N - mg = -ma \Rightarrow N = mg - ma$.

$$N = 80\,\text{kg}(9.8 - 3)\text{m/s}^2 = 544\,N.$$

That is, the normal force is less than normal by about 240 $N \approx$ 50 lb.

[9.11] Answer, no.

[9.13] Answers: **b)** $a = g \sin\theta$ **c)** $0.7\,g$, no.

[9.15] Answers: **b)** $\Sigma F_t = mg \sin\theta$, $\Sigma F_r = T - mg\cos\theta$ **c)** $a_t = g\sin\theta$, $a_r = \dfrac{v^2}{L}$.

[9.17] Answer, $v = \dfrac{mg}{b}$.

CHAPTER 10

[10.1] **a)** $3\vec{A} = 3(3\hat{i} + 1\hat{j}) = 9\hat{i} + 3\hat{j}$, $|3\vec{A}| = \sqrt{9^2 + 3^2} = 9.5$ $\theta = \tan^{-1}\dfrac{A_y}{A_x} = \tan^{-1}\dfrac{3}{9} =$ 18.4°. **b)** $\theta_A = 18.4°$, and $\theta_B = \tan^{-1}\dfrac{2}{-2} = -45°$ up from the negative x-axis. The angle between the vectors is $\theta_B - \theta_A = 135° - 18.4° = 116.6°$. $\vec{A}\cdot\vec{B} = AB\cos\theta = \sqrt{3^2 + 1^2}\,(2\sqrt{2})\cos 116.6° = -4$. Note this is the same as $A_xB_x + A_yB_y = -6 + 2 = -4$. **c)** $|\vec{A}\times\vec{B}| = AB\sin\theta = \sqrt{3^2 + 1^2}\,(2\sqrt{2})\sin 116.6° = 8.0$. By the right hand rule, lay your fingers along $\vec{A}$ with your palm facing $\vec{B}$. Your thumb will point out of the page. This is the same direction as $\hat{i}\times\hat{j} = \hat{k}$.

[10.3] $\vec{A}\times\vec{B}$ is perpendicular to $\vec{A}$; thus, the angle between $\vec{A}$ and $\vec{A}\times\vec{B}$ is $\theta = 90°$ so $\vec{A}\cdot(\vec{A}\times\vec{B}) = |\vec{A}||\vec{A}\times\vec{B}|\cos 90° = 0$, i.e., $\vec{A}\times\vec{B}$ can have no overlap with $\vec{A}$.

[10.5] $|\vec{L}\times\vec{w}| = Lw\sin\theta$. The angle between the vectors is given by the subtraction of the individual angles. $\theta_L = \tan^{-1}\dfrac{2}{4} = 26.6°$ $\theta_w = \tan^{-1}\dfrac{-1}{3} = -18.4°$ $\theta = 26.6° - (-18.4°) = 45°$. The area has a magnitude of $|\vec{L}\times\vec{w}| = \sqrt{4^2 + 2^2}\times\sqrt{3^2 + (-1)^2}\sin 45° = 10$, and points in the negative z direction, that is, the surface of the parallelogram defined by the two vectors faces in the negative z direction (into the page).

[10.7] The minimum potential energy is when U has its least value which in this case is when it is most negative. This occurs when the angle between $\vec{p}$ and $\vec{E}$ is $\theta = 0$. Then, $U = -pE\cos 0 = -pE$. If $\theta = 0$ the dipole moment $\vec{p}$ is parallel to the electric field $\vec{E}$.

[10.9] $\hat{i}\cdot\hat{i} = 1, \hat{i}\cdot\hat{j} = 0, \hat{i}\cdot\hat{k} = 0, \hat{j}\cdot\hat{i} = 0, \hat{j}\cdot\hat{j} = 1, \hat{j}\cdot\hat{k} = 0, \hat{k}\cdot\hat{i} = 0, \hat{k}\cdot\hat{j} = 0, \hat{k}\cdot\hat{k} = 1.$

[10.11] Answer, 613 J.

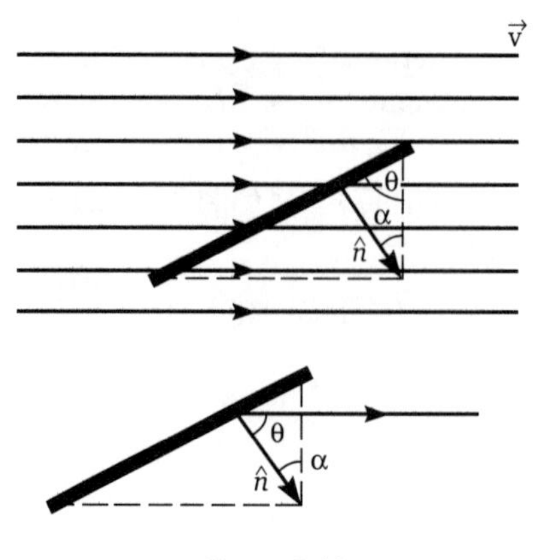

Figure B.28

[10.13] On Figure 10.15 draw the unit normal to A, such that it forms a triangle with the vertical side as shown in Figure B.28. Note that θ and α are complementary $\Rightarrow \alpha = 90° - \theta$. Look at the angle $\hat{n}$ makes with the flow. That angle is also complementary with α so it must be θ.

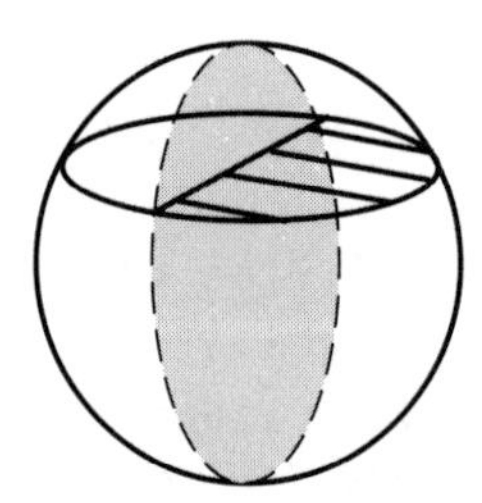

Figure B.29

[10.15] Consider an imaginary plane perpendicular to the rays coming from the sun as shown in Figure B.29. If this plane intersects the earth and goes through its center, the intersection with the earth would form a circle. The area of that circle is projection of the earth's surface area that's perpendicular to the rays of the sun, that is, each point on the surface of the earth facing the sun would project down into a unique point on the surface of the disk. That area is

$$\pi R_e^2 = \pi(6.4 \times 10^6 \text{ m}) = 1.3 \times 10^{14} \text{ m}^2.$$

[10.17] **a)** 88.2 J. **b)** 0. **c)** 7.7 m/s

Index

Conversions

Length	1 m = 3.23 ft 1 km = 0.62 mi 1 in. = 2.54 cm (exact) 1 mi = 5280 ft
Velocity	1 m/s = 2.24 mi/hr 30 mi/hr = 44 ft/s
Mass and Weight	1 kg = 0.0685 slugs 1 kg weighs 2.2 lb at sea level 1 N = 0.22 lb

Algebra

The Quadratic Formula

If $ax^2 + bx + c = 0$,

then
$$x = \frac{-b \pm \sqrt{b^2 - 4ac}}{2a}$$

Properties of Logarithms and Exponentials

Logarithms	Exponents
$\log a^x = x \log a$	$(a^x)^y = a^{x \cdot y}$
$\log_a a = 1$	$a^1 = a$
$\log (a \cdot b) = \log a + \log b$	$a^x a^y = a^{x+y}$
$\log \frac{a}{b} = \log a - \log b$	$\frac{a^x}{a^y} = a^{x-y}$
$\log_a 1 = \log_a a^0 = 0$	$a^0 = 1$
$\log \frac{1}{a} = -\log a$	$a^{-x} = \frac{1}{a^x}$
$\log \sqrt{a} = ½ \log a$	$\sqrt{a} = a½$

Series Expansion

$$e^x = 1 + x + \frac{x^2}{1 \cdot 2} + \frac{x^3}{1 \cdot 2 \cdot 3} + \frac{x^4}{1 \cdot 2 \cdot 3 \cdot 4} + \ldots$$

$$ln(1 + x) = x - \frac{1}{2}x^2 + \frac{1}{3}x^3 - \frac{1}{4}x^4 + \ldots$$